On-Site Power Generation:
A Reference Book
SECOND EDITION

Edited by
Gordon S. Johnson

NOTE: By publishing this reference book, The Electrical Generating Systems Association does not ensure anyone using the information it contains against any liability arising from that use. Users of the information in this reference book should make independent investigation of the validity of that information for their particular use of any items referred to herein.

This reference book is subject to revision at any time by The Electrical Generating Systems Association. Comments (recommendations, additions, or deletions) and any pertinent data that may be of use in improving this reference book are requested and should be addressed to EGSA Headquarters, 10251 West Sample Road, Suite B, Coral Springs, Florida 33065-3939. Such comments will receive careful consideration by EGSA and the author of the comments will be informed of the response to the comments.

No part of this publication may be reproduced in any form, in an electronic system or otherwise, without the prior written permission of the publisher.

© 1993, Electrical Generating Systems Association.
Printed in U.S.A.   All rights reserved.   ISBN 0-9625949-1-1

Preface to the Second Edition

I again had the privilege of editing this reference book. It was just three years ago that we ventured into the unknown with the first edition of *On-Site Power Generation*. EGSA had never published a book, but we felt the need was there. We were right. The on-site power industry reacted to the book as we had hoped it would. We received grateful praise from engineers, service men and salesmen in every part of the industry. Our stock is virtually exhausted.

The entire book has been carefully reviewed for errors and obsolete material. Chapters 10, Liquid-Cooled Diesel Engines, and 20, Vibration Isolation, have completely new texts. We appreciate the efforts of Roman Gawlowski and Francis Andrews in rewriting these chapters. Chapter 2 has a major addition covering non-linear loading of generators. Our faithful generator mentor, Jim Wright, worked very hard to produce this very useful section. Chapter 4 has new material and figures to bring us up to date on solid state circuit breaker progress. That work was done by new author, Ray Queen. Herbert Daugherty took over Chapter 6, Transfer Switches, edited the whole chapter and added new material, particularly on no-break load transfer. Marjorie Bernahl and her staff reviewed and updated Chapter 14, Diesel Fuel Systems. Dave Fredlake added a section on digital governing to Chapter 13, Governor Fundamentals.

Summit Technical Associates did such excellent work on the first edition that we returned to them for the book format and arrangement. As before, professionals James Malwitz, Barbara Riley and Hermine Scragg have been of inestimable value in producing a coordinated, attractive and readable book.

Welcome, readers of our second edition. We know that you will find it useful and even enjoyable.

Gordon Johnson, Editor

Table of Contents

Chapter	Page

1 Electrical Fundamentals 1
  1.1 Definitions 1
  1.2 Direct current (dc) 2
  1.3 Alternating current (ac) 2
  1.4 Rectification 3
      1.4.1 Diodes 3
  1.5 Magnetism 5
      1.5.1 Permanent magnets 5
      1.5.2 Electromagnetism and electromagnets 5
  1.6 Electromagnetic induction 6
  1.7 Elements of an alternating current circuit 7
      1.7.1 Resistance 7
      1.7.2 Inductors and inductance 10
      1.7.3 Capacitors and capacitance 12
      1.7.4 Impedance 13
  1.8 Power in an alternating current electrical circuit (kVA, kW, power factor) 14
      1.8.1 Power 14
      1.8.2 Alternating current power 15
      1.8.3 Power factor 17

2 Alternators (Synchronous Generators) 21
  2.1 Configurations 21
      2.1.1 Definitions 21
      2.1.2 Arrangements of alternators 22
      2.1.3 Types of alternator fields 23
      2.1.4 Types of exciters used with revolving field alternators 23
      2.1.5 Generation of the ac voltage 26
  2.2 Single and three phase armatures 27
      2.2.1 Basic principles of single and three phase alternating current 28
      2.2.2 Single and three phase armatures 28
      2.2.3 Basic features of an armature winding 29
      2.2.4 Connecting coils in an armature winding 29
      2.2.5 Interconnecting armature windings 31
      2.2.6 One and two circuit armatures 31
      2.2.7 Output load lead identification 32
  2.3 Connections for single and three phase alternators 35
      2.3.1 Single phase alternators 35
      2.3.2 Three phase alternators 37
      2.3.3 Delta connected three phase alternators 39
      2.3.4 Three phase six and twelve lead reconnectable alternators 40
  2.4 Frequency, frequency regulation, voltage and voltage regulation 43
      2.4.1 Frequency and frequency regulation 44
      2.4.2 Voltage and voltage regulation 45
  2.5 Temperature and environmental considerations of alternators 47

Chapter **Page**

   2.5.1 Temperature and classes of insulation materials 48
   2.5.2 Temperature considerations of alternator operation 48
   2.5.3 Effects of environment on alternators 50
   2.5.4 Thermal devices embedded in stator windings 50
  2.6 Alternator loading considerations 52
   2.6.1 Conditions of loading 52
   2.6.2 Electrical loads 54
   2.6.3 Steady state (constant) load analysis 56
   2.6.4 Alternator load bank testing 57
  2.7 Considerations of polyphase induction motor starting on engine-generator sets 58
   2.7.1 Three phase induction motor starting 58
   2.7.2 Alternator sizing for induction motor starting 63
  2.8 Application considerations of synchronous ac generators to non-linear electrical loads 67
   2.8.1 Definitions unique to non-linear loading 67
   2.8.2 Gas discharge lighting 68
   2.8.3 Adjustable speed motor drives 68
   2.8.4 Uninterruptible power supplies (UPS systems) 69
   2.8.5 Thyristor/SCR bridge circuits 70
   2.8.6 Why so much concern? 71
   2.8.7 Impact of static power converters upon generator voltage waveform 72
   2.8.8 Effects of voltage harmonics upon automatic voltage regulators 74
   2.8.9 Generator performance data 74
   2.8.10 Total load considerations 75
   2.8.11 Generator sizing guidelines for static power converter loads 75

3 Induction Generators 79
  3.1 Operating principles 79
   3.1.1 Basic operation 79
   3.1.2 Synchronous speed 79
   3.1.3 Speed-torque characteristics 79
   3.1.4 Performance 79
   3.1.5 Construction 79
   3.1.6 Excitation 80
   3.1.7 Frequency and voltage 80
  3.2 Applications 80
   3.2.1 Prime movers 80
   3.2.2 Generator protection 80
   3.2.3 Power factor correction 81
   3.2.4 Starting 81
  3.3 Application considerations 81
  3.4 Advantages and disadvantages of induction generators 81

4 Automatic Voltage Regulation 83
  4.1 Generator characteristics 83
  4.2 Manual excitation control — battery source 84
  4.3 Manual excitation control — self excitation 86
  4.4 Automatic excitation control 87
  4.5 Voltage regulator stability 89
  4.6 Types of sensing circuits 90
  4.7 Power input circuit 92
  4.8 Frequency compensation 93
  4.9 Fault current support 94
  4.10 Motor starting capability 95

Chapter   **Page**

    4.11 Parallel compensation ... 95
    4.12 Summary ... 96

5 Circuit Breakers ... 97
    5.1 Definition ... 97
        5.1.1 Function ... 97
        5.1.2 Standards ... 97
    5.2 Circuit breaker components ... 98
        5.2.1 Molded case (frame) ... 98
        5.2.2 Operating mechanism ... 98
        5.2.3 Arc extinguishers ... 98
        5.2.4 Trip elements ... 99
        5.2.5 Terminal connectors ... 102
    5.3 Accessories and modifications ... 102
        5.3.1 Shunt trip ... 102
        5.3.2 Undervoltage release ... 103
        5.3.3 Auxiliary switches (extra contacts) ... 103
        5.3.4 Alarm switches ... 103
        5.3.5 Motor operators ... 104
        5.3.6 Mechanical interlocks ... 104
    5.4 How to select a circuit breaker ... 105
        5.4.1 Voltage ... 105
        5.4.2 Frequency ... 105
        5.4.3 Interrupting capacity ... 105
        5.4.4 Continuous current rating ... 105
        5.4.5 Unusual operating conditions ... 105
        5.4.6 Testing and maintenance (UL testing) ... 106

6 Automatic Transfer Switches ... 107
    6.1 Supplying emergency power ... 108
    6.2 Transferring power ... 109
    6.3 Manual devices ... 109
    6.4 Automatic devices ... 110
    6.5 Factors to consider when choosing an automatic transfer switch ... 110
        6.5.1 Close against high inrush currents ... 110
        6.5.2 Interrupt current ... 112
        6.5.3 Carry full rated current continuously ... 114
        6.5.4 Withstand fault currents ... 114
        6.5.5 Prevent simultaneous closure of both the normal and emergency sources ... 116
        6.5.6 Protect main contacts ... 117
        6.5.7 Powered from the live source ... 117
        6.5.8 Convenient to maintain ... 117
    6.6 Controlling automatic transfer switches ... 118
        6.6.1 Voltage and frequency sensing ... 119
        6.6.2 Time delays ... 119
        6.6.3 Engine control contacts ... 120
        6.6.4 Test switch ... 120
    6.7 Transferring motor loads with automatic transfer switches ... 120
        6.7.1 Avoiding damage to motors ... 120
        6.7.2 Motor load shedding delayed reconnection ... 123
        6.7.3 Electronic variable frequency drives ... 124
    6.8 Ground-fault protection ... 124

Chapter   **Page**

    6.9   Sizing automatic transfer switches ... 126
          6.9.1   Types of loads ... 127
          6.9.2   Voltage ratings ... 127
          6.9.3   Continuous current rating ... 127
          6.9.4   Overload and fault current withstand ratings ... 128
          6.9.5   Protective device ahead of transfer switch ... 131
          6.9.6   UL long-time fault current, withstand, and closing ratings ... 132
          6.9.7   Special considerations ... 132
          6.9.8   Selection summary ... 132
    6.10 Closed transition transfer switches (CTTS) ... 132
          6.10.1 Applications ... 133
          6.10.2 Utility approval ... 133
    6.11 Maintaining emergency power transfer systems ... 133
          6.11.1 Installation precautions ... 133
          6.11.2 Drawings and manuals ... 134
          6.11.3 Testing ... 134
          6.11.4 Check list ... 135
          6.11.5 Record keeping ... 135
          6.11.6 Inspection ... 135
          6.11.7 Detailed inspection suggestions ... 135
          6.11.8 Preventive maintenance and repair ... 136
          6.11.9 Automatic transfer and bypass-isolation switches for testing and service ... 137

7 **Generator Switchgear** ... 143
    7.1   Voltage classifications ... 143
    7.2   Switchgear types ... 143
          7.2.1   Low voltage metal-enclosed switchgear ... 143
          7.2.2   Metal-clad switchgear ... 144
          7.2.3   Metal-enclosed interrupter switchgear ... 145
    7.3   Applications ... 145
          7.3.1   Generator control panel ... 145
          7.3.2   Automatic transfer systems ... 146
          7.3.3   Paralleling switchgear ... 147
          7.3.4   Paralleling with the utility ... 151

8 **Liquid-Cooled Spark-Ignited Engines** ... 155
    8.1   Cooling system ... 156
    8.2   Lubricating oil system ... 157
    8.3   Fuel system ... 157
    8.4   Air intake system ... 157
    8.5   Ignition system ... 157
    8.6   Exhaust system ... 158
    8.7   Electrical ... 158
    8.8   Governors ... 158
    8.9   Safety ... 158

9 **Air-Cooled Spark-Ignited Engines** ... 159
    9.1   Advantages ... 159
    9.2   Disadvantages ... 159
    9.3   Two versus four stroke cycle ... 160
    9.4   Aluminum versus cast iron ... 161
    9.5   Cooling ... 161
    9.6   Lubricating oil ... 161

Chapter	Page
9.7 Fuel	162
9.8 Ignition	162
9.9 Horizontal versus vertical shaft	162
9.10 Cylinder configurations	162
9.11 Summary	162
10 Liquid-Cooled Diesel Engines	**165**
10.1 Diesel principle	167
10.2 Two- and four-stroke designs	167
10.2.1 Four-stroke cycle	167
10.2.2 Two-stroke cycle	167
10.3 Combustion chamber design	168
10.3.1 Direct injection	168
10.3.2 Pre-combustion chamber	168
10.4 Diesel fuel injection system	168
10.5 Engine performance ratings	169
10.6 Cooling system	171
10.6.1 Set-mounted radiator	171
10.6.2 Remote radiator	171
10.6.3 Cooling air requirement	172
10.6.4 Cooling system design	172
10.6.5 Temperature control	172
10.6.6 Quantity of coolant	173
10.6.7 Heat rejection to coolant	174
10.7 Special application: cogeneration	175
11 Air-Cooled Diesel Engines	**177**
11.1 Advantages	177
11.2 Disadvantages	178
11.3 Aluminum versus cast iron	179
11.4 Cooling	179
11.5 Lubrication	181
11.5.1 Lubricating system	181
11.5.2 Extended run lubricating oil system	181
11.5.3 Lubricating oil	181
11.6 Fuel systems	181
11.7 Air intake system	182
11.8 Exhaust system	182
11.9 Starting systems	184
11.10 Governors	185
11.11 Ratings	185
11.12 Safety	185
11.13 Summary	185
12 Gas Turbines	**187**
12.1 Principle of operation	187
12.1.1 Compressor	187
12.1.2 Combustion	187
12.1.3 Power turbine	187
12.2 Advantages	187
12.2.1 Performance	187
12.2.2 Reliability	188
12.2.3 Size and weight	188

Chapter   **Page**

        12.2.4   Multi-fuel and dual fuel capability ... 188
        12.2.5   Recoverable heat ... 188
        12.2.6   Low emissions and noise ... 188
        12.2.7   Low maintenance ... 189
  12.3   Disadvantages ... 189
        12.3.1   Cost ... 189
        12.3.2   Specific fuel consumption (SFC) ... 189
        12.3.3   Sensitivity to ambient conditions ... 189
        12.3.4   Training of service personnel ... 190
  12.4   Single shaft and split shaft ... 190
  12.5   Aero-derivative versus industrial ... 190
  12.6   Cooling ... 190
  12.7   Engine lubrication ... 190
  12.8   Fuels and combustion ... 190
  12.9   Air inlet and exhaust systems ... 190
  12.10   Generator set related improvements ... 191
  12.11   Summary ... 191

13 Governor Fundamentals ... 193
  13.1   Speed sensing techniques ... 193
        13.1.1   Centrifugal pump ... 193
        13.1.2   Constant displacement pump ... 193
        13.1.3   DC generator ... 193
        13.1.4   Permanent magnet alternator, rectified ... 193
        13.1.5   Permanent magnet alternator, frequency sensitive ... 194
        13.1.6   Centrifugal ballhead ... 194
  13.2   Mechanical governor basics ... 196
  13.3   Hydromechanical governor basics ... 197
  13.4   Servomotors ... 198
  13.5   Hydraulic governor ... 199
  13.6   Speed droop governor ... 200
  13.7   Isochronous governor ... 202
  13.8   Proportional type electric governor systems — basic electric speed governor ... 203
        13.8.1   Loop dynamics ... 205
  13.9   Electrical versus mechanical ... 205
  13.10   Load sharing ... 206
  13.11   Isochronous speed control ... 206
  13.12   Droop speed control ... 206
  13.13   Isochronous-droop parallel operation ... 207
  13.14   Droop-droop parallel operation ... 207
  13.15   Isochronous load sharing ... 207
  13.16   Digital governing ... 211

14 Fuel Systems, Diesel ... 213
  14.1   Transfer tank ... 213
        14.1.1   Sizing ... 213
        14.1.2   Connections ... 213
        14.1.3   Construction ... 213
  14.2   Pumping system ... 214
        14.2.1   Lifting system ... 214
        14.2.2   Pump lift ... 214
        14.2.3   Pump head ... 216

Chapter **Page**

      14.2.4   Pump prime 216
      14.2.5   Pumping notes 216
      14.2.6   Gravity system 218
  14.3 Electrical control 218
      14.3.1   Single system 218
      14.3.2   Duplex system 218
      14.3.3   Testing 218

15 Fuel Systems, Gaseous 219
  15.1 Natural gas engines 219
      15.1.1   Pressure regulators 219
      15.1.2   Carburetors 219
      15.1.3   Piping 220
      15.1.4   Fuel treatment 221
      15.1.5   Gas mixing 221
  15.2 HD-5 propane fuel systems 221
      15.2.1   Filter 221
      15.2.2   LPG vaporizer 221
      15.2.3   Fuel tank 221
      15.2.4   Piping 222
      15.2.5   Carburetors 222
  15.3 Digester (sewage) gas fuel systems 222

16 Cooling Systems, Liquid 223
  16.1 Water cooled systems 223
  16.2 Fan cooled systems 225
      16.2.1   Construction 225
      16.2.2   Remote radiators 225
      16.2.3   Engine mounted 226
  16.3 Radiator system considerations 226
  16.4 Additional systems 228
  16.5 Exhaust gas heat exchangers 229
  16.6 Fuel cooler 229
  16.7 Deaeration 230
      16.7.1   Why dearation? 230
      16.7.2   How to deaerate 230
      16.7.3   How much deaeration? 230
  16.8 Summary 232

17 Engine Exhaust Systems 233
  17.1 Exhaust piping 233
      17.1.1   Empirical formula 233
      17.1.2   Nomograph 234
  17.2 Dry exhaust systems 234
  17.3 Pressure tight pipe 234
  17.4 General 235

18 Exhaust Silencers 239
  18.1 Sound definitions 239
  18.2 Sound pressure 239
  18.3 Addition of levels 239
  18.4 Why distance from the sound source affects sound levels......................................240

Chapter   **Page**

- 18.5 System evaluation ... 241
- 18.6 Back pressure ... 242
- 18.7 Material ... 242

19 Engine Protective Controls ... 245
- 19.1 Lubrication ... 246
  - 19.1.1 Full pressure lubrication ... 246
  - 19.1.2 Oil level ... 246
  - 19.1.3 Lube level maintaining devices ... 246
- 19.2 Cooling ... 246
  - 19.2.1 Liquid cooling ... 246
  - 19.2.2 Air cooling ... 247
- 19.3 Overspeed ... 248
  - 19.3.1 Magnetic sensor ... 248
  - 19.3.2 Mechanical sensor ... 248
  - 19.3.3 Override during startup ... 249
  - 19.3.4 Manifold vacuum ... 249
- 19.4 Miscellaneous ... 249
- 19.5 Alarms ... 249
  - 19.5.1 Function and general ... 249
  - 19.5.2 Visual ... 249
- 19.6 Shutdown ... 249
  - 19.6.1 Method ... 249
  - 19.6.2 Grounding ignition and fuel valve ... 250
  - 19.6.3 Diesel engines ... 250
- 19.7 Summary ... 251

20 Vibration Isolation ... 253
- 20.1 Sources of vibration ... 253
- 20.2 Vibration theory ... 254
- 20.3 Practical considerations ... 260
- 20.4 Types of isolators ... 260
  - 20.4.1 General ... 260
  - 20.4.2 Stiffness characteristics ... 260
  - 20.4.3 Isolator types ... 261
    - 20.4.3.1 Spring isolators ... 262
    - 20.4.3.2 Elastomeric isolators ... 263
    - 20.4.3.3 Wire rope isolators ... 265
    - 20.4.3.4 Pnuematic springs (air springs) ... 265

21 Vibration Analysis for a Sound Generator Set Design ... 273
- 21.1 Types of generator sets ... 273
- 21.2 Generator anatomy ... 273
  - 21.2.1 Mass-elastic shaft system ... 273
  - 21.2.2 Skid and mountings ... 273
  - 21.2.3 Responsibility ... 275
- 21.3 Calculations to avoid lateral vibration failures ... 275
  - 21.3.1 Natural frequency of a six-point mount gen-set ... 275
  - 21.3.2 Turbine generator shaft ... 276
  - 21.3.3 Shaft axial vibration ... 276
- 21.4 Calculations to avoid torsional failures ... 277
  - 21.4.1 Torsional excitation input ... 277

Chapter	Page
21.4.2 The determination of tangential effort	280
21.4.3 A review of the classical Holzer forced vibration torsional analysis	283
21.4.4 Gen-set crankshaft evaluation criteria	288
21.5 Harmonic synthesis method	290
21.5.1 Torsional simulation code	290
21.6 Torsional dampers	291
21.6.1 Elastomeric damper	291
21.6.2 Viscous damper	292
21.6.3 Pendulum absorber	292
21.7 Procedures to avoid shaft bending failure and abnormal vibration	293
21.7.1 Bending failures	293
21.7.2 Shaft misalignment	295
21.7.3 Imbalance	296
21.7.4 Overhang and web deflection	296
21.7.5 Main bearing distress	296
21.7.6 Abnormal vibration	296
21.8 Sound skid design and foundation requirements	297
21.8.1 Skid design	297
21.8.2 Engine flange load and SAE flywheel designs	298
21.8.3 Front engine mounts	299
21.8.4 The selection of mounting system	299
21.8.5 Vibration isolator	301
21.8.6 Foundation considerations and bearing capacity of the soil	301
21.9 Troubleshooting for vibration	302
21.10 Summary	302
22 Enclosure Design	305
22.1 Design considerations	305
22.2 Materials	305
22.3 Operating environment considerations	306
22.4 Noise control	306
22.5 Temperature control	306
22.6 Ancillary equipment	307
22.7 Future considerations	309
22.8 Summary	309
23 Sound Attenuation	311
23.1 Regulations, the numbers game	311
23.2 Basics of sound	312
23.2.1 Sound power level	312
23.2.2 Sound pressure level	312
23.2.3 The decibel	312
23.2.4 Response to sound	313
23.2.5 Octave bands	313
23.2.6 Adding noise levels	313
23.2.7 Ambient noise	313
23.2.8 The mass law	314
23.2.9 Resonance	315
23.3 Sound transmission (propagation)	315
23.4 Sound absorption	315
23.5 Transduction	316
23.6 Refraction	316

Chapter	Page
23.7 Diffraction	316
23.8 Straight-line transmission	316
23.9 Measuring sound working in the real world	317
23.10 Summary	318
24 Batteries and Battery Chargers	**321**
24.1 Battery characteristics	321
24.2 Battery charging	322
24.2.1 Float voltage	322
24.2.2 Equalize voltage	322
24.3 Design of float systems	323
24.4 Battery selection	323
24.5 Types of batteries	323
24.5.1 Starting, lighting and ignition (SLI)	323
24.5.2 Stationary	324
24.5.3 Valve regulated, recombinate	324
24.5.4 Pocket plate nickel-cadmium	324
24.5.5 Fiber plate nickel-cadmium	324
24.6 Battery charger selection	324
24.7 Types of chargers	324
24.7.1 Silicon controlled rectifier (SCR) type	324
24.7.2 Mag-amp type	324
24.7.3 Ferroresonant type	324
24.8 Battery charger requirements	326
24.8.1 Float/equalize output	326
24.8.2 Automatic ac line compensation	326
24.8.3 Battery discharge protection	326
24.8.4 Charger protection	326
24.8.5 Isolation	326
24.8.6 Output indication	326
24.8.7 Interference	326
24.9 Optional charger accessories	326
24.9.1 0–24 hour equalize timer	326
24.9.2 Automatic equalize timer	326
24.9.3 Low voltage alarm	326
24.9.4 High voltage alarm	326
24.9.5 Current failure relay	326
24.9.6 AC input power failure relay	326
24.9.7 Ground fault relay	326
24.9.8 Diagnostic lights	327
24.10 Maintenance	327
24.10.1 Water additions	327
24.10.2 Connections	327
24.10.3 Cleaning	327
24.10.4 Cell balance	327
24.10.5 Capacity	327
24.11 Other considerations	327
24.11.1 Equipment operating voltage	327
24.11.2 Cable resistance	327
24.11.3 Point of charger supply	327
24.11.4 Codes and standards	327
24.11.5 Cold temperatures	327

Chapter	Page

Appendix 329
   A1. SI conversions 329
   A2. Power conversion factors 330
   A3. Pressure conversion chart 330
   A4. BTU energy equivalents 331
   A5. Barrel of oil equivalent energy 331
   A6. Liquid measure equivalent volumes 331
   A7. Approximate weights of various liquids 331
   A8. International standard number prefixes 331
   A9. Propane bulk tank table 332
   A10. Useful equations 332
      A10.1 Brake mean effective pressure (BMEP) 332
      A10.2 Torque 332
      A10.3 Peak of sine wave 332
      A10.4 Rectification formulae 332
   A11. Exhaust back pressure nomographs 332
   A12. Rules of thumb 333
      A12.1 Engine horsepower 333
      A12.2 Motor starting 333
      A12.3 Total motor load 333
      A12.4 Engine derating 334
      A12.5 Generator heat rejection to cooling air 334
      A12.6 Engine heat radiated to cooling air, liquid cooled engines .. 334
      A12.7 Engine energy distribution at rated load 334
      A12.8 Elevator feedback 334
      A12.9 Sound attenuation 334
   A13. Effect of voltage and frequency variation on induction motors ... 334
   A14. Ampere charts 335
      A14.1 kVA 335
      A14.2 kW at 0.80 power factor 335
   A15. Motor ratings 335
   A16. Power factor improvement 335

Acknowledgements 339

Index 341

List of Figures

Figure		Page
1-1	Conventional Current	1
1-2	Electrical Circuit (Closed)	2
1-3	Electrical Circuit (Open)	2
1-4	Direct Current	2
1-5	Alternating Current Wave Shape	2
1-6	Diode	3
1-7	Diode Type	3
1-8	Half Wave Recitification	4
1-9	Full Wave Rectification	4
1-10	Magnetic Flux	5
1-11	Right Hand Rule	6
1-12	Electromagnetism in a Coil	6
1-13	Electromagnet	6
1-14	Magnetic Field Held Stationary	7
1-15	Conductor Held Stationary	7
1-16	Ohm's Law	7
1-17	Ohm's Law Currents	8
1-18	Voltage and Current Waves	8
1-19	Voltage and Current Relationships in a Resistive Circuit	9
1-20	Battery Powered DC Circuit with an Induction Coil	10
1-21	Circuit with an Alternator Supplying an Inductor	11
1-22	Plot of an AC Current Sine Wave	11
1-23	Plot of Voltage in an AC Inductive Circuit	11
1-24	Plot of Voltage Versus Current in a Purely Inductive AC Circuit	11
1-25	DC Circuit with Capacitor and Switch	12
1-26	Plot of Voltage Versus Current in a Purely Capacitive AC Circuit	13
1-27	Relationships of Voltage, Current, and Power in a Resistive AC Circuit	16
1-28	Relationships of Voltage, Current, and Power in an AC Inductive Circuit	17
1-29	Relationships of Voltage, Current, and Power in a Capacitive AC Circuit	18
1-30	Power Right Triangle	19
2-1	Amortisseur Winding with Shorting Rings	21
2-2	Amortisseur Winding with Shorting Laminations	21
2-3	Simple Revolving Armature Alternator	22
2-4	Simple Revolving Field Alternator	23
2-5	Brush Type Alternator with External Belt Driven DC Exciter	24
2-6	Rotor Assembly of Brush Type Alternator with Integral Brush Type Exciter	24
2-7	Typical Externally Excited Brush Type Revolving Field Alternator	25
2-8	Typical Self Excited, Externally Regulated, Brushless Alternator	25
2-9	Basic Revolving Armature Alternator	26
2-10	Generation of an AC Sine Wave by a Two Field Pole Alternator	26
2-11	Generation of an AC Sine Wave by a Four Field Pole Alternator	27
2-12	Three Phase Voltage Sine Waves	28

Figure   Page

Figure	Title	Page
2-13	Rotating Armature Core with Winding	28
2-14	Stationary Armature Core with Winding	28
2-15	Form Wound Coil	29
2-16	Mush Wound Coil	29
2-17	Cross Section of a Typical Stationary Armature Winding	30
2-18	Coils Connected in Series	30
2-19	Parallel Connected Coils	31
2-20	Center-Tapped Coil	31
2-21	Single Circuit Single Phase Armature	31
2-22	Single Circuit Three Phase Armature	32
2-23	Single Phase Two Circuit Armature	32
2-24	Three Phase Two Circuit Armature	32
2-25	NEMA Single Phase Single Circuit Armature	33
2-26	NEMA Single Phase Two Circuit Armature	33
2-27	NEMA Three Phase Single Circuit Armature Lead Wire Numbering System	33
2-28	NEMA Three Phase Two Circuit Armature Lead Wire Numbering System	34
2-29	IEC/BS Three Phase Single Circuit Armature Lead Wire Numbering System	34
2-30	IEC/BS Three Phase Two Circuit Armature Lead Wire Numbering System	35
2-31	Single Phase, Two Load Lead Alternator Armature	35
2-32	Single Phase, Three Load Lead Armature	36
2-33	Parallel and Series Connections	36
2-34	Parallel, 2 Wire 120 Volt Service	36
2-35	Series 2 Wire 240 Volt Service or Series 3 Wire 120/240 Volt Service	37
2-36	Three Wire WYE Connection	37
2-37	Four Wire WYE Connection	38
2-38	Voltages in a Three Phase WYE (Star) Connected Alternator	38
2-39	Four Lead WYE Connected Alternator with Two Phases Center Tapped	39
2-40	Ten Lead WYE Connected Alternator Connections	39
2-41	Delta Connected Three Phase Armature	40
2-42	Voltages in a Four Wire, Three Phase Delta Connected Alternator	40
2-43	Six Lead WYE Connection	41
2-44	Six Lead Delta Connection	41
2-45	Low Voltage Delta (Parallel Connected)	41
2-46	Low Voltage WYE (Parallel Connected)	41
2-47	High Voltage Delta (Series Connected)	42
2-48	High Voltage WYE (Series Connected)	42
2-49	Single Circuit, Six Lead Zig-Zag Connection	43
2-50	Two Circuit, Twelve Lead Zig-Zag Connection	43
2-51	Twelve Wire Double Delta Single Phase Connection	43
2-52	Light Beam Voltage Oscillograph Strip Chart	53
2-53	Typical Speed/Start kVA Curve for NEMA Design B Squirrel Cage Induction Motor	61
2-54	Typical Motor Starting Curve 400 kW Generator	64
2-55	Load Analysis Form	65
2-56	Harmonics of a Sine Wave	68
2-57	Typical Saturation Curve	68
2-58	Silicon Controlled Rectifier	69
2-59	Typical Voltage and Current Waveforms Associated with Saturated Coils	69
2-60	Static UPS System	70
2-61	Hybrid Static-Rotary UPS System	70
2-62	Phase Control of Rectified DC Output with Silicon Controlled Bridge Circuit	71
2-63	DC Output from Three Phase Half Wave Bridge Rectifier Circuit with Full Conduction of Rectifiers	72
2-64	SCR Firing Notches on a Three Phase Voltage Waveform	72

List of Figures

Figure		Page
2-65	Notches on AC Voltage Wave	73
2-66	Distorted Voltage Waveform with SCR Converter Load	73
2-67	Generator Performance Data Worksheet	76
3-1	Typical Speed-Torque Characteristics of Induction Motor and Generator	79
3-2	Typical Performance Curves for a Two-Pole Induction Generator	80
4-1	Typical Generator Saturation Curve	83
4-2	Model of AC Generator	84
4-3	Effect of Increasing Load on Model Generator	84
4-4	Generator Voltage Versus Frequency (Speed)	84
4-5	Example Generator Saturation Curve	85
4-6	Manual Excitation — Battery	85
4-7	Generator Performance Manual Excitation — Battery	85
4-8	Generator Performance Manual Excitation Increase by Steps	85
4-9	Generator Performance Manual Excitation Increase in One Step	86
4-10	Generator Performance Manual Excitation One Step Increase and Decrease	86
4-11	Manually Controlled Self-excited Generator	86
4-12	Self-excited Generator No Excitation Control	87
4-13	Self-excited Generator Manual Control	87
4-14	Automatic Voltage Regulator	88
4-15	Generator Performance with Automatic Voltage Regulator	88
4-16	Generator Hunting	89
4-17	AVR Block Diagram	89
4-18	Stability Control	90
4-19	Voltage Sensing Single Phase Generator	90
4-20	Voltage Sensing Three Phase Generator	91
4-21	Fusing Power and Sensing Circuit	91
4-22	Fuse Failure Detection	91
4-23	Automatic Voltage Regulator Performance, Too Low Input Voltage	92
4-24	Generator Voltages Available	92
4-25	Power Input Fusing	92
4-26	Power Input Isolation	93
4-27	One Step Load Performance, No Recovery	93
4-28	One Step Load with Frequency Compensation	93
4-29	Voltage Performance with Frequency Compensation	94
4-30	Simple Battery Boost	94
4-31	Current Boost	94
4-32	Voltage/Curent Boost	95
4-33	PMG Generator	95
4-34	Parallel Droop Compensation	96
5-1	Molded Case Circuit Breaker	98
5-2	De-Ion Arc Quenchers Extinguishing Arc	98
5-3	Bi-Metal Type Protection	99
5-4	Thermal Action	99
5-5	Typical Deflection Curve for 100-Amp Thermal Element of a Breaker	99
5-6	Magnetic Action	100
5-7	Typical Trip Curve for Fixed Magnetic Action	100
5-8	Thermal Magnetic Breaker with Adjustable Magnetic Trip	100
5-9	Typical Trip Curve for Adjustable Magnetic Action	100
5-10	Thermal Magnetic Action	101
5-11	Thermal Magnetic Curve	101

List of Figures

Figure		Page
5-12	Interchangeable Trip Unit	101
5-13	Electronic Trip Components	102
5-14	Seltronic Test Kit	102
5-15	Typical Shunt Trip Installation	103
5-16	Undervoltage Release	103
5-17	Typical Installation of Auxiliary Switch	103
5-18	Alarm Switch	104
5-19	Sliding Bar Type Interlock	104
6-1	Shopping Mall	107
6-2	Airport	107
6-3	Two Utility Sources	108
6-4	On-site Generator Emergency Source	108
6-5	Emergency System with Transfer Switch	109
6-6	Manual Transfer Switch with Remote Control	110
6-7	Manually Operable Transfer Switch	110
6-8	Emergency System with Automatic Transfer Switch	111
6-9	Arc While Interrupting Circuit	112
6-10	Arc in Transfer Switch	112
6-11	Transfer Switch Arc Gap	112
6-12	Arc Splitters	113
6-13	Arc Interruption	113
6-14	Movable Arm Action	114
6-15	Low X/R Ratio	115
6-16	Medium X/R Ratio	115
6-17	High X/R Ratio	115
6-18	Single-solenoid Operator in Normal Postion	116
6-19	Single-solenoid Operator in Emergency Postion	116
6-20	Arc Runner	117
6-21	Arcing Contacts	117
6-22	Front Connected Transfer Switch	118
6-23	Control Panel	119
6-24	Motor Transfer with Inphase Monitor	120
6-25	Phase Angle Advance Versus Time	120
6-26	Motor Transfer with Motor Disconnect Circuit	122
6-27	Motor Transfer with Timed Center-off Position	122
6-28	Closed Transition	123
6-29	Improper Sensing Due to Multiple Ground Connections	125
6-30	Transfer Switch with Overlapping Neutral Contacts	126
6-31	Overlapping Neutral Contacts	126
6-32	System with Overlapping Neutral Contacts	127
6-33	Main and Arcing Contact Operation	136
6-34	Automatic Transfer and Bypass-Isolation Switch	137
6-35	Schematic of Automatic Transfer and Bypass-Isolation Switch, Normal Position	138
6-36	Schematic of Automatic Transfer and Bypass-Isolation Switch, Emergency Position	139
6-37	Schematic of Automatic Transfer and Bypass-Isolation Switch, Bypass to Normal Position	139
6-38	Schematic of Automatic Transfer and Bypass-Isolation Switch, Bypass to Emergency Position	140
6-39	Removing Automatic Transfer Switch	141
7-1	Metal-Enclosed Switchgear	144
7-2	Metal-Clad Switchgear	145
7-3	Generator Control Panel	146
7-4	Starting Control	146

Figure / Page

Figure		Page
7-5	Two Source Transfer System	146
7-6	Three Source Transfer System	147
7-7	Three Source Priority Load System	147
7-8	Sequential System	148
7-9	Random Access System	148
7-10	Prioritized Loads	149
7-11	Load Control with Automatic Transfer Switches	149
7-12	Reverse Power	150
7-13	Paralleling Conditions	150
7-14	Instrumentation	151
7-15	Peak Demand Reduction System	152
7-16	Parallel with Utility	152
7-17	Cogeneration System	153
8-1	Otto Four Stroke Cycle	155
8-2	Cross Section Sunbeam Farm-Lite Plant	156
8-3	Comparison of SI Engine to CI Engine with Same cu. in. Displacement	157
9-1	Portable Set	160
9-2	Recreational Vehicle Set	160
9-3	Reverse Cooling	161
9-4	Vertical Shaft Set	162
9-5	Two Cylinder Opposed Configuration	163
10-1	Modern Multi-Cylinder Turbocharged, Intercooled, Diesel Engine	166
10-2	Four-Stroke Cycle	167
10-3	Two-Stroke Cycle	167
10-4	Direct Injection System	168
10-5	Pre-combustion Chamber in Cylinder Head	168
10-6	Engine Performance Curve	170
10-7	Arrangement Pusher-Type Fan and Air Flow Through Radiator	172
10-8	Water Pump Flow Rate	173
10-9	Hypothetical Figures for Engine Heat Rejection	174
10-10	Jacket Water Heat Rejection at Varied Load and Speed	175
10-11	Cogeneration Flow Diagram of a Typical Prepackaged Module	175
11-1	Two-Cylinder Air-cooled Diesel	177
11-2	20 kW Generator Set	178
11-3	Engine with Thin Wall Iron Crankcase	179
11-4	Axial Belt Driven Fan and Flywheel Fan Cooled Engines	180
11-5	Cooling Data	180
11-6	Lubricating Oil System	182
11-7	Self Regulating Oil System	182
11-8	Long Run Oil Systems	184
11-9	Fuel System	184
12-1	Power Section Air Flow Schematic	188
12-2	Double Shaft Turbine	189
13-1	Air Vane Governor	193
13-2	Centrifugal Pump Govenor	194
13-3	Orifice Governor	194
13-4	Permanent Magnet DC Generator	194
13-5	Permanent Magnet AC Generator	194
13-6	Frequency Sensing Governor	195

Figure   **Page**

Figure	Title	Page
13-7	Centrifugal Ballhead	195
13-8	Torque Balance	195
13-9	Modern Ballhead	196
13-10	Thrust Point	196
13-11	Control by Pilot Valve	197
13-12	Spring Comparison	198
13-13	Double Acting Piston	198
13-14	Single Acting Piston	199
13-15	Differential Piston	199
13-16	Servo System	200
13-17	Simple Droop Governor	201
13-18	Droop Curve	201
13-19	Floating Lever	201
13-20	Moveable Bushing	202
13-21	Common Hydraulic Governor	202
13-22	Use of Buffer Piston	202
13-23	Droop System	203
13-24	Single Unit Speed Control	204
13-25	Closed Control Loop	204
13-26	Summing Point	205
13-27	Engine Performance Curve	206
13-28	Isochronous Speed Control	206
13-29	Droop Speed Control	207
13-30	Droop Parallel Operation with an Infinite Bus	207
13-31	Paralleled Generator Sets (one in Isochronous and one in Droop)	208
13-32	Paralleling Operation	208
13-33	Paralleling Sequence	209
13-34	Isochronous Load Sharing Governor System	210
13-35	Load Sharing System Schematic	211
14-1	Day Tank System — Above Main Tank	214
14-2	Day Tank System — Below Main Tank	217
15-1	Gas Fuel System Installation	220
15-2	Pressure Regulator	220
16-1	Freezing Points of Aqueous Ethylene Glycol Antifreeze Solutions	223
16-2	Heat Exchanger	224
16-3	Heat Exchanger Installation	224
16-4	Remote Radiator	225
16-5	Engine-mounted Radiator	226
16-6	Water Boiling Temperature At Altitudes Above Sea Level	227
16-7	Remote Radiator Installation	228
16-8	Hot Well System	229
16-9	Diagram of Deaeration System	231
16-10	Diagram of Deaeration System with Surge Tank	231
17-1	Nomograph for Exhaust Back Pressure	234
17-2	Dry Exhaust System	235
17-3	Factory-Built Pressure-Tight Exhaust System	236
18-1	Incremental Sound Level, Two Machines	240
18-2	Typical Attenuation Curve	241
18-3	Typical Attenuation Curve	241

Figure | **Page**

18-4	Silencer Back Pressure	242
19-1	Engine Temperatures	247
19-2	Oil Temperature Versus Cylinder Head Temperature	248
20-1	Schematic of Simple Mounting System	254
20-2	Transmissibility Versus Natural Frequency and Critical Damping Ratio	256
20-3	Static Deflection Versus Natural Frequency	257
20-4	Vibration Transmissibility Chart	258
20-5	Transmissibility Versus Frequency and Critical Damping Ratio, Showing Effect of Rocking Modes	259
20-6	Static Load Deflection Curves for Several Types of Vibration Isolators	261
20-7	Helical Coil Spring Isolators	263
20-8	Seismic Helical Coil Spring Isolator	264
20-9	Elastomeric Isolators	266
20-10	Elastomeric Isolators	267
20-11	Elastomeric Isolators	268
20-12	Fully Bonded Buckling Mount	269
20-13	Wire Rope Isolators	270
20-14	Pneumatic Springs	271
21-1	Different Types of Generator Sets	274
21-2	A Gen-set Exposed	274
21-3	Shaft System	274
21-4	Cylinder Gas Pressure	274
21-5	Bearing System	274
21-6	Crankline Alignment	275
21-7	Force and Torque Reaction	275
21-8	Simple Model for Lateral Natural Frequency	275
21-9	Bearing Flexibility	275
21-10	Flexible Disc Coupled Gen-set	276
21-11	Long Turbine Generator Shaft	276
21-12	Axial Vibration	276
21-13	Cracks from Radial Hole	277
21-14	Four-Cycle Engine Cylinder Gas Pressure	278
21-15	Single Cylinder Gas Torque on the Shaft	278
21-16	Rated Load T_{gas} and Its Harmonics	279
21-17	Gas Torque T_{gas} at Part Load	279
21-18	Lloyd Tangential Effort Table (Ref. 3)	279
21-19	Output Torque	280
21-20	Four-Cycle In-line Engine Vector Sum	281
21-21	Two-Cycle Three-Cylinder In-line Engine Vector Sum	282
21-22	Non-Elastic Shift	283
21-23	Elastic Shaft	283
21-24	Generator Crankline and Mass Elasticity Determination	284
21-25	Normal Elastic Curve	285
21-26	Amplitude for Mass m and Order n	286
21-27	Amplitude at Resonance	286
21-28	Dynamic Multiplier	287
21-29	Actual Normal Elastic Curve	287
21-30	Mass-Elasticity System	288
21-31	Resonance Curve	288
21-32	No-Break Gen-set	289
21-33	Comparison of Data	290

List of Figures

Figure		Page
21-34	Model of Eight-Cylinder System	290
21-35	Mass-Elastic Simulation, Eight-Cylinder System	290
21-36	Torque Excitation, Mass i	290
21-37	Rubber Damper/Pulley	292
21-38	Pure Viscous Type Torsional Vibration Damper	292
21-39	NuLastic Rubber-viscous Torsional Vibration Damper	292
21-40	Damping with Viscous Damper	293
21-41	Pendulum Absorber	293
21-42	10 Cylinder 8.125 × 10 OP Diesel	293
21-43	Crank Shaft Bending	294
21-44	Filleted Shaft	295
21-45	Misaligned Generator Shaft	295
21-46	Crank Web Deflection	296
21-47	Generator Sets on Skids	297
21-48	Method of Limiting Static Deflection During Lifting	297
21-49	Couples Acting on the Skid	298
21-50	Overhung Load	298
21-51	Four-Point Mounting and Minimal Flange Load	299
21-52	Four-Point Mounting and Increased Flange Load	299
21-53	Simple Base with Cradle Mount	299
21-54	CEBRA Generator with Front Foot Mount to Support the Engine Weight	299
21-55	Rubber Sandwich Front Mount	300
21-56	Four- and Six-Point Mounting	300
21-57	Eight- and Twelve-Point Mounting	300
21-58	Gas Turbine Driver and Two-Bearing Generator	301
21-59	Shock Mount Between the Skid and the Foundation	301
21-60	Natural Frequency of Foundations of Two Types of Soil	302
21-61	Generator-Set Linear Vibration Test Log	303
22-1	Enclosure with Intake Hood	307
22-2	Intake Louvers	308
23-1	Calculating dBA From an Octave-Band Analysis	314
23-2	Calculating dBA From an Octave-Band Analysis	314
24-1	Battery Discharge Rates	322
24-2	Battery Capacity Versus Temperature	322
24-3	Float System	323
24-4	SCR Charger	325
24-5	Controlled Mag Amp Charger	325
24-6	Ferroresonant Charger	326
A-1	Back Pressure Nomograph	333
A-2	Back Pressure Nomograph	333
A-3	General Effect of Voltage Variation on Induction Motor Characteristics	334
A-4	General Effect of Frequency Variation on Induction Motor Characteristics	334

List of Tables

Table		Page
1-1	Calculations for Figure 1-27	16
1-2	Calculations for Figure 1-28	17
2-1	Voltages of Three Phase Wye Connected Alternators	38
2-2	Synchronous Speeds for 2, 4, 6, and 8 Pole Alternators	44
2-3	Table of Insulation Temperature Limits	49
2-4	Steady State Load List	57
2-5	NEMA Motor Starting Code Letters	58
2-6	Table of Methods and Types of Motor Starting	62
2-7	Table of Three Phase Design B Motor Characteristics	66
2-8	Per Unit Practical Harmonic Currents for Six and Twelve Pulse Converters	74
4-1	Generator Specifications	84
6-1	Inrush Current, Tungsten Lamps	111
6-2	Typical Transfer Switch Available Fault Current Ratings	115
6-3	Generator Output Current Ratings	129
6-4	UL 1008 Minimum Withstand Current Ratings	129
6-5	UL 1008 Preferred Withstand Current Ratings	129
6-6	UL 1008 Test Power Factor Ranges	130
6-7	Power Factor Versus X/R Ratio	130
6-8	Circuit Breakers Versus Test Power Factor	131
6-9	Fuses Versus Test Power Factor	131
6-10	Transfer Switch Rating Versus Available Symmetrical Short Circuit Amperes	131
14-1	Frictional Head Loss for 100 Feet of Standard Weight Pipe at 60°F at Sea Level — Diesel Fuel	215
14-2	Frictional Loss in Pipe Fittings in Terms of Equivalent Feet of Straight Pipe	215
14-3	Lifting Capacities at Various Elevations	215
14-4	Pump Discharge Pressure (psi)	216
15-1	Fuel Specifications	219
18-1	Converting Octave Level to an Overall dBA Level	239
18-2	Octaves	240
18-3	Calculated Exhaust Flow Rate for a Silencer with Pressure Drop of 1.0 in. Hg or Less	242
21-1	Design and Performance Responsibilities	275
21-2	Calculated Lateral Vibration Frequencies	276
21-3	Significant Orders of Gas Pressure Torque	282
21-4	Multiplier BAM, vee angle 45 degrees	282
21-5	First Mode Amplitudes and Natural Frequencies	285
21-6	Torsional Stress Calculations Versus Test	291
21-7	Soil Load Characteristics	301
23-1	Example of Logarithmic Addition of L_wA Octave Band Value Using Table 1 Factors	314
24-1	Voltage Ranges Per Cell at 77°F (25°C)	323

List of Tables

Table		Page
A-1	Conversions to SI	329
A-2	Approximate Power Conversion Factors	330
A-3	Pressure Conversion Chart	330
A-4	Energy Equivalents	331
A-5	Barrel of Oil Equivalent Energy	331
A-6	Liquid Measure Equivalent Volumes	331
A-7	Weights of Liquids	331
A-8	International Standard (SI) Numerical Prefixes	331
A-9	Propane Bulk Tank Table (Tank kept at least 1/2 full)	332
A-10	Multipliers to Convert Three Phase kVA to Amperes and Amperes to Three Phase kVA	335
A-11	Generator Set Ampere Ratings — Three Phase 0.80 P.F.	336
A-12	Typical Three Phase Motor Characteristics for NEMA Design B, C and D Motors	337
A-13	kW Multipliers for Determining Capacitor Kilovars	338

Electrical Fundamentals

James Wright

CHAPTER 1

INTRODUCTION

The intent of this chapter is to discuss the various electrical concepts necessary to the understanding of on-site electric generation equipment. We will not present the theory for an in-depth understanding of the engineering design of these machines.

1.1 DEFINITIONS

Before launching into the subject of electricity and electric current it would be useful to define some of the basic terms we will be using to describe or explain the various aspects of the subject. Sources for most of these definitions are the Electrical Generating Systems Association Glossary of Standard Industry Terminology and Definitions-Electrical, EGSA 101E-1984, and IEEE Standard Dictionary of Electrical and Electronics Terms, ANSI/IEEE Std 100-1992. Additional terms and expressions will be defined as necessary throughout the rest of this and subsequent chapters.

Electron. An electron is a constituent of an atom, with one or more electrons orbiting about the nucleus of the atom, and is the most elemental charge of electricity. Electrons are essentially spinning masses of electromagnetic energy.

Electric current. Electric current has been defined as the movement of free electrons within an electrical conductor caused by a difference of electric pressure between the ends of the conductor. The electron has a negative charge, and will be attracted to the positive end of the conductor that is the point of higher potential. **Therefore, the flow of electrons is from negative to positive, N to P.** However, electric current, sometimes termed conventional current, is considered to flow from positive to negative, P to N, as shown in Figure 1-1. The use of conventional current stems from the work of Benjamin Franklin who theorized that electric current was the flow of positively charged particles from a point of high potential to a point of lower potential, that is, from positive to negative (P to N). This theory became the convention in all of the texts on physics, electricity and electric equipment and survives today in the form of the accepted direction of current flow.

All standard electrical symbols, diagrams, and descriptions of electric current phenomena used in this, and almost all other texts, deal with conventional current. The direction of electric current flow is from positive to negative.

Ampere. The ampere is the measure of electric current. A quantity of 6.24×10^{18} electrons flowing past a point per second equals one ampere.

Electromotive force (EMF). Electromotive force is the force that causes current to flow within a conductor, and can be thought of as electric pressure. This force is also referred to as electric potential or just potential.

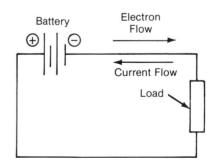

Figure 1-1. Conventional Current

Volt. The volt is the unit of measure of electromotive force or potential. The difference in potential between two charges, or two points in a circuit, is called voltage.

Ohm. The ohm is the unit used to measure the opposition or resistance to flow of electric current in a circuit. A way of thinking could be to compare electrical resistance to mechanical friction. Like friction, it tends to hinder the flow of current just as friction tends to hinder the flow of a fluid in a pipe. As with friction, resistance to the flow of current through a conductor creates heat within the conductor.

Conductor. A conductor is any material, such as copper or aluminum, which has a very low resistance to the flow of electric current. The term conductor is also used to refer to a wire, cable, or bus bar used to carry electric current.

Insulator, insulation. An insulator is any material or device that, under normal conditions, will not allow the flow or passage of electric current. As an example, insulation is used to electrically isolate one conductor from another.

Circuit. An electric circuit is a conductor or a series of conductors through which an electric current is intended to flow. A circuit may have one or more electric components within it which are called circuit elements, or circuit components. Electric current will only flow in a closed or continuous circuit. There must be a continuous, non-interrupted path from the source through the entire circuit and back to the source. Any open or interruption in this continuous path will cause the flow of electrical current to stop. See Figures 1-2 and 1-3.

1.2 DIRECT CURRENT (DC)

Direct current is an electric current that flows in one direction only. A pure direct current is one that will continuously flow at a constant rate. If we imagine a device that would allow us to observe the actual flow of the individual electrons in a circuit with a battery as a power source, we would see that the electrons were passing our point of observation in a single direction, and at a steady rate of flow. See Figure 1-4.

1.3 ALTERNATING CURRENT (AC)

Alternating current is an electric current which flows first in one direction for a given period of time, and then in the reverse direction for an equal period of time, constantly changing in magnitude. See Figure 1-5. Note that ac current builds from zero to maximum in the positive or forward direction, decays to zero, builds to maximum in the negative or reverse direction, and again decays to zero.

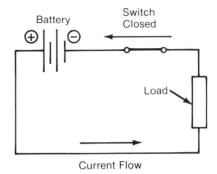

Figure 1-2. Electrical Circuit (Closed)

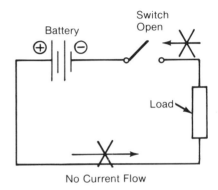

Figure 1-3. Electrical Circuit (Open)

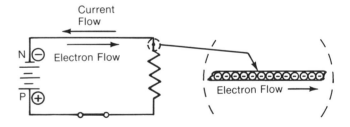

Figure 1-4. Direct Current

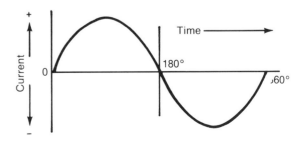

Figure 1-5. Alternating Current Wave Shape

From Figure 1-5 it can be seen that the shape of alternating current (ac) flow is that of a wave. The common and most important alternating current wave shape is the sine wave, named for the graphic curve of the trigonometric function it follows. The complete pattern of zero to maximum positive, back to zero, down to maximum negative, and back to zero is termed a cycle. The length of time necessary to complete one cycle is termed the period of the wave. The number of cycles completed in one second is termed the frequency of the alternating current. The unit of frequency measurement is hertz. For example, if 60 complete cycles are being produced in one second, the frequency of the alternating current being produced would be 60 hertz (60 Hz).

1.4 RECTIFICATION

Alternating current has become the most common form of electricity available throughout the world. However, there are many applications where only ac electric power is present, but the systems or devices being used either require only dc power, or require both forms of power in varying amounts. What is needed to service these applications is a means to convert ac power to dc. Commonly there are two methods to accomplish this task. First, for applications with a rather large dc power requirement, an ac electric motor can be used to drive a dc generator, whose output is then used to provide the dc power the electrical load requires. The second, and by far the most frequently used method of converting alternating current to direct current, is to *rectify* the alternating current.

Rectification can be defined as the process of converting alternating current into direct current by means of a rectifier. A rectifier will permit current flow in one direction, and block the flow of current that attempts to flow in the opposite direction. An analogy can be drawn that a rectifier is an electrical check valve. As a check valve in a fluid system prohibits reverse flow of a fluid, a rectifier readily allows the flow of electricity in one direction, but blocks its flow in the opposite direction.

1.4.1 Diodes. Modern rectifiers are most often solid state devices called diodes, and are usually constructed of silicon. Figure 1-6 illustrates the electrical symbol for a diode. Note that the arrow head points in the direction of current flow, and that the bar at the point of the arrow head indicates that current will be blocked from entering from the opposite direction.

Physically, diodes (Figure 1-7) may be small devices with a lead wire on either end, or larger power rectifiers of the press-in or stud type having a lead wire or connection terminal on only one end. With these larger devices, the diode body or mounting stud is used as the second terminal.

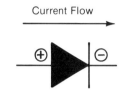

Figure 1-6. Diode

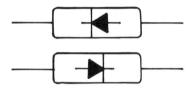

Pigtail Diode

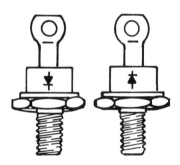

Stud Type Diode

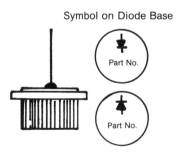

Press In Diode

Figure 1-7. Diode Type

Electrical Fundamentals

Since the small pigtail diodes have identical lead wires on both ends, they offer the convenience of being a universal device: the diode may be inserted into the circuit to block flow in either direction, merely by installing the device in the electric circuit with the diode arrow symbol pointing in the desired direction of current flow. On the other hand, due to the limitations of their physical construction, press-in or stud type diodes may be installed in only one way: before installing, select the proper polarity, either forward (normal) or reverse (R), as indicated by the diode symbol on the case. Further, these larger power rectifiers are usually mounted on an aluminum, copper or brass heat sink to aid in dissipating heat. In many designs the heat sink, being constructed of conductive material, also serves as the second rectifier terminal.

a. *Half wave.* Rectification of ac power to dc power may take one of two forms, depending upon whether one half or both halves of the ac sine wave is used. Figure 1-8A illustrates in schematic form a single phase half wave rectification circuit with diode CR1, load R1, and an ac generator as the power source. When current is flowing from generator terminal A, diode CR1 permits the current to freely flow through it to the load, and back to the generator via terminal B. However, when the direction of current flow is reversed during the bottom half of the sine wave as shown in Figure 1-8B, with current attempting to flow from generator terminal B to load R1, diode CR1 blocks the flow of current and the negative or bottom half of the ac current wave is clipped. This action is illustrated by the dotted line in Figure 1-8C. The result of this is pulsating dc. The current *does flow* in only one direction, but in the form of regular pulses rather than in pure dc. This form of ac to dc rectification is termed **half wave rectification** since only half of the ac sine wave is used.

b. *Full wave.* Figure 1-9A illustrates in schematic form a single phase, full wave rectification circuit using four diodes: CR1, CR2, CR3 and CR4, load R1 and an ac generator as a power source. This rectifier arrangement is called a **bridge**. Current flows from generator terminal A to the rectifier bridge circuit input point A1 and must flow through diode CR2 and out to load R1 through point C, since diodes CR1 and CR3 prevent current flow in any other direction. Flow is then through load R2, back into the bridge through bridge terminal D through diode CR4, and returns to the generator through bridge terminal B1. Then, when current flow is reversed during the bottom half of the sine wave (Figure 1-9B), current flow is from generator terminal B, into the bridge at B1, through diode CR1, and out to the load through point C as before. The point to note here is that while current circulates in opposite directions through the rectifier bridge during each half of the sine wave, current flow to and through the load is always in the same direction, i.e., direct current. This method of rectification is called **full wave rectification** as the bottom half of the current wave has been flip-flopped so that it is in the same direction as the top half of the wave, instead of being clipped as in half wave rectification. See Figure 1-9C. However, since current flow drops to zero at the end of each half cycle, the result is still pulsating dc, and not the smooth flow of pure dc.

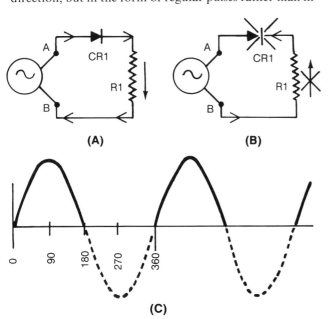

Figure 1-8. Half Wave Rectification

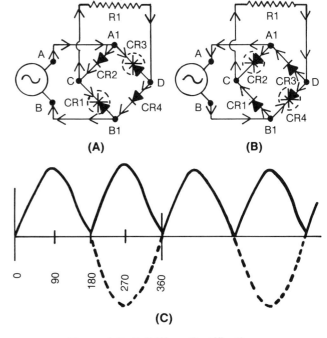

Figure 1-9. Full Wave Rectification

1.5 MAGNETISM

A magnet is said to establish a magnetic field or field about itself which is represented pictorially by directed lines termed **magnetic lines of force**, or **magnetic flux**. If a magnet has been freely suspended, it will align itself in a north-south direction. The end or pole of the magnet that points in the northern direction is designated the north pole (N) of the magnet, and the other pole pointing in the southern direction is then designated the south pole (S). The surface areas on each end of the magnet are termed the **pole faces** of the magnet. The lines of force or magnetic flux are considered to leave the magnet's north (N) pole, travel externally, and re-enter the magnet at its south (S) pole (see Figure 1-10). The strength of a magnet is dictated by the number of lines of force it contains per unit of area of its pole face. A good permanent magnet may have as many as 40,000 or more lines of force per square inch of pole face area. The number of lines of force per unit area of the pole face is referred to as the magnet's **flux density**. Therefore, the greater the flux density of a magnet, the stronger the magnet.

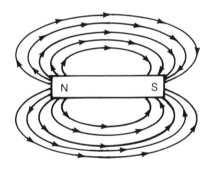

Figure 1-10. Magnetic Flux

1.5.1 Permanent magnets. Permanent magnets were originally manufactured of hardened steel which has the property of becoming magnetized when exposed to a strong magnetic field, and retaining this magnetism after the magnetic field has been removed. Today, many other alloys and materials have been developed which display the properties of a permanent magnet. Once a permanent magnet has been magnetized, its flux density becomes a fixed value and cannot readily be changed. A permanent magnet may be weakened by exposing it to abuse such as heat, vibration, being struck sharply, or exposure to another magnetic field. After such abuse it may be re-magnetized, and it will assume the same strength of field it had originally. The advantage of a permanent magnet is that once it has been magnetized, no outside force or source of energy is needed to maintain its magnetism. The disadvantage of a permanent magnet is that its magnetic strength or flux density cannot readily be controlled or altered. As an example, a permanent magnet would be of little use to a salvage yard for moving scrap iron, for while it will pick up the material, it cannot be turned off to release the material.

1.5.2 Electromagnetism and electromagnets. In 1819 Hans Oersted, a Danish physicist, discovered that when an electric current passes through a conductor, a magnetic field surrounds the conductor. The direction of the magnetic lines of flux which form concentric circles around the conductor depends upon the direction of current flow in the conductor. The actual direction of these lines of flux can be determined by the **right hand rule**, which states that if a current-carrying conductor is grasped with the right hand with the thumb pointing in the direction of current flow, the fingers will circle the conductor in the direction of the lines of flux, as shown in Figure 1-11.

The strength of the magnetic field shown in Figure 1-11 may be increased or decreased by raising or lowering the amount of current flowing through the conductor. The field formed around a single conductor is relatively weak; if, however, the conductor is formed or wound into a coil, the field around each turn in the coil interacts with the other turns. The net result is a large number of long parallel lines of flux running down the axis of the coil, creating a large field around the entire coil. See Figure 1-12.

The net effect of this is that the coil has become a magnet, and since the field has been created by electromagnetism, the coil may be referred to as an electromagnet. Further, if a piece of magnetic material such as soft iron is used as a core, with the coil wound around it as in Figure 1-13, the magnetic properties of the coil are greatly increased. This increase in magnetic strength is due to greater **permeability**, the ease with which flux lines pass through a material of soft iron as opposed to air.

The ideal material to be used to fabricate the core of an electromagnet is one which, when current flow in the coil stops, no magnetic flux will remain in the core. Any magnetism retained in the core material after current flow has ceased is called **residual magnetism**. Later on in this text we will discuss applications where the core material is specifically selected to assure a given degree of residual magnetism. For the vast majority of applications a high degree of residual magnetism in the core is undesirable, thus the cores of most electromagnets have little **retentivity**, ability to retain magnetism.

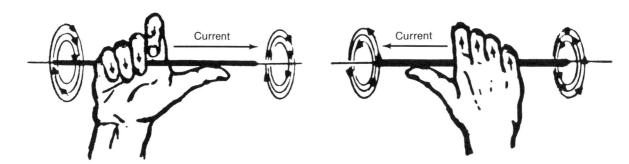

Figure 1-11. Right Hand Rule

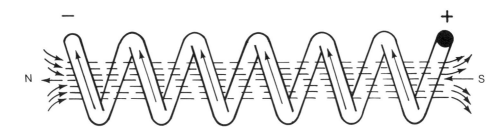

Figure 1-12. Electromagnetism in a Coil

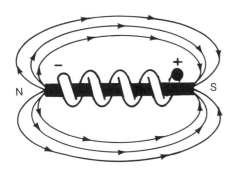

Figure 1-13. Electromagnet

1.6 ELECTROMAGNETIC INDUCTION

Michael Faraday has been credited with what, in many circles, is considered to be one of the most important scientific discoveries of all time. In 1831, Faraday discovered that when a conductor or coil was moved through a magnetic field, a voltage was produced across the conductor or coil. This phenomenon is called **electromagnetic induction**. The voltage produced in the conductor or coil is said to have been induced. All electric generator, motor, and transformer action is based upon this principal. Faraday further discovered that the magnitude of the voltage induced was dependent upon the following factors:

a. *The velocity and number of lines of flux cut by the conductor.* If the flux density of the field is held constant, any movement of the conductor relative to the field will change the value of the voltage induced in the conductor in direct proportion to velocity.

b. *The change of number of lines of flux affecting the conductor.* Changing the flux density of the field will change the value of the voltage induced in the conductor in direct proportion to the change in flux density per unit time.

c. In either a. or b., the voltage induced is in direct proportion to the total number of turns.

It is important to note at this point that we have been discussing the relative velocity of the conductor coil to the field. The significant and required action is the cutting of lines of flux. The magnetic field may be held stationary and the conductor moved through the field, as illustrated in Figure 1-14, or the conductor may be held stationary and the field moved, as shown in Figure 1-15.

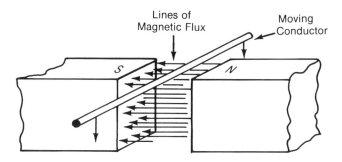

Figure 1-14. Magnetic Field Held Stationary

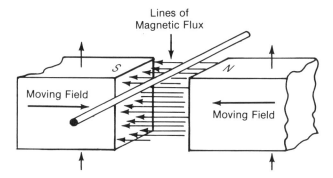

Figure 1-15. Conductor Held Stationary

1.7 ELEMENTS OF AN ALTERNATING CURRENT CIRCUIT

In this section we will discuss the three basic elements, or types of load that comprise an alternating current electrical circuit. These three elements are:

- resistance,
- inductance, and
- capacitance.

Obviously there is one additional component which must be present in order to have an electric circuit, and that is the conductors or load leads (cables) that carry the current from the source to the circuit. In most cases, when considering one or more of these three basic circuit elements in a given circuit, the effects of the load leads are considered to be so small as to be negligible, but under certain circumstances of load current, voltage, and frequency, the conductors themselves may well exhibit characteristics of one, two, or all three of these elements. The effects of this will be brought out in subsequent sections dealing with each of these specific circuit elements.

1.7.1 Resistance

a. *Conductors and resistors.* In Section 1.1 we discussed the fact that when an electric current is caused to flow through a material or substance, there is a resistance to this current flow. This resistance to current flow may be thought of as **electrical friction**. Resistance, like friction, generates heat when current flows through the substance, which may be either an advantage or disadvantage, depending upon the specific use of the electricity.

A conductor is any material, such as copper, which has a very low value of resistance to current flow, and when properly sized to the load, will exhibit a minimum of heating. A resistor, on the other hand, is any material that will conduct current, but which has a significant value of resistance to current flow. Thus, a characteristic of a resistor, is that when a current is caused to flow through it, a considerable amount of heat is generated. In the process, electric energy is being converted into heat energy. This conversion into heat energy is an advantage in those devices wherein we wish to create heat, such as in water heaters, ovens, space heaters and lights. In the case of the incandescent lamp, heat is used to raise the temperature of the filament of the lamp to the point at which it glows, or emits a visible light.

b. *Ohm's law.* The unit of measurement of resistance to current flow in a circuit is the **ohm**, named after the man who formulated one of the most important of all basic ideas of electricity, that of the relationship between current flow, voltage, and resistance to current flow in an electrical circuit. Ohm's Law, simply stated, is that voltage in a circuit is equal to the current flow times the resistance in the circuit. Mathematically this is stated:

$$E = IR \qquad \text{(Equation 1-1)}$$

Where:

E = EMF in volts
I = Current in amperes
R = Resistance in ohms

For those of our readers who may be a bit rusty in algebra, this equation is often written as illustrated in Figure 1-16.

Figure 1-16. Ohm's Law

To solve for any given element of the equation, place your finger on the element, and perform the arithmetic indicated with the remaining two elements. For example, to solve for voltage (E), placing a finger on E indicates that we should multiply I by R.

Ohm's law may be illustrated by considering a circuit with a 10 ohm resistor in a dc circuit powered by a battery. See Figure 1-17.

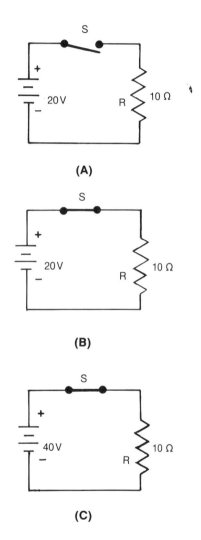

Figure 1-17. Ohm's Law Currents

If the voltage across the resistor R is zero, as in Figure 1-17A with switch open, the current through resistor R is zero. When switch S is closed and the battery is producing 20 volts, as in Figure 1-17B, the current through resistor R is calculated by:

$$I = \frac{E}{R} = \frac{20}{10} = 2 \text{ amperes}$$

If we increase the voltage in the circuit to 40 volts, as in Figure 1-17C, the current will be:

$$I = \frac{E}{R} = \frac{40}{10} = 4 \text{ amperes}$$

We need not be concerned with the length of time involved in these voltage changes from zero to 20 volts, and then to 40 volts, but rather with the fact that when the voltage was changed the amount of current flow changed in direct proportion to the amount of voltage change. Voltage is equal to the product of current times the resistance, or in this case, current is equal to the voltage divided by the resistance. If we supply the voltage to the circuit in the form of a sine wave, as in alternating current, with the period of one cycle equal to one second, we can plot the voltage and current sine waves for the above examples as shown in Figures 1-18A and 1-18B. Figure 1-18A represents the point when the voltage is zero as A, the point when voltage is 20 as B, and 40 as point C. In Figure 1-18B point A will represent zero current flow, B, 2 amperes, and C, 4 amperes. We could,

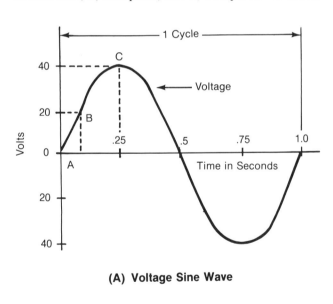

(A) Voltage Sine Wave

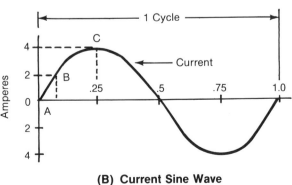

(B) Current Sine Wave

Figure 1-18. Voltage and Current Waves

of course, make similar calculations to determine current value against voltage for as many points as we want, and plot these points to illustrate from instant to instant. For purposes of clarity we will only plot these three points.

Since we have plotted both the voltage and current sine waves with a period of one second, we can superimpose one plot over the other on a common plot and achieve a clearer representation, as in Figure 1-19.

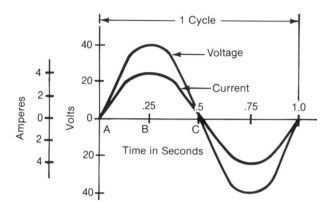

Figure 1-19. Voltage and Current Relationships in a Resistive Circuit

From this common plot, several aspects of an ac electrical circuit containing only resistance can be seen:

1. The zero (0) values of both waves occur at the same instant. The peak points (both positive and negative) of both waves occur at the same instant.

2. Under this condition the waves are said to be in phase with each other.

3. The effects of resistance in an ac circuit are independent of the frequency of the alternating current, and the only hindrance to current flow is the resistance of the circuit.

c. *Line voltage drop.* In the beginning, we pointed out that when considering any specific circuit, the effects of resistance of the load leads are usually ignored as having too low a value for any serious consequences. However, load leads (conductors) do have a specific value of resistance which is usually given in ohms per unit length for the size and material of the conductor. Thus, the resistance of the conductor can be calculated by multiplying the length of the conductor in feet by the resistance per unit length. In actual installations where the alternator used to power a load may be remote from the load, the conductor cables connecting the alternator to the load may extend for a considerable distance. In some cases, such as a rather large construction site, this may be as much as a mile or more. The resistance of these long leads cannot be neglected. The reduction in voltage between the voltage measured at the alternator's terminals, and the terminals of the load is termed **line voltage drop**, or simply **line drop**, and could have a very real effect upon the performance of the various electrical devices comprising the load. Further, since Ohm's law (E = IR) shows us that the actual drop in voltage due to the resistance of the conductors is directly proportional to current flow, the amount of voltage drop will be greatest at full load, and least at no load.

d. *Line voltage drop compensation.* In installations having rather long conductor runs such as the example of a construction site given above, some form of compensation may be required to ensure proper performance of the devices comprising the load. This compensation, when the power source is an externally regulated alternator with an automatic voltage regulator, can be accomplished in two ways:

1. In situations where all, or most of the load will be on line during periods of operation, a simple form of compensation is to adjust the terminal voltage of the alternator upward so that the voltage at the load terminals at full load is at the desired level or value. We will discuss alternator sizing problems this type action may present in Chapter 2.

2. For installations where there is the possibility that a very minor amount of load may be on line for a period of time, compensations as in paragraph 1 above could cause the devices comprising this light load to have too high a voltage with possible injurious effects. Here a device such as a line drop compensator could be used in conjunction with the voltage regulator. This device consists of a controller and a current transformer. The current transformer sends a signal which is proportional to load current to the controller, which then sends a signal to the voltage regulator via the voltage adjust circuit that will raise the alternator's terminal voltage automatically in proportion to current flow, in order to maintain voltage at the load terminals at a constant value.

e. *Resistance as a current limiter.* In our study of Ohm's law we saw that current (I) in a circuit is equal to the voltage (E) divided by the resistance (R). Therefore, current flow in the circuit is inversely proportional to resistance. Thus by increasing the value of any resistance in a circuit, we will reduce the amount of current that can flow in the circuit with a given fixed value of voltage supplied to the circuit. When we transpose our Ohm's law formula to solve for resistance (R), it becomes R = E divided by I. Knowing the circuit voltage, and the maximum current we wish to flow in the circuit, using this formula we can calculate the exact value of resistance to insert into a circuit to limit current to the desired level.

Manufacturers of automatic voltage regulators used with brushless exciters take advantage of the current-limiting ability of resistance to protect their product from overloading, by specifying a minimum value of resistance of the exciter field windings to be used with their regulator. While the dc output voltage of the voltage regulator will vary with loading on the alternator, the regulator does have a maximum dc voltage it can produce when turned full on. Knowing this maximum output voltage, and the maximum current handling capability of the voltage regulator, the manufacturer can readily compute the minimum value of resistance in the exciter field circuit to ensure overcurrent protection. Since current flow is inversely proportional to resistance, any value of resistance in the exciter field circuit in excess of this minimum value will only further inhibit current flow.

From the above, we can see that the manufacturers of alternators will design the exciter fields of their units to have a minimum resistance slightly above the regulator's minimum stated value to compensate for any manufacturing tolerances. However, the resistance of the exciter field will usually be designed low enough to enable sufficient current to flow in the winding to afford adequate excitation during conditions of high excitation demands such as during starting of large motors.

1.7.2 Inductors and inductance

a. *Electromagnetic induction and inductor coils.* In Section 1.6 we discussed Faraday's laws of electromagnetic induction. The second of our three circuit elements, the inductor or induction coil, is based upon these laws. When a conductor is wound or formed into a coil, current flowing through the coil will produce a magnetic field around the coil, as shown in Figure 1-20. When switch S is closed, a current starts to flow in the coil, and the lines of magnetic flux start to go out, forming or establishing a magnetic field around the coil. Since these flux lines pass through the turns of conductor forming the coil on their way out to form the field, they are actually being cut by these turns of conductor. Thus, during the brief period of time necessary for the field to become completely established, a voltage is being induced in the coil itself. Once current flow has reached a stable full load value, the field will be completely established, and no further cutting of flux lines will occur. Thus the induction of voltage into the coil itself will stop. When switch S is opened, the opposite action will occur. Current will cease to flow in the coil, thus for a brief period of time while the field is in the process of collapsing, the flux lines will once again cut the turns of conductor in the coil, and induce a voltage in the coil. When the field has fully collapsed, all induction of voltage in the coil will cease. In a direct current circuit, such as is illustrated in Figure 1-20, induction of voltage in the coil will occur only when power is being applied or removed. During the time when current flow is constant, there is no change of current flow either increasing or decreasing, so there will be no increase or decrease in the magnetic field of the unit to cause induction of voltage in the coil. For this reason, induction in a constant current dc circuit is of little consequence, except when closing or opening the switch.

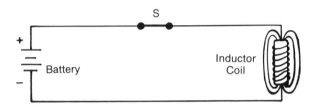

Figure 1-20. Battery Powered DC Circuit with an Induction Coil

b. *Inductance in ac circuits.* The presence of an inductor or induction coil can be of great significance in an alternating current circuit. The reason for this is that in an ac circuit the flow of current is in the form of wave, and is constantly changing, both in magnitude and direction. We have seen in the paragraph above that inductance, or the act of inducing voltage in an inductor coil, will occur only when a magnetic field is in the process of forming or collapsing due to a change in current flow. Therefore, when a coil is present in an ac circuit, a voltage is constantly being induced in the coil as long as the circuit receives power.

c. *Factors affecting induction of voltage in a coil.* In our study of Faraday's laws, we learned that the magnitude of a voltage induced into a coil is dependent upon two basic factors:

1. *The number of turns of conductor wire in the coil.* As the number of turns comprising a coil is fixed once it has been wound, the effect of the number of turns comprising any specific coil or inductor upon inductance will be a constant for that particular coil, regardless of any other changes in the ac power supplied to the coil.

2. *The number of lines of magnetic flux being cut per unit time*, or as was pointed out above, the rate of change of current flow in amperes per second. This rate of change in current flow will be directly affected by two factors:

(a) *The peak value of current demanded by the coil or conductor.* The greater the value of peak current, the greater will be the rate of change in current flow from zero to peak value.

(b) *The frequency of the alternating current applied to the device.* In alternating current the length of time to complete one cycle is termed **the period of the wave.** The higher the frequency, the shorter will be the period of the wave, and thus the shorter the length of time for current to change from peak value in one direction to peak value in the opposite direction. From this we can see that any change in the frequency of the alternating current supplied to the inductor will directly affect the rate of change in current flow. The unit for frequency is hertz. One hertz equals one cycle per second, 60 hertz equals 60 cycles per second.

d. *Effects of inductance on current flow in an ac circuit.* It can be shown that the direction or polarity of the voltage induced into a coil is such as to oppose current flow. If we consider a circuit with an alternator supplying an ac sine wave current to a coil with an inductance L only, ignoring any resistance of the conductors (Figure 1-21), we could represent a plot of this current sine wave, as shown in Figure 1-22.

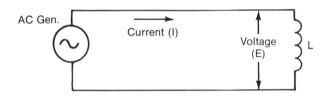

Figure 1-21. Circuit with an Alternator Supplying an Inductor

A study of Figure 1-22 will show that the ampere change per second will be maximum at points A, C, and E, and zero at points B and D. The voltage in the circuit will be maximum at points A, C, and E, and be zero at points B and D. Figure 1-23 illustrates a plot of the voltage as it would appear in this circuit.

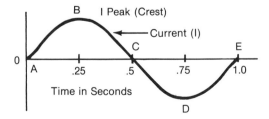

Figure 1-22. Plot of an AC Current Sine Wave

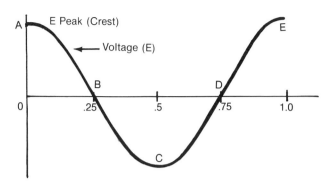

Figure 1-23. Plot of Voltage in an AC Inductive Circuit

If we superimpose the plots of current (Figure 1-22) and voltage (Figure 1-23), on a common plot, and use a base of degrees instead of time, the new plot will be as shown in Figure 1-24, where we see several aspects of an ac circuit having a purely inductive (disregarding resistance of the coil and load leads) load imposed on it.

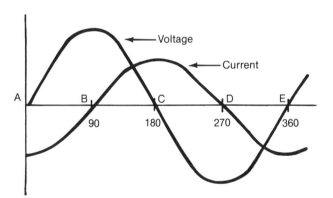

Figure 1-24. Plot of Voltage Versus Current in a Purely Inductive AC Circuit

1. When voltage is at peak value (points B and D), current is zero.

2. When voltage is at zero (points A, C, D), current is at peak value.

3. During 2 of the 4 quarters of cycle (A to B, and C to D) the voltage and current waves are of opposite sign or polarity.

4. The zero point of current (B) occurs 90° after the zero point (A) of voltage, thus the current wave lags the voltage wave by exactly 90°.

In summary, we can say that the effect of pure inductance in an ac circuit is to retard the current wave in relation to the voltage wave by exactly 90°. These two waves are out of phase by 90° and the current is said to lag the voltage. In the industry, inductive loads are referred to as **lagging loads**.

Without going into the physics of inductance, we can say that the presence of an inductor in a circuit creates a hindrance to current flow, termed **inductive reactance**, by the induced voltage opposing the flow of current. Inductive reactance is given the symbol XL. Just as the voltage across a resistance is equal to the current times the resistance (E = I × R), the voltage measured across an inductor is equal to the current times the inductive reactance, or E = I × XL. Note that these two equations are very similar, and it is because of this similarity that inductive reactance is measured in ohms. Inductive reactance, like resistance, tends to limit the magnitude of current flow in an ac circuit. For any given voltage across an inductance, the higher the inductive reactance, the lower the amount of current flow. However, unlike resistance, which is unaffected by frequency, inductive reactance will vary directly with any change in frequency. Increasing frequency will increase the inductive reactance in direct proportion, and current flow will be reduced. For example, if we hold the voltage across an inductor at a constant level, and change frequency from 60 hertz to 400 hertz, the current flow at 400 hertz will be only 15% of its magnitude at 60 hertz. In a similar fashion, if system voltage is held constant and the frequency is reduced from 60 hertz to 50 hertz, current flow will increase to 120% of the value at 60 hertz.

Effects of inductance in load leads. In Section 1.1 we discussed the fact that the magnitude of flux around a single conductor, even at high levels of current flow, is relatively weak. For this reason, inductance in load leads is usually considered to be of such low value (especially for frequencies of 50 and 60 hertz) as to be safely ignored.

1.7.3 Capacitors and capacitance

a. *Capacitors as a circuit element.* In Section 1 of this chapter we discussed the resistor or resistance as a circuit element which inhibits the flow of electric current, and which generates heat when a current is caused to flow though it. In Section 1.7.2 we covered the second of the three circuit elements, the inductor, which inhibits current flow by an opposing voltage induced into the inductor. In this section, we will discuss the capacitor, sometimes called a condenser, the third alternating current circuit element, which exhibits the ability to store up an electric charge.

b. *Capacitors.* The capacitor is a simple electric device comprising three separate parts: two plates of metallic conductor material, and an insulator (non-conductor) inserted between the two conductor plates. The two metallic plates of a capacitor form the **terminals**. Figure 1-25 illustrates a simple dc circuit incorporating a battery as a power source, switch S, and a capacitor C. With switch S in the closed position, at first glance one would be inclined to say that the presence of the insulator would prohibit current flow. However, experiment would show that at the instant switch S is closed there will be (for a very brief time) current flow, as charges from the right hand plate of the capacitor are moved to the left hand plate through the battery. As charges have been removed from the right hand plate, it is left with a negative charge; as charges have been added, the left hand plate has assumed a positive charge. Once the voltage across the insulator plate from A to B is exactly equal to the battery's voltage, current will cease to flow, and will not flow as long as the battery voltage is held constant. If, however, we were to raise the voltage, current would again flow for a brief time as before from plate B through the battery to plate A until stable voltage is achieved. Conversely, if we were to lower the voltage, current would again flow, but from plate A back through the battery to plate B for a brief time until a stable voltage condition exists. Note the similarity to an inductance in a dc circuit where voltage is induced only when there is a change in current flow; here with a capacitor in a dc circuit, current will only flow when there is a change in voltage.

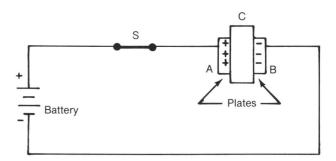

Figure 1-25. DC Circuit with Capacitor and Switch

c. *Capacitance in an ac circuit.* Capacitive current, as with inductance, occurs in a circuit only when there is change. In the case of inductance, a change of rate of current flow is present; in capacitance there must be a change in voltage. As with inductance, capacitance is of great significance in ac circuits where the value of voltage, being of sine wave form, is constantly undergoing change. Another similarity is that, as is the case of inductance in an ac circuit, capacitance has the effect of hindering current flow. But unlike inductance, which hinders flow by opposing it, capacitance has the effect of hindering current flow by storing it. This hindrance to current flow caused by the presence of capacitance in an ac circuit is termed **capacitive reactance**, and is given the symbol XC. Capacitive reactance, like resistance and inductive reactance being a hindrance to current flow, is measured in ohms. The voltage across a capacitor in an ac circuit is equal to the current times the capacitive reactance, or $E = I \times XC$.

The voltage current relationship can be shown by analysis of voltage and current waves in a purely capacitive circuit. This storing of current causes the current sine wave to precede or lead the voltage sine wave by 90°; see Figure 1-26.

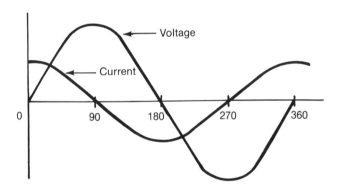

Figure 1-26. Plot of Voltage Versus Current in a Purely Capacitive AC Circuit

Capacitance is dependent upon the rate of voltage change in volts per second. Therefore, just as we saw with inductance, frequency and change in frequency will affect capacitance. Capacitive reactance, like inductive reactance, tends to limit the magnitude of current flow in an ac circuit. The higher the value of capacitive reactance in the circuit, the lower the magnitude of current flow. However, where inductive reactance is directly proportional to frequency, capacitive reactance is inversely proportional to frequency. Therefore, by increasing frequency, the value of capacitive reactance is reduced. This is just the opposite of the example we saw in the section on inductive reactance. If we hold voltage constant in an ac circuit containing capacitive reactance, and raise the frequency from 60 hertz to 400 hertz, the magnitude of current flow will increase by 667% of the 60 hertz value.

d. *Effects of capacitance in load leads.* There can exist a very small, but measurable amount of capacitance between a load lead to ground, or between two adjacent load leads since both are conductors, and the separating air space acts as an insulator. For 50 or 60 hertz systems at or below 600 volts, this effect is so small as to be safely ignored. However, when considering high frequency circuits such as 400 hertz, or systems with voltages in excess of 600 volts, this capacitive effect in the load leads may have to be taken into account.

1.7.4 Impedance. Throughout this section we have been considering each of the three basic ac circuit elements individually, and totally disregarding the possible existence of either of the other elements in the circuit being studied. For practical purposes, there will always be some amount of resistance in any electrical circuit due to load lead resistance, and the resistance of the various conductors used in the fabrication of any electrical device that may be in the circuit. Depending upon the nature of the devices comprising the load in an ac circuit, there is a very real possibility that at least two, if not all three elements will be present in a circuit at any time.

Total hindrance to current flow. From our study of resistance, inductive reactance, and capacitive reactance we saw that E, the voltage across a circuit, could be calculated by one of three equations depending upon the circuit element involved:

(1) For resistive elements: $E = I \times R$ (Equation 1-2)
(2) For inductive elements: $E = I \times XL$ (Equation 1-3)
(3) For capacitive elements: $E = I \times XC$ (Equation 1-4)

If we transpose these equations to solve for hindrance to current flow they would be:

$$R = \frac{E}{I} \quad XL = \frac{E}{I} \quad XC = \frac{E}{I}$$

Since all of these individual hindrances are equal to the voltage across the circuit divided by the current flowing in the circuit, you would think that to obtain the total hindrance to current flow in a circuit containing two or more of these elements, we should simply add them together. However, our study of inductive reactance showed that it has the effect of causing the current wave to lag the voltage wave, while our study of capacitive reactance showed it has the effect of causing the current wave to lead the voltage wave. It would follow then, that in an ac circuit containing both inductive reactance and capacitive reactance, these two factors would tend to offset each other, which in fact they do. So to obtain the net total of these two reactances, we must subtract one from the other, and then add this difference vectorially to any resistance to obtain the total hindrance to current flow.

Total hindrance to current flow in an ac circuit is called **impedance**, and is given the symbol Z. Since it is the measure of total hindrance to current flow it also is measured in Ohms. The equation for impedance is:

$$Z = \sqrt{R^2 + (X_L - X_C)^2} \qquad \text{(Equation 1-5)}$$

Impedance, the expression of total hindrance to current flow in an ac circuit, is the ratio of the voltage across the circuit to the total current flow. The equation for this ratio is:

$$Z = \frac{E}{I} \qquad \text{(Equation 1-6)}$$

Ohm's law, voltage is equal to current flow times resistance to current flow (E = IR), is valid in dc circuits where the values of voltage and current remain constant. For Ohm's law to be valid in ac circuits where resistance is only one of three possible forms of hindrance to current flow, we must substitute impedance (Z) for resistance (R) and the formula becomes:

$$E = IZ \qquad \text{(Equation 1-7)}$$

Where:

  E = voltage across the circuit in volts
  I = total current flow in the circuit in amperes
  Z = the impedance of the circuit in ohms

1.8 POWER IN AN ALTERNATING CURRENT ELECTRICAL CIRCUIT (kVA, kW, POWER FACTOR)

In Section 1.7 we discussed the three basic elements of an alternating current electric circuit: resistance, inductance, and capacitance, and their effects upon voltage and current flow. In this section we will discuss the principles of electric power, and how these basic circuit elements affect the performance and output of an alternator.

1.8.1 Power. Power is defined as the rate at which work is done, or heat released. This rate of doing work can be measured in any of several different units. Since these various units of measurement are simply different expressions of the same thing, the rate at which work is done, or heat is released, it follows that they can be used interchangeably given the appropriate equivalents. For example, the most common measurement unit of work done mechanically is horsepower. An electric motor develops one **horsepower** by lifting a weight of 550 pounds through a distance of one foot in one second, the product of force and rate of motion.

The most common unit of measurement of electrical power is the **watt**. It can be shown by experiment that the electric motor in the example above will consume 746 watts of electric power, discounting efficiency losses, to accomplish one horsepower of mechanical work. Therefore, one horsepower of mechanical power is equal to 746 watts of electric power. In fact, the International System of Units, SI Units, (Systeme International D'Unites) now rates mechanical power in watts, or kilowatts. It is now becoming quite common to see manufacturers of stationary industrial diesel engines rate the power output capabilities of their engines in kilowatts rather than in horsepower.

Electric power. We have seen that power, the rate of doing work, is the product of force times rate of motion. Electrically, the product of force (EMF in volts) times rate of motion (current flow in amperes) equals power (watts). The equation for electric power is expressed by the equation:

$$P = EI \qquad \text{(Equation 1-8)}$$

Where:

  P = power in watts
  E = voltage in volts
  I = current flow in amperes

For example, if an electric hot plate has 120 volts across it, and the current flowing through it is 10 amperes, the power consumed by the hot plate is:

$$P = EI \quad P = 120 \times 10 \quad P = 1{,}200 \text{ watts}$$

In a dc circuit, Ohm's law tells us that voltage is the product of current times resistance; E = IR. Since voltage is equal to IR, we could substitute the expression IR for E in Equation 1-8 above, and then the equation for power in an electrical circuit would be:

$$P = (IR)I \quad \text{or}$$
$$P = IRI \quad \text{or}$$
$$P = I^2R \qquad \text{(Equation 1-9)}$$

Therefore, from our example of the hot plate above, if the resistance of the hot plate is 12 ohms, and the current through it is 10 amperes, the power consumed by the hot plate in producing heat is:

$$P = I^2R \quad P = (10 \times 10) \times 12 \quad P = 100 \times 12;$$
$$P = 1{,}200 \text{ watts.}$$

The watt is a satisfactory unit of measurement of power in small appliances, lamps, etc.; it is, however, a rather small unit when dealing with any large amount of power usage. As a result, the electric power generating industry frequently uses the larger unit of a kilowatt, or 1,000 watts, usually referred to just as kW.

$$kW = \frac{WATTS}{1000} \qquad \text{(Equation 1-10)}$$

Thus the hot plate in our example used 1,200 watts divided by 1,000 or 1.2 kW of electric power.

1.8.2 Alternating current power

(a) *Volt-amperes (apparent power)*. Up to now, if the reader had no knowledge of alternating current circuit elements, he could well assume that the power delivered by, or taken from an alternator would be the product of the alternator's voltage (measured at its terminals) times the current flowing at its terminals. In the case of a circuit containing an electric motor as the total load on an alternator having a voltage measured at its terminals of 240 volts, and 25 amperes of current flowing at its terminals, he would use P = EI, multiply 240 volts times 25 amperes, and consider the motor to be taking 6,000 watts (6 kW) from the line. However, since the stator of the motor is an armature winding consisting of coils of conductor wire, the motor stator winding is in fact an inductor. From our study of inductors in Section 1.7.2, we know that the inductive reactance of the motor's stator windings will create an angle of separation between the sine wave of the voltage and the current sine wave. The product of voltage and current in an ac electric circuit, without regard to the angle of separation between them is called apparent power, and is expressed as volt-amperes, or kilovolt-amperes, VA, or kVA. The equations for apparent power VA, or kVA are:

$$VA = EI \qquad \text{(Equation 1-11)}$$

$$kVA = \frac{EI}{1000} \qquad \text{(Equation 1-12)}$$

Where:

    VA = apparent power in volt-amperes
   kVA = apparent power in kilovolt-amperes
     E = voltage in volts
     I = current flow in amperes

Therefore, in the example given above of an alternator with a motor as the total load, we can now correctly state that the apparent power taken from the alternator is:

VA = EI   VA = 240 volts × 25 amperes
VA = 6,000 volt amperes

or:

$$kVA = \frac{EI}{1,000} \qquad kVA = \frac{240 \times 25}{1,000} \qquad kVA = \frac{6,000}{1,000}$$

kVA = 6 kilovolt-amperes

b. *Power in ac resistive circuits*. In direct current circuits where, for any fixed load, the values of voltage and current are constant, the product of voltage and current at any instant of time will be the same as that of any other instant of time. However, as we saw from the common plot of voltage and current in a resistive circuit, Figure 1-19, the values of both voltage and current are constantly changing, so the product of these two values will constantly be changing from one instant to the next.

Figure 1-27 is a replot of the relationship of voltage and current in a purely resistive circuit, with the base line graduation given letter designations rather than degrees for easier reference. Also, since we are going to add a curve to represent power in this plot, we have added a third vertical scale graduated in watts. Table 1-1 gives calculations for Figure 1-27.

In our analysis of power in this figure we will examine the curves on a point to point basis. As a refresher on the rules of multiplication they are:

1. When multiplying two positive quantities, the product is positive.

2. When multiplying two negative quantities, the product is positive.

3. When multiplying a positive quantity by a negative quantity, or vice versa, the product is negative.

In our calculations of power we will be using Equation 1-8, P = EI.

Note that in every case where voltage (E) is negative, current (I) is also negative, but the product of these two negative quantities is a positive quantity of power. This is right in line with our study of current through a resistor. It does not matter in which direction current flows through a resistor, heat is liberated by current flow regardless of its direction. From this we can see that, regardless of which direction current flows in a resistive ac circuit, power flows from the source to the resistive load.

It is interesting to note that the power being delivered to the resistor is ever-changing from one instant to the next in a manner that suggests a pulsation of two times the base frequency. The value of power taken over the entire cycle may be shown as: P = E (effective) times I (effective), or usually written just as P = EI. The power equation may be used with either instantaneous values of voltage and current to calculate the instantaneous value of power at any point of time within the cycle, or with the effective values of voltage and current to determine the active power.

Another fact of great importance to us here, is that in a circuit containing only resistance, the waves of both voltage and current are in phase, having a 0° angle of separation, and that the product of voltage times current at the alternator's terminals (VA or kVA) is the same as the real power delivered to the load (watts or kilowatts).

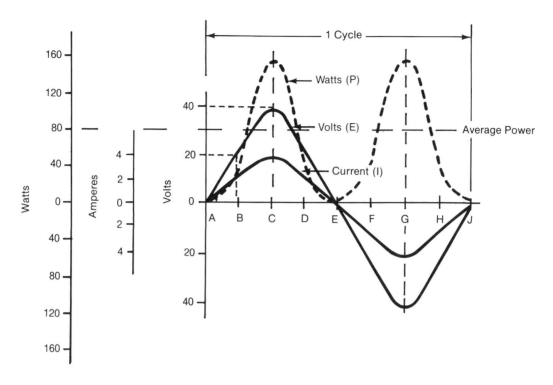

Figure 1-27. Relationships of Voltage, Current, and Power in a Resistive AC Circuit

Table 1-1. Calculations for Figure 1-27

A.	E =	0;	I =	0;	P =	0 × 0;	P = 0 watts
B.	E =	20;	I =	2;	P =	20 × 2;	P = 40 watts (+)
C.	E =	40;	I =	4;	P =	40 × 4;	P = 160 watts (+)
D.	E =	20;	I =	2;	P =	20 × 2;	P = 40 watts (+)
E.	E =	0;	I =	0;	P =	0 × 0;	P = 0 watts
F.	E =	(−)20;	I =	(−)2;	P =	(−)20 × (−)2;	P = 40 watts (+)
G.	E =	(−)40;	I =	(−)4;	P =	(−)40 × (−)4;	P = 160 watts (+)
H.	E =	(−)20;	I =	(−)2;	P =	(−)20 × (−)2;	P = 40 watts (+)
J.	E =	0;	I =	0;	P =	0 × 0;	P = 0 watts

c. *Power in ac inductive circuits.* To illustrate the power in an alternating current circuit having only inductance (here again we will ignore any resistance) we will replot Figure 1-24 from Section 1.7 using the same form and base we used to construct the plot in Figure 1-27. Again, in Figure 1-28, we will analyze these curves on a point to point basis, and compute the power at each of these points with the power equation P = EI. Calculations for Figure 1-28 are in Table 1-2.

To analyze these points: at point A no power is being delivered to the load. At B, power is positive, and 50 watts are being delivered to the load. At C, voltage is zero, and once again no power is being delivered to the load. At point D, we see negative 50 watts indicating that power is being transferred from the load back to the source. From points E through J, we see the power curve repeating as from A through E. From this we can see that from points A to C, and from E to G, power is positive and being delivered to the load. But, from points C to E, and G to J, power is negative and being returned to the source.

A study of this data indicates that, during the two periods of positive power, power was being stored in the inductor, and then this power is returned to the source during the periods of negative power. Note also, that the peaks of the power curve, both positive and negative, are identical (50 watts). This leads to the conclusion that the build up, and subsequent collapse of a magnetic field does not dissipate heat, so that no power is consumed in the process. Remember, we are considering here a pure inductance. In reality, all of the conductors in the coil and the circuit do have resistance, and some amount of heat will be created by this resistance, but none by the inductance of voltage into the inductor coil. We will discuss this aspect of resistance in an inductive circuit in a later section. For our purposes here where we are only considering pure inductance in a circuit, when the current lags the voltage by 90°, no power is consumed by the inductor. Therefore, in this case, the product of the voltage at the alternator's terminals times the current flowing at the terminals will have some value of VA or kVA indicating that the alternator was apparently delivering power to the load, even though the load was consuming no real power.

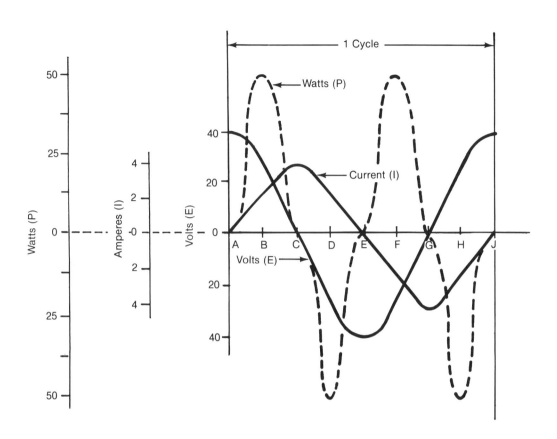

Figure 1-28. Relationships of Voltage, Current, and Power in an AC Inductive Circuit

Table 1-2. Calculations for Figure 1-28

A.	E = +40;	I = 0;	P = +40 × 0,	P =	0 watts
B.	E = +25;	I = +2;	P = +25 × +2,	P =	+50 watts
C.	E = 0;	I = +4;	P = 0 × +4,	P =	0 watts
D.	E = -25;	I = +2;	P = -25 × +2,	P =	-50 watts
E.	E = -40;	I = 0;	P = -40 × 0,	P =	0 watts
F.	E = -25;	I = -2;	P = -25 × -2,	P =	+50 watts
G.	E = 0;	I = -4;	P = 0 × -4,	P =	0 watts
H.	E = +25;	I = -2;	P = +25 × -2,	P =	-50 watts
J.	E = +40;	I = 0;	P = +40 × 0,	P =	0 watts

d. *Power in ac capacitive circuits.* To illustrate power in a purely capacitive circuit, we will replot Figure 1-26 from Section 1.7 in the same manner as in Figures 1-27 and 1-28; see Figure 1-29.

When we analyze these points, we see that the power curve in Figure 1-29 is identical to the power curve in Figure 1-28. From this we can draw the same conclusions in a capacitive circuit as we did in an inductive circuit. The 90° separation of voltage and current causes electric power to be delivered to the capacitor, and then returned to the source from the capacitor without power being consumed in the process. Here again we will make the same disclaimer as to resistance in the circuit.

From this analysis and the analysis of a purely inductive circuit, we can now state: in an alternating current circuit, where current is displaced from voltage, either lagging or leading, by 90° (one quarter cycle), no real power is consumed by the load. It is also important to remember that there is a voltage at the alternator terminals and current is flowing through these terminals. This voltage and current represent a load on the alternator, in this case all VA or kVA, that is very real to it, even though the load is not consuming any real power. This apparent power load upon the alternator resulting in no real power being consumed by the load is termed **reactive power**, or sometimes it is called **wattless power**.

1.8.3 Power factor. To this point in Section 1.8 we have been considering power in ac circuits containing only one of our three basic circuit elements, either resistance, inductance, or capacitance. This has suited our purpose in developing the theory of exactly how each of these elements affect power in a load. We saw in Section 1.7.4 in our discussion of impedance that in reality there may well be at least two, if not all three of these basic circuit elements present in an ac circuit. Certainly some degree of resistance will always be present in an electric circuit since the circuit must contain conductors.

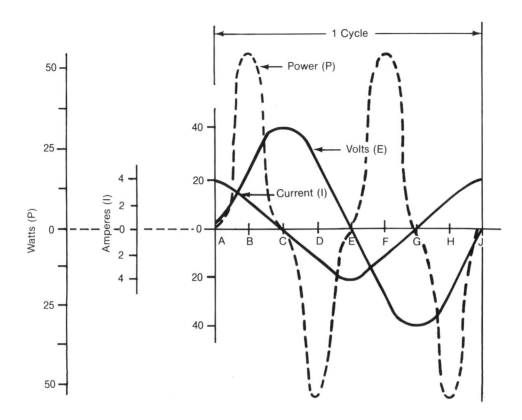

Figure 1-29. Relationships of Voltage, Current, and Power in a Capacitive AC Circuit

In keeping with the purpose of this text, we will not undertake the rather involved vector analysis, and corresponding circuit analysis to compute the exact angle of separation between the voltage and current waves in a complex circuit containing resistance, inductive reactance and capacitive reactance. Rather, we will state that a combination of resistance with inductive reactance and/or capacitive reactance in a circuit will result in an angle of displacement, either lagging or leading, between the voltage and current waves. This angle of displacement will be greater than the zero angle of a purely resistive circuit, and less than the 90° of a purely inductive or capacitive circuit. The predominant reactance, whether inductive or capacitive, will determine whether this angular displacement of current versus voltage will be either lagging or leading. Electrical engineers have given this angle of separation the term **Theta** (θ) from the Greek alphabet.

a. *Apparent power, real power, and reactive power.* We have discussed the fact that the product of voltage and current in an ac electric circuit without regard to the angle of separation between the waves of voltage and current is called VA or kVA, and that VA and kVA is termed **apparent power**. The product of voltage and the current flowing through a reactance, either inductive, capacitive, or a combination of both, is referred to as volt-amperes reactive (var), or kilovolt-amperes reactive (kvar).

From a study of power in resistive, inductive and capacitive circuits it can be shown vectorially that the real power (watts or kilowatts) is the product of voltage times the component of current that is in phase (of the same polarity — either positive or negative) with the voltage. The product of voltage times the current that is out of phase (of the opposite sign: current + with voltage –, or vise versa) with the voltage is reactive power (var or kvar). The product of voltage times the total current flow is the apparent power (VA or kVA).

b. *Power factor.* The ratio of real power (watts or kilowatts) used in a circuit to the apparent power delivered to the circuit is termed **power factor** (PF) and is expressed:

$$\text{POWER FACTOR} = \frac{\text{REAL POWER}}{\text{APPARENT POWER}}$$

$$PF = \frac{kW}{kVA} \qquad \text{(Equation 1-13)}$$

c. *Power right triangle.* The relationship of kilowatts (kW), kilovolt-amperes (kVA), kilovolt-amperes reactive (kvar), and power factor (PF) are often represented in a **power right triangle**. The power right triangle is a triangle where the horizontal leg is drawn proportional to kilowatts; the vertical leg is drawn proportional to kilovolt-amperes reactive; and the hypotenuse is drawn proportional to kilovolt-amperes as illustrated in Figure 1-30. The angle between the hypotenuse (kVA) and the horizontal leg (kW) is made equal to the angle of separation between voltage and current, and is labeled θ (theta).

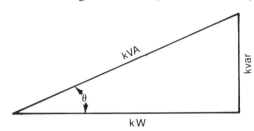

Figure 1-30. Power Right Triangle

In trigonometry, the cosine of an angle of a right triangle is equal to the side adjacent to the angle divided by the hypotenuse. In our power right triangle (Figure 1-30), the cosine of the angle theta (θ) would be: the adjacent side (kW) divided by the hypotenuse (kVA). When written in a formula, the term cosine is cos. The equation for the cosine of theta would be:

$$\cos \theta = \frac{kW}{kVA} \qquad \text{(Equation 1-14)}$$

From the paragraph above we saw that in Equation 1-13, power factor is equal to kilowatts divided by kilovolt-amperes, just as cosine theta is equal to kilowatts divided by kilovolt-amperes. Therefore, power factor (PF) is equal to the cosine of the angle of separation between the waves of voltage and current (cos θ), or PF = cos θ.

d. *Power factor in a purely resistive circuit.* We have seen that in an ac circuit containing only resistance, there is no angle of separation between the waves of voltage and current. Thus, in a purely resistive circuit, theta equals zero. The cosine of zero is 1. This agrees with our study of power in a resistive circuit, where we saw that kW and kVA are equal. If we transpose Equation 1-13 to solve for kW it would be:

$$kW = kVA \times PF \qquad \text{(Equation 1-15)}$$

From this, if PF equals 1; kW = kVA × 1, kW = kVA. A power factor of 1 is referred to as **unity power factor**. Electrical loads that are purely resistive are frequently referred to as unity power factor loads.

e. *Power factor in an ac circuit containing only inductance or capacitance.* Our studies of purely inductive or capacitive circuits have shown us that the waves of voltage and current are separated by exactly 90°. The cosine of 90° is zero (0). In a circuit where the waves of voltage and current are separated by exactly 90°, no real power is consumed in the circuit (Section 1.8.2), therefore kW will equal zero. This is reaffirmed by the equation kW = kVA × PF, or kW = kVA × cos θ, where cos θ is zero; kW = kVA × cos θ, kW = kVA × 0, kW = 0.

f. *Summary of power factor*

1. Power factor (PF) is the ratio of the real power (watts or kilowatts) consumed by a load to the apparent power (volt-amperes or kilovolt-amperes) delivered to the load. As a ratio, power factor can be expressed as a decimal from 0 to 1.0, i.e., 0.4, 0.8, or multiplied by 100 and expressed as a percent, i.e., 40%, 80%, 100%. However, since the real power consumed by a load can never exceed the apparent power delivered to it, 1.0 or 100% is the highest possible value of power factor.

2. The equations for expressing the relationships of power factor, real power, and apparent power in an ac electric circuit are:

$$PF = \frac{kW}{kVA} \qquad kVA = \frac{kW}{PF} \qquad kW = kVA \times PF.$$

3. Power factor is equal to the cosine of the angle of separation between the waves of voltage and current. Therefore, the expression cos θ can be substituted for PF in all of the above power factor equations.

4. The power factor of ac circuits containing only inductive reactance or capacitive reactance is zero. This type load is referred to as **zero power factor lagging**, or **zero power factor leading**, depending upon the type of reactance of the circuit. Inductive reactance will cause a lagging power factor, while capacitive reactance will cause a leading power factor.

5. In ac circuits containing only resistance, the power factor will be 1 (unity). Purely resistive loads are referred to as **unity power factor loads**.

6. For the effects of power factor loads upon alternators see Chapter 2, Section 2.4.2, Voltage and voltage regulation.

REFERENCES

[1] Glossary of Standard Industry Terminology-Electrical, EGSA 101E-1984, Electrical Generating Systems Association

[2] IEEE Standard Dictionary of Electrical and Electronics Terms, ANSI/IEEE Standard 100-1992.

BIBLIOGRAPHY

Eaton, J. R. Beginning Electricity, The McMillian Co.

G. L. Oscarson, ABC of Power Factor, E-M Synchronizer 200 SYN-50, Electric Machinery Mfg. Co.

Oscarson, G. L. ABC of Semiconductors, E-M Synchronizer 200 SYN-61, Electric Machinery Mfg. Co.

Permanent Magnet Guidelines, Magnetic Materials Producer Association, January 1972.

Wallace, J. H. Magnetism, What You Need to Know Series — Part 5, Power Transmission Design, November 1968.

Alternators (Synchronous Generators)

James Wright

CHAPTER 2

An alternator, or ac generator is a rotating electrical machine designed to convert rotating mechanical energy into alternating current (ac). As is frequently the case with mechanical-electrical devices, depending upon the nature of the application, an alternator may be a very simple machine or it can become rather complex.

2.1 CONFIGURATIONS

The purpose of this section is to define and describe the various types, styles, and physical configurations of alternators.

2.1.1 Definitions. The various types and configurations of alternators are often given descriptive names that incorporate the configuration of one or more of their component parts. While no single alternator will incorporate all of the below listed component parts, these component parts will all be used in one or more types or models of alternators or alternator sub-assemblies.

Air gap. The separating space between the rotor and stator of an alternator.

Amortisseur bars. Also termed amortisseur winding, damper winding, or damper bars. A squirrel cage winding consisting of bars or rods of a conductor material, usually copper or aluminum, embedded in slots or holes in each of the pole faces of a revolving field core. These bars run the entire length of the core and are electrically connected on each end to the bars of the other pole/poles of the core by means of either shorting rings, Figure 2-1, or shorting laminations, Figure 2-2. Several manufacturers also use die-cast aluminum damper windings. The purpose of the amortisseur or damper winding is to dampen rotor oscillations or hunting during load changes, thereby improving transient stability. This winding is especially helpful when two or more alternators are being operated in parallel.

Figure 2-1. Amortisseur Winding with Shorting Rings

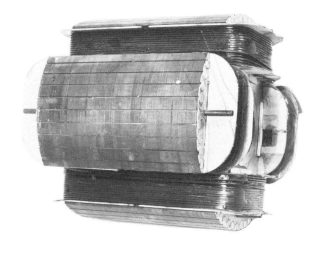

Figure 2-2. Amortisseur Winding with Shorting Laminations

Alternators (Synchronous Generators)

Armature. An assembly of coils placed in slots of a laminated steel core. The interconnected coils make up the armature winding. The alternator output voltage and current are generated in this winding.

Bearing bracket. Sometimes referred to as a bearing carrier, or an end bell. A bearing bracket is the structural member of an alternator that houses or supports a rotor bearing. In some designs, a bearing carrier is also used to house or support other stationary component parts such as brushes, brush holders, bearing temperature detectors and exciter stators.

Bearing carrier. See **bearing bracket**.

Brush. A conducting element constructed of carbon, graphite, or in some cases, copper. This element is used to maintain sliding electrical contact between rotating and stationary components.

Brush holder. A device used to house and support a brush or brushes enabling it to maintain contact with a rotating surface such as a collector ring or commutator.

Collector ring. Often referred to as a slip ring. A conductor ring usually constructed of an alloy of copper, and mounted on the shaft of an alternator. Brushes on the ring maintain sliding contact, affording electrical continuity between stationary and rotating components.

Commutator. A device mounted on the shaft of a brush type exciter (dc generator) that mechanically converts the alternating current generated in an exciter armature winding to direct current. Typically, a commutator is constructed of wedge shaped segments of hard drawn copper that are insulated from each other by thin strips of mica insulation. See Figure 2-6.

Core. An assembly of thin magnetic steel laminations, tightly compressed, and then either riveted, bolted, or welded together to form a magnetic path. It is this core around which, or into which, the coils of an armature or field winding are wound or inserted.

End bell. See **bearing bracket**.

Exciter. A device used to supply excitation (dc current) to the field windings of an alternator. Exciters are classed in one of two general categories: static exciters (constructed with no rotating components) and rotating exciters (constructed with a rotating component) and may be either of the brush, brushless, or permanent magnet type. See Section 2.1.4 of this chapter for further information on the various forms and types of exciters.

Exciter armature. The output winding of a rotating type exciter.

Exciter field. The dc field winding of a rotating type exciter.

Exciter rotor. The rotating component of a rotating exciter.

Exciter stator. The stationary wound component of a rotating exciter.

Field coil. A suitably insulated winding to be mounted on a field pole to magnetize it.

Field pole. The part of the magnetic structure of a rotating electrical machine, usually a laminated core, onto which the field coils are wound, formed, or placed. See **core**.

Frame. The structural member of an alternator used to house and support the stationary winding or windings of an alternator.

Laminated core. See **core**.

Rectifier assembly. An assembly consisting of rectifiers (diodes), heat sinks, terminal lugs or strips mounted on an insulated board, plate or ring. A rectifier assembly used with an alternator may be stationary, or mounted on the alternator shaft and rotated. Some rectifier assemblies may incorporate one or more surge suppressors in the assembly.

Rotor. Any rotating winding or element of an alternator.

Stator. Any stationary winding of an alternator.

NOTE: It is important to note at this point that when referring to wound components of an alternator, the terms rotor and stator are locational in nature, and may be used to refer to either an armature or field element. Armature winding or field winding are functional terms and may refer to either a stationary or rotating element.

2.1.2 Arrangements of alternators. There are two basic physical arrangements for alternators:

a. *Revolving armature alternators.* In a revolving armature alternator the field is held stationary and an ac voltage is induced/generated in the revolving armature winding. AC power is then supplied to an external load via slip rings and brushes, as illustrated in Figure 2-3. As

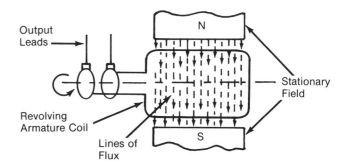

Figure 2-3. Simple Revolving Armature Alternator

a general rule, revolving armature alternators are small units in the range of 500 to 5,000 watts capacity.

b. *Revolving field alternators.* The armature windings of a revolving field alternator are held stationary and the field is rotated. Thus an ac voltage is induced/generated in the stationary winding of the machine, as shown in Figure 2-4. The advantage of this type design is that power is taken directly from insulated load leads coming directly out of the armature windings. Most modern alternators are of the revolving field type.

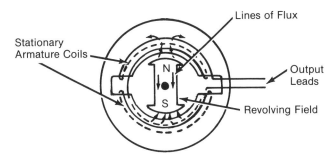

Figure 2-4. Simple Revolving Field Alternator

2.1.3 Types of alternator fields

a. *Permanent magnet field.* One or more permanent magnets may be used to provide the magnetic field necessary for the induction of voltage in the armature windings. An alternator constructed with a permanent magnet field has the advantage of eliminating the need for an exciter, thereby becoming very cost effective in the smaller sizes. One disadvantage is that its field strength (flux density) is maintained at a single constant, and voltage regulation no load to full load tends to be poor. However, in applications where the connected load will be of a fixed value, with only minor variations, permanent magnet alternators can be very effective.

b. *Electromagnetic field.* A field using an electrical winding used to provide the necessary magnetic field. The advantage of this type field is that the flux density of the field can be controlled quite accurately by varying the magnitude of the dc voltage imposed on the field windings. The disadvantage of this type field is that it does require an external source of dc current to initiate and maintain the magnetic field. Output voltage control of an alternator constructed with an electromagnetic field is obtained by coordinating the amount of excitation current provided to the field windings with load demand, or changes in load demand. As load demand is increased, a corresponding increase in excitation current is required to maintain output voltage at a given level. Conversely, as the load demand is reduced, a reduction in excitation is necessary to maintain a fixed value of output voltage.

2.1.4 Types of exciters used with revolving field alternators

a. *Brush type exciter.* A small dc generator termed a brush type exciter, since the construction of dc generators includes commutators and brushes. These exciter/generators (see Figure 2-5) are either mounted externally to the main alternator or constructed as an integral component of the alternator with the armature and commutator assembled on the shaft with the main field. See Figure 2-6. The advent of both static and brushless exciters have rendered this type of exciter virtually obsolete.

b. *Static exciter.* Static exciters are so named since they have no rotating component parts. This type of exciter is a solid state electronic device using rectifiers, transformers, reactors and the like. In many cases an automatic voltage regulator is incorporated as an integral part of the unit, and then it is referred to as a static exciter-regulator. While any source of ac power may be used as power input for these exciters, it is usually obtained directly from the output of the parent alternator. As with any brush type unit, the direct current excitation output of these exciters is applied directly to the main rotating field of the alternator through brushes and slip rings. See Figure 2-7.

When static exciters are powered by alternating current supplied by the main alternator, an external source of dc power is usually required to flash the field for initial field excitation and voltage buildup. In a typical engine generator set with battery powered engine starting, the engine cranking batteries are used to supply this flashing dc power. For those applications where the engine employs some other method of starting such as air, steam, or a small cranking engine, some system of a battery and possibly battery charger may be required to provide the dc flashing current. In addition, a field discharge resistor is frequently required with these exciters to prevent damage to the exciter during shutdown.

c. *Brushless exciter.* A three phase revolving armature alternator with its armature mounted on the shaft of the main alternator. Its stationary field windings are housed either in the main frame or the rear bearing bracket exactly the same as the integral brush type exciter discussed above. The only difference between these two types of exciters is that the commutator and brushes of the brush type exciter are eliminated, and a three phase rotating rectifier assembly is substituted to rectify the exciter armature's ac output to dc, which is then applied to the main rotating field windings via two directly connected main field lead wires. Since excitation is directly applied to the main field without the use of brushes and slip rings, the main alternator is termed a brushless alternator, or brushless ac generator. See Figure 2-8.

Alternators (Synchronous Generators)

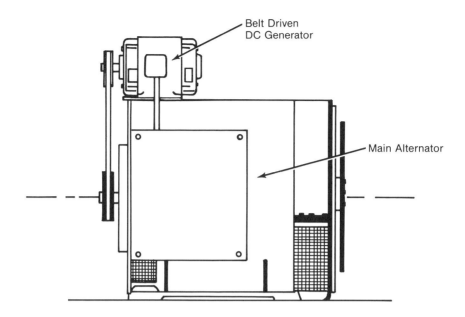

Figure 2-5. Brush Type Alternator with External Belt Driven DC Exciter

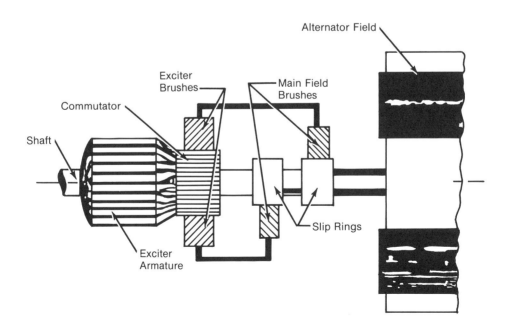

Figure 2-6. Rotor Assembly of Brush Type Alternator with Integral Brush Type Exciter

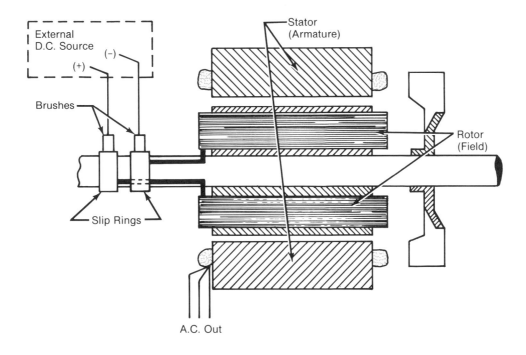

Figure 2-7. Typical Externally Excited Brush Type Revolving Field Alternator

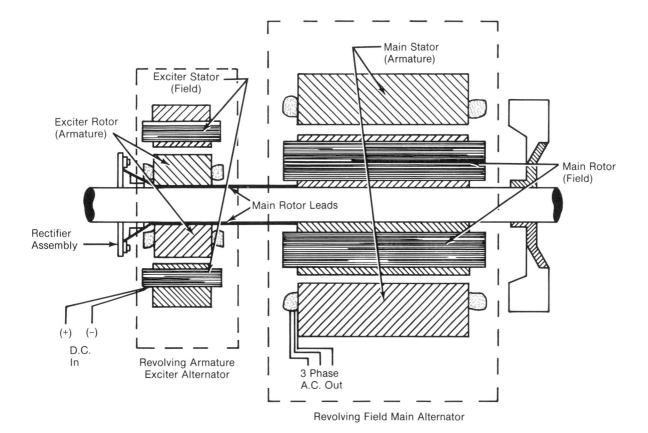

Figure 2-8. Typical Self Excited, Externally Regulated, Brushless Alternator

The material for the core of a brushless exciter field is usually selected to retain a degree of residual magnetism sufficient to generate an adequate residual ac voltage in the main stator to assure initial voltage buildup without the need for field flashing. Automatic voltage regulators used with brushless alternators usually incorporate some additional features to assist in initial voltage buildup. This will be discussed in the chapter on voltage regulators.

Another feature of a brushless exciter is that since its output ac voltage is rectified to dc, there is no need to hold the exciter frequency to that of the main alternator. To give a smoother rectified dc output and to take advantage of the shorter time constant, most alternator manufacturers design their exciter fields with at least two more poles than the main rotor to generate a higher exciter output frequency than that being produced by the alternator.

d. *Permanent magnet (pilot) exciter.* This is not actually an alternator exciter in the true sense of the term. A pilot exciter is a small revolving permanent field, single phase alternator whose output voltage is used to provide a single phase ac voltage as a power input for the unit's automatic voltage regulator. These devices fall into the category of excitation support systems which will be covered in Chapter 4 dealing with automatic voltage regulators and their accessories. Pilot exciters are usually mounted on the extreme end of the alternator shaft opposite the drive end outboard to the rear bearing. The stationary winding is housed either in the rear bearing bracket or in its own frame bolted to the bearing bracket.

2.1.5 Generation of the ac voltage.

Figure 2-9 is an illustration of a basic revolving armature alternator. For purposes of clarity, winding is illustrated as being composed of a single coil, with the start and finish leads of the coil affixed to slip rings and connected to a voltmeter via brushes.

Figures 2-10A and B illustrate the generation of the ac sine wave during one complete revolution of the armature winding through the magnetic field. In order to reduce the complexity of Figure 2-10A, we show a cross section of the leading side of the coil in increments of 45° of arc. Figure 2-10B shows the corresponding value and sign of the voltage in the generated sine wave.

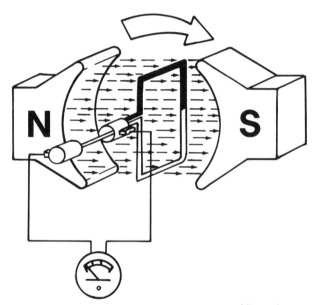

Figure 2-9. Basic Revolving Armature Alternator

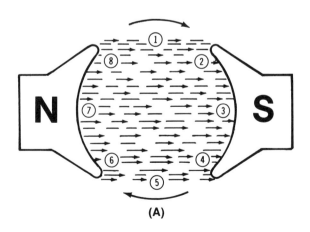

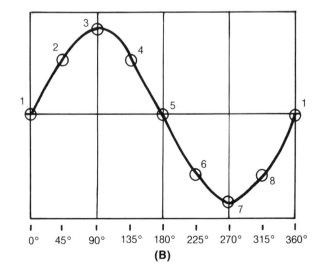

Figure 2-10. Generation of an AC Sine Wave by a Two Field Pole Alternator

At position 1, the conductor is traveling parallel to the field and not cutting any lines of flux; therefore, the induced voltage is zero. As the conductor rotates through positions 2 and 3 it begins to cut the lines of flux at an ever increasing rate of lines per unit time. Therefore, the voltage being induced is constantly increasing in value until it reaches position 3, where it is cutting the maximum number of lines per unit time and the induced voltage reaches its peak value. As the armature continues to rotate past position 3, the moving conductor begins to cut the lines of flux at an ever decreasing rate of lines cut per unit time. This causes the induced voltage to decrease in value until it reaches position 5, where again it is parallel to the field and voltage has decayed to zero. As the conductor begins the second half of its revolution, the direction of the flux lines in relationship to the conductor is reversed, and the induced voltage is of the opposite sign or direction. Again, as during the first half revolution, the conductor begins to cut more and more lines of flux per unit time as it moves from position 5 to position 7. The point of peak voltage is in the opposite direction, and during the final quarter of the revolution, moves from position 7 to position 1. A steady decrease in the number of lines per unit time being cut causes the induced voltage once again to decay to zero, completing the generation of one cycle of the voltage sine wave.

It is interesting to note here that two separate and distinct units of angular measurement are involved in the generation of the ac sine wave. First, the magnetic field in Figure 2-10 is composed of two magnetic poles. The shaft supporting these poles has made one complete revolution, or has physically rotated through a complete 360° circle. Second, the sine wave is 360 electrical degrees in length, and was generated by the conductor moving from a position mid-way between a north and south magnetic pole, traveling through the field past both a south pole and then a north pole and finishing in a position midway between a north and south pole. If the number of magnetic poles in the alternator's field is increased to four poles, as illustrated in Figure 2-11A, the 360° sine wave will be generated by only 180° of the shaft's rotation, as shown in Figure 2-11B. Section 2.4 will discuss in detail the effect the number of pairs of magnetic poles comprising the field of an alternator has upon the frequency generated by the unit.

2.2 SINGLE AND THREE PHASE ARMATURES

Alternating current is induced or generated in the armature windings of an alternator, and in general, the electrical characteristics of the unit's output power are determined by the design and construction of its armature windings. This section will cover the basics of single and three phase alternating current, general features of design and construction of alternator armatures, and output load lead identification. The basic principles of construction are the same for both stationary and rotating armatures. However, since stationary armature units are the most common, illustrations in this section will deal primarily with stationary armatures.

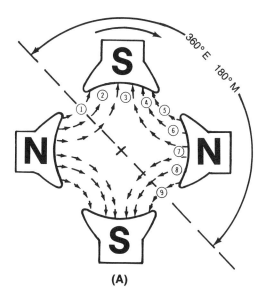
(A)

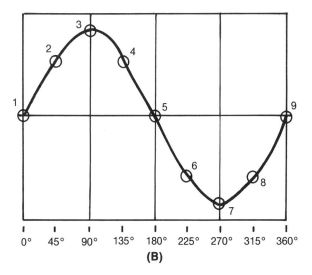
(B)

Figure 2-11. Generation of an AC Sine Wave by a Four Field Pole Alternator

Alternators (Synchronous Generators)

2.2.1 Basic principles of single and three phase alternating current

a. *Single phase ac.* Single phase ac consists of either a single voltage or two voltages in series with exactly the same phase relationship. Single phase ac supplies power over two or three lines.

b. *Three phase ac.* Three phase ac power consists of three separate voltages spaced 120 electrical degrees apart, utilizing three lines. The phases are usually given letter designations as, Phase A, Phase B, and Phase C. Figure 2-12 illustrates the three sine waves and points out several interesting aspects of the balance achieved in the system by the 120 electrical degree separation of the individual phases.

1. No two phases are ever at zero voltage at the same time.

2. No two phases are ever at peak voltage at the same time.

3. When any one phase is at zero (0) voltage, the remaining two phases are of opposite sign or polarity, and are at 86.6% of peak voltage.

2.2.2 Single and three phase armatures.
A complete armature assembly consists of two major components, a steel core and a winding, as defined in 2.1.1.c.

a. *Armature core.* The armature core consists of a number of laminations of electrical grade steel plate tightly compressed, and then either welded, riveted or bolted together to form a cylinder. Each of the individual laminations contains slots to accommodate the coils of the armature winding, as illustrated in Figures 2-13 and 2-14. Among the design considerations given to an armature winding are the length and diameter of the core along with the number, shape and spacing or separation of the individual slots.

Figure 2-13. Rotating Armature Core with Winding

Figure 2-14. Stationary Armature Core with Winding

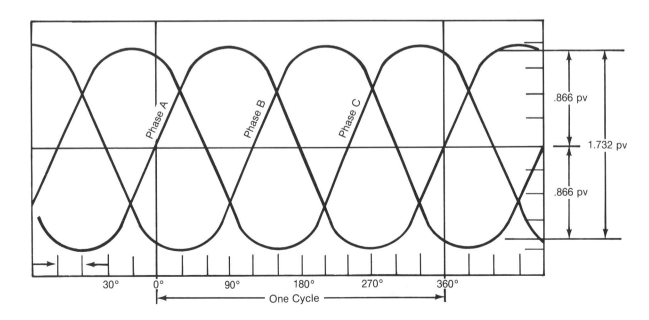

Figure 2-12. Three Phase Voltage Sine Waves

b. *Armature winding.* An armature winding consists of a number of groups of individual coils comprised of one or more turns of conductor wire, termed magnet wire, that has been precoated with an insulating enamel. These coils may be either form wound, with one or more turns of rectangular wire wound in layers and formed into shape, or mush wound, consisting of multiple turns of random wound round conductor which may, or may not be formed into shape. However constructed, the finished coils have two straight sides and end turns that are either rounded or angular, as shown in Figures 2-15 and 2-16. The coil sides are then inserted in the core slots with the end turns extending beyond the core and spanning the slots.

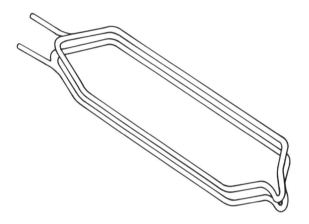

Figure 2-15. Form Wound Coil

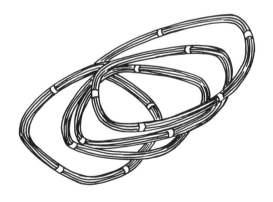

Figure 2-16. Mush Wound Coil

2.2.3 Basic features of an armature winding.
All of the coils used to make up a specific armature winding usually have three basic common features: they are wound of copper magnet wire of the same size or diameter, they contain the same number of individual strands of wire, and have the same number of turns. Since the current carrying capacity of a conductor is basically dependent upon its cross sectional area, all coils of equal construction will have the same current carrying capacity. Further, since the magnitude of the voltage induced in a coil is dependent on the number of turns of conductor comprising the coil, all of the individual coils of the winding will have the same magnitude of voltage induced into them.

Figure 2-17 is a sketch of a portion of a typical stationary armature core, with instructions to the winder for the proper placement of the coils, and the use and placement of the various insulating materials. This sketch also illustrates several aspects of the actual construction of an armature winding.

1. The slot cell insulation serves to separate and insulate the coils from the stator core iron.

2. An armature winding can be constructed with either one or more coils occupying the same slot. When two coils are inserted in the same slot it is called a **double layered winding**, and the coils are both separated and insulated from each other by the center stick.

3. The slot wedge locks the coils in the slot.

4. The two sides of each coil are inserted in separate slots with a designed space or number slots separating them.

5. Coils of different phases may be inserted into adjacent slots. The end turns of adjacent coils of different phases are separated and insulated from each other by phase insulation inserted between the end turns.

2.2.4 Connecting coils in an armature winding.
After all of the individual coils have been inserted in the core, the winder proceeds to group and interconnect the various coils in accordance with instructions provided by the winding data. Every coil has what we will refer here to as a start end and a finish end, and it is these ends or wires that are used to connect two or more individual coils into pairs or groups. There are two methods of connecting coils together.

a. *Series connection.* To connect two or more coils in series, the finish end or lead of the first coil is connected to the start end or lead of the next coil, and so forth, so that the end result resembles a chain with the individual coils being the links. See Figure 2-18. The voltage induced into the entire series of coils, as measured from the start lead of the first coil to the finish lead of the last coil in the series, will be the sum of the voltages induced into each coil. However, since electrical current must flow through the entire length of all the coils, the maximum current carrying capacity of a series of coils will be that of the coil having the least capacity, or in the case of coils of equal capacity, that of any individual coil in the series.

Alternators (Synchronous Generators)

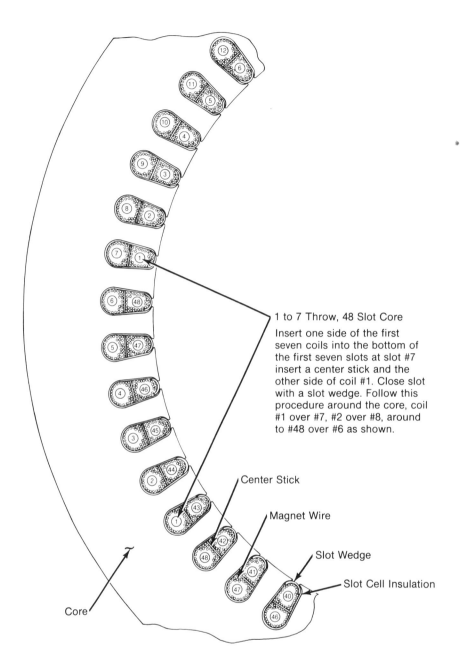

Figure 2-17. Cross Section of a Typical Stationary Armature Winding

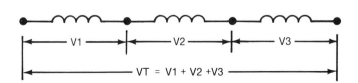

Figure 2-18. Coils Connected in Series

b. *Parallel connection.* To connect two or more coils in parallel, the start leads of all coils are tied together, and the finish leads of all coils are tied together, as shown in Figure 2-19. If all the coils connected in parallel are identical, the induced voltage across the entire group of paralleled coils will be that of any one coil. However, since each coil in the group is capable of carrying its full capacity of current, the total current carrying capacity of a group of coils in parallel will be the sum of the capacities of each coil in the group.

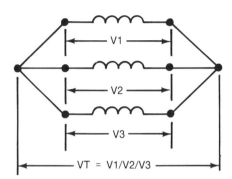

Figure 2-19. Parallel Connected Coils

Sometimes a particular design will require that an individual coil be center-tapped. This is accomplished by affixing a load lead to the middle or center turn of the coil. See Figure 2-20. The voltage measured from the center-tap lead to either end of the coil will be equal to one-half the voltage, end to end, of the entire coil.

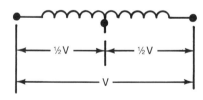

Figure 2-20. Center-Tapped Coil

2.2.5 Interconnecting armature windings. After all the individual coils have been inserted into the core slots, and the various slot and end turn insulating materials have been installed, the individual coils are grouped and connected in accordance with the winding design. At this point, the output load leads are secured to the start and finish ends of the major coil groups and the end turns are laced with lacing tape to provide mechanical strength.

The entire winding, less the load leads, are then given a number of coatings of a liquid insulating material such as varnish or epoxy compound applied by one of several methods. Each coat or layer of the insulating material is usually cured or dried in an oven prior to the application of the next coat. Depending upon the nature of the primary insulating material, a final overcoat of a special material may be required or specified to render the winding more impervious to adverse environmental conditions such as excessive moisture, corrosive atmospheres, abrasive dust, etc. As a general rule, Class F and H insulating materials are non-nutrient for most known forms of fungi, and therefore, by their basic nature will not require the addition of fungus inhibitors as is the case for Class A and B materials. See Section 2.5 for further information on the classes and nature of the various insulating materials used in alternator construction.

2.2.6 One and two circuit armatures. Sections 2.1.2–2.1.5 have covered the differences between single and three phase armatures, and in general, the basics of armature design and construction. Once the individual coils of an armature winding have been interconnected into major coil groups, the entire coil group assumes the properties of, and may be treated as though it were a single large coil, mechanically and electrically located in a single position on the core. Figures 2-21 and 2-22 illustrate a single phase armature with one coil centered at 12 o'clock (0°); and a three phase armature with three separate coils 120° apart, located at 12 o'clock; 4 o'clock; and 8 o'clock, or 0°, 120°, and 240°. It is important to bear in mind that this method of illustrating armature windings is a graphic depiction of the various aspects of single- and three phase phenomena, and not the way in which an armature winding is actually designed and constructed.

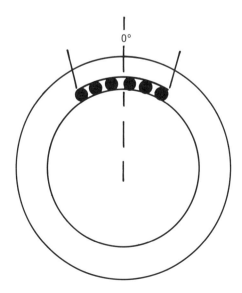

Figure 2-21. Single Circuit Single Phase Armature

Alternators (Synchronous Generators)

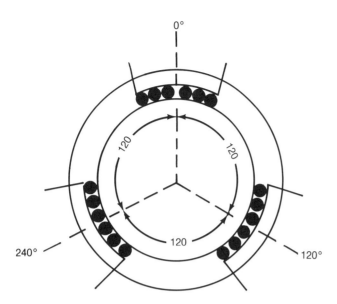

Figure 2-22. Single Circuit Three Phase Armature

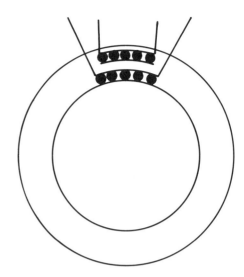

Figure 2-23. Single Phase Two Circuit Armature

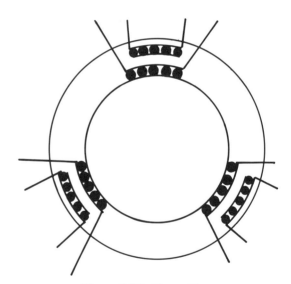

Figure 2-24. Three Phase Two Circuit Armature

a. *Single circuit armatures.* Armature windings such as illustrated in Figures 2-21 and 2-22 are termed **single circuit windings** since they are comprised of only one major coil group per phase, and have two load leads, one start and one finish per phase, available to be brought out for external connection.

b. *Two circuit armatures.* Frequently specifications will require alternators to be reconnectable, or have a high voltage range and a low voltage range, for example: a 40 KW, 50 KVA, 60 Hz alternator to be reconnectable for either 240 volts or 480 volts three phase. With a single circuit armature winding, this high-low voltage range reconnection is not possible. The voltage phase-to-phase may be either 240 or 480 volts, but not both. The answer to this requirement is to provide a unit with a two circuit winding (which in effect splits each of the three phase major coil groups into two equal groups as shown in Figures 2-23 and 2-24), and bring out all load leads, two start and two finish, from each phase for external connection. The low range voltage requirement is then met by connecting the two coils of each phase in parallel, while the high range voltage is achieved by connecting the two phase coil groups in series.

An important application note must be emphasized here. Having a unit with a two circuit stator simply means that the unit may be connected to produce either a low voltage range or a high voltage range, depending upon whether the windings are connected in parallel or in series. It does not mean that the machine is capable of producing both ranges of voltage simultaneously. Methods of single and three phase armature connections will be addressed in depth in Section 2.3.

2.2.7 Output load lead identification. Worldwide two standard systems, National Electrical Manufacturers Association (NEMA) and International Electrotechnical Commission (IEC), are basically used for identifying alternator load leads brought out for external (user) connection.

a. Alternators manufactured in the United States will usually follow the NEMA standard NEMA MG1 Part 2 (Ref. 1), where applicable, for marking load leads. Armature leads are given a number and a prefix letter T, that is: T1, T2, T3,...etc. In three phase systems, the individual phases are given letter designations (A, B, C), and are referred to as Phase A, Phase B, and Phase C.

Figures 2-25 and 2-26 illustrate both graphically and schematically the NEMA single phase lead identification system for both single and two circuit armatures.

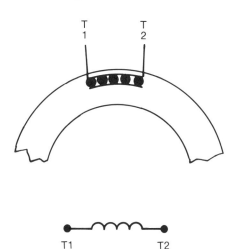

Figure 2-25. NEMA Single Phase Single Circuit Armature

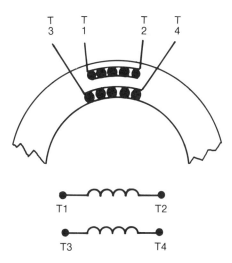

Figure 2-26. NEMA Single Phase Two Circuit Armature

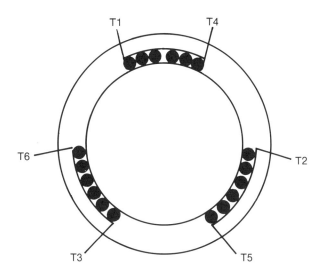

Figure 2-27. NEMA Three Phase Single Circuit Armature Lead Wire Numbering System

Similarly, Figures 2-27 and 2-28 illustrate the NEMA three phase single and two circuit lead terminal designations. It may be helpful to consider leads T1, T2, T3, T7, T8, and T9 as the start leads of their respective coil groups, while leads T4, T5, T6, T10, T11, and T12 represent the finish leads of these individual coil groups.

b. The second system of alternator armature lead numbering is outlined in IEC Standard 34, Part 8 (Ref. 2), and also British Standard BS4999 Part 3 (Ref. 3). In this system, the three phases are designated U, V and W. Armature leads are given a prefix letter and a number (as with the NEMA system). However, unlike the NEMA system, the prefix letter is used to indicate a particular phase, as illustrated in Figures 2-29 and 2-30. A study of both the NEMA and the IEC systems of lead wire identification, Figures 2-28 and 2-30, can lead to an alternate configuration for a three-phase two circuit armature winding by considering coil groups T1-T4; T2-T5; T3-T6 and U1-U2; V1-V2; W1-W3 as outer coil groups, and groups T7-T10; T8-T11; T9-T12; and U5- U6; V5-V6; W5-W6 as inner coil groups of a two circuit winding. Our studies of the method of alternator armature construction earlier in this chapter clearly indicate that there can be no such distinction of outer and inner coils in an actual armature winding. However, the author has found this type of thinking to be useful both

Alternators (Synchronous Generators)

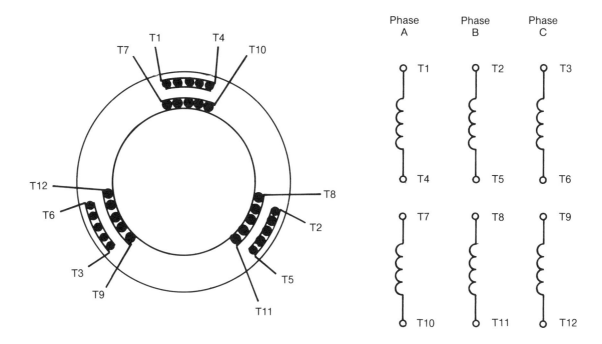

Figure 2-28. NEMA Three Phase Two Circuit Armature Lead Wire Numbering System

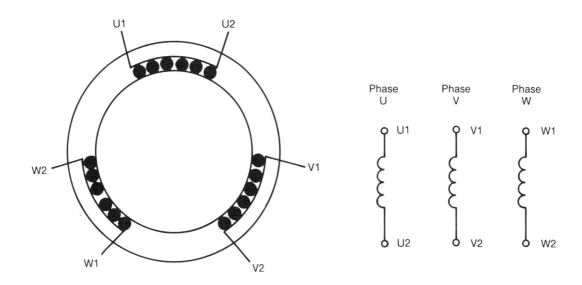

Figure 2-29. IEC/BS Three Phase Single Circuit Armature Lead Wire Numbering System

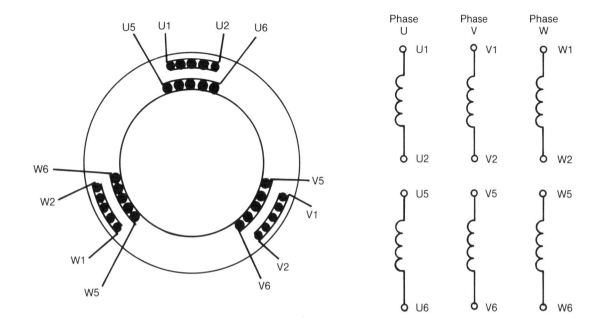

Figure 2-30. IEC/BS Three Phase Two Circuit Armature Lead Wire Numbering System

in analysis of wiring diagrams and in re-identification of load leads whose number tags have been lost or obliterated in service.

Alternator field leads are given the prefix F. Both the NEMA and IEC systems specify that the field leads of synchronous generators shall be designated F1 and F2, with F1 being the positive lead, and F2 being the negative lead. However, many automatic voltage regulator manufacturers are now designating the Automatic Voltage Regulator's (AVR) exciter output terminals as F+ and F−, not F1 and F2. Therefore, many brushless alternator manufacturers have deviated from strict compliance with NEMA and IEC standards to not only conform with voltage regulator terminal markings, but to avoid confusing operating and maintenance personnel with conflicting marking systems when flashing exciter fields with batteries whose terminals are either marked (+) and (−) or POS and NEG.

2.3 CONNECTIONS FOR SINGLE AND THREE PHASE ALTERNATORS

In Section 2.2 we covered the basics of single and three phase alternator armatures and the NEMA load lead identification system. In this section we will cover in depth the various methods of external connection for single and three phase one and two circuit armatures, the output voltages available from the various connections, and reconnection of one and two circuit three phase alternators for total single phase output.

2.3.1 Single phase alternators. NEMA lists 120 volts, 240 volts, or a combination of 120 and 240 volts as standard single phase voltages for 60 hertz systems.

a. *Single phase, single voltage, two lead alternators.* This type unit is usually constructed having a single circuit armature and two load leads brought out. It is designed to produce a single output voltage, usually either 120 volts or 240 volts (Figures 2-31A and 2-31B). Units built to produce a specific non-standard voltage such as 480 volts or 600 volts single phase are usually designed as two load lead machines.

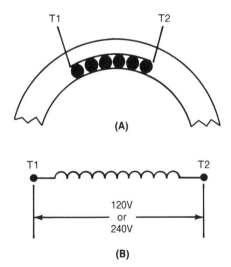

Figure 2-31. Single Phase, Two Load Lead Alternator Armature

Alternators (Synchronous Generators)

b. *Single phase, three load lead, dual voltage alternators.* This type of single phase alternator is essentially identical to a two load lead design. The difference is that the phase coil has been center-tapped and the center tap lead T0 brought out with start and finish leads T1 and T2 for external connection. Thus, this configuration will allow a dual single phase voltage output, with the voltage from either T1 or T2 to T0 one-half the designed line to line (T1 to T2) voltage of the machine. See Figures 2-32A and 2-32B.

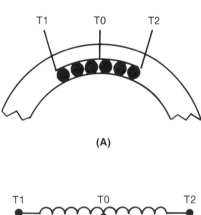

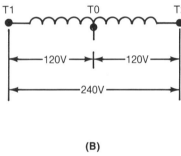

Figure 2-32. Single Phase, Three Load Lead Armature

c. *Single phase, four load lead reconnectable alternators.* Due to their versatility, these alternators are the most commonly produced type of single phase units. They have a two circuit armature winding, with each of the two individual coil groups usually designed for 120 volts output. As stated in Section 2.2.6, a low voltage range is accomplished by connecting the coils in parallel, and a high voltage range by connecting the coils in series as shown in Figures 2-33A and 2-33B. With the unit in parallel, it is used for 120 volt two wire service, as shown in Figure 2-34. When connected in series the unit may be used for 240 volt two wire service. With a third output load lead L0 attached to, and brought out from the junction of leads T2 and T3 (series connection point), the unit may be used for three wire 120/240 volt service, as shown in Figure 2-35.

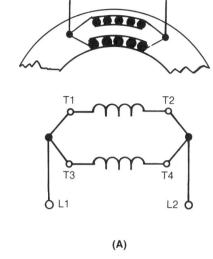

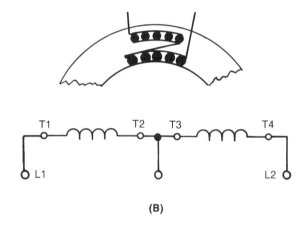

Figure 2-33. Parallel and Series Connections

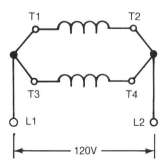

Figure 2-34. Parallel, 2 Wire 120 Volt Service

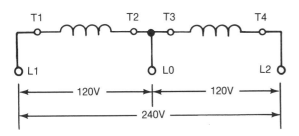

Figure 2-35. Series 2 Wire 240 Volt Service or Series 3 Wire 120/240 Volt Service

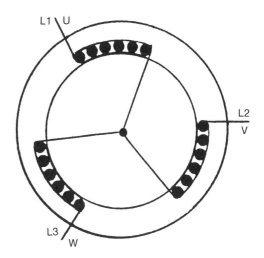

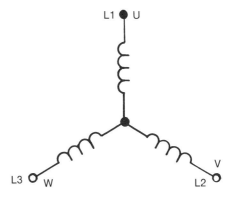

Figure 2-36. Three Wire WYE Connection

2.3.2 Three phase alternators. In Section 2.2.1 we learned that a three phase alternator armature is constructed so there are three separate independent phase coil groups spaced 120 electrical degrees apart, and that this 120 electrical degrees of separation between the three phase coil groups provides the balance in a three phase power generation system. We also learned that each phase is given a letter designation; Phase A, Phase B, and Phase C (NEMA Standard designation), or Phase U, Phase V, Phase W (IEC Standard designation), and that each phase may either be comprised of one major coil group (single circuit design), or that each major phase coil group may be divided into two equal coil groups (two circuit design). In this section we will cover the two methods of connection for three phase armatures, use of a three phase unit for single phase loading, and the methods of externally reconnecting both single and two circuit three phase armature windings for total single phase output.

a. *Wye connected three phase alternators.* General: The majority of three phase alternators presently in service throughout the world are wye (star) connected. In this method of three phase armature connection, the finish leads of each of the three phase coil groups are tied together to form a neutral point, sometimes called the **wye point** or **star point**, and the start leads become the three output load leads. See Figure 2-36. Frequently, a fourth or neutral lead is affixed to the neutral point and brought out for external connection. See Figure 2-37.

b. *Grounding (earthing) the neutral in a wye connected alternator.* In the majority of fixed installations, the neutral of a wye connected alternator is connected to ground. Grounding of an alternator should be done in accord with the National Electrical Code NFPA 70, Article 250 (Ref. 4), or the code or standard having jurisdiction at the site of the installation. Several grounding methods are approved. One method involves driving a copper ground rod eight feet or more into the earth, and connecting the lead to be grounded to this rod. If the neutral is not brought out, or if the neutral lead is insulated and left ungrounded, it is referred to as being a **floating neutral**.

c. *Voltages available in a three phase wye connected alternator.* The voltage generated in each phase of a wye connected alternator, measured from an output load lead to neutral, is termed **line to neutral** (L to N), or **phase voltage**. The voltage measured between any two output load leads in a wye connected unit is termed **line to line** or **phase to phase**. See Figure 2-38.

d. In Section 2.2.1 we discussed the results of the 120 electrical degree separation between each of the three phases of a three phase alternator. Figure 2-12 illustrates graphically that the voltage measured between any two lines (L to L) of a wye connected unit is equal to the voltage measured from line to neutral (L to N) times 1.732, the square root of 3 ($\sqrt{3}$). Thus, the voltages

Alternators (Synchronous Generators)

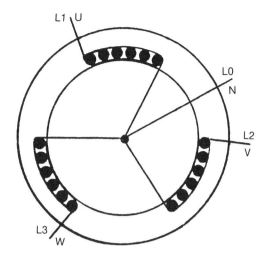

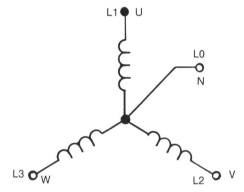

Figure 2-37. Four Wire WYE Connection

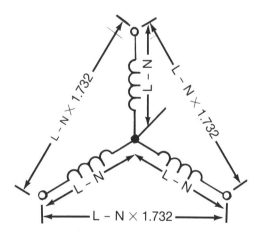

Figure 2-38. Voltages in a Three Phase WYE (Star) Connected Alternator

present in a wye connected alternator may be calculated by the following equations:

Voltage LL = Voltage LN × 1.732   (Equation 2-1)

Voltage LN = $\dfrac{\text{Voltage LL}}{1.732}$ (Equation 2-2)

Since the reciprocal of the square root of 3 (1 divided by 1.732) is 0.577, Equation 2-2 above may be modified to be:

Voltage LN = Voltage LL × 0.577   (Equation 2-3)

Three phase voltages are usually specified as the line to line voltage desired. However, frequently both the line to line and the line to neutral voltages will be called out. In these cases the line to line voltage is listed, then a slant line, and then the line to neutral voltage, i.e., 208/120 volts where L-L is 208 volts and L-N voltage is 120 volts.

Table 2-1 lists the line to line and line to neutral voltages for a number of commonly seen three phase system voltages, 600 volts and lower.

Table 2-1. Voltages of Three Phase Wye Connected Alternators

Voltage L-L	Voltage L-N	Voltage L-L	Voltage L-N
208	120	416	240
220	127	440	254
225	130	450	260
230	133	460	266
240	139	480	277
380	220	575	332
400	230	600	346

A study of Table 2-1 shows that when the three phase line to line system voltage is above 208 volts (low wye), or 416 volts (high wye), the voltage available from line to neutral will be higher than standard 120 or 240 volt single phase system voltage. For example, when the three phase system voltage is either 240 volts or 480 volts, the line to neutral voltage available will be either 139 volts or 277 volts, a value that is 15% higher than normal nameplate voltage for most single phase equipment. For this reason, care must be taken when using wye connected units for both three phase and single phase loads that the line to neutral voltage is not too high for the connected load.

e. *Three phase, four lead, wye connected alternators.* These machines are internally connected at the manufacturer's facility in a wye configuration, with the fourth or neutral lead brought out for external connection. In some instances, a 480 volt wye connected unit may have two of its three phase coil groups center tapped to provide a single phase 240 volt source for automatic voltage regulator input power. Refer to Figure 2-39. In instances where these center taps are not made, or in the case of 600 volt alternators where 120 volts or 240 volts single phase is not readily available, a power isolation transformer is used to supply the unit's automatic voltage regulator with the required voltage. Additional discussion on use of power isolation transformers is provided in Chapter 4, Section 4.7.

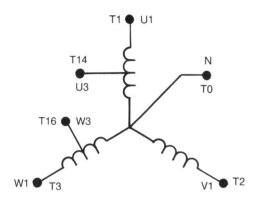

Figure 2-39. Four Lead WYE Connected Alternator with Two Phases Center Tapped

f. *Three phase, 10 lead, wye connected alternators.* These units are constructed as two circuit machines with the finish leads of the inner coil groups tied together, and a tenth or neutral lead affixed to this junction and brought out for external connection. These units are then reconnectable for low range voltages in a parallel connection (Figure 2-40A), or high range voltages in a series connection (Figure 2-40B).

2.3.3 Delta connected three phase alternators.

On delta connected alternators, the phase coil groups are arranged in a triangular configuration, as illustrated in Figure 2-41. As stated earlier, the symbol for a delta connection is a triangle. The connection derives its name from the Greek letter Delta, which is also represented by a triangle.

A study of Figure 2-41 shows that the delta connection is made by joining the start lead of one phase coil group to the finish lead of the phase coil group immediately to its

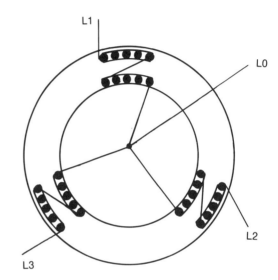

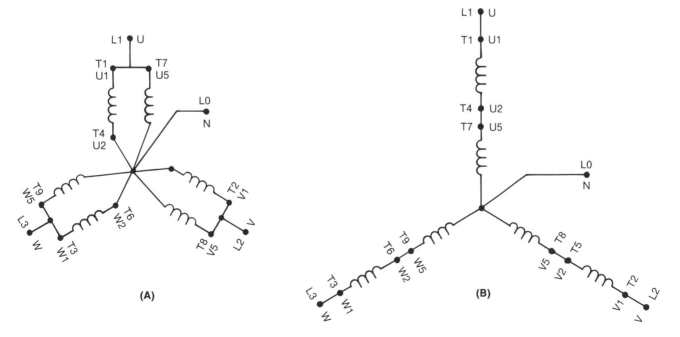

Figure 2-40. Ten Lead WYE Connected Alternator Connections

left, and so on, from phase-to-phase around the triangle. An output lead is then affixed to each corner of the triangle, creating a three wire delta connected alternator.

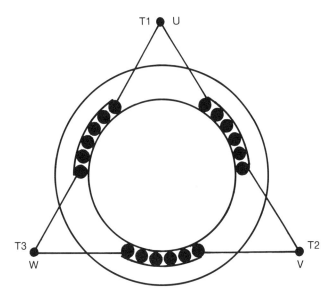

Figure 2-41. Delta Connected Three Phase Armature

a. *Three phase, four lead delta connected alternators.* On these units one of the phase coil groups is center tapped, and the center tap lead brought out as lead L0 or N. If, for example, the designed voltage of each phase coil group is 240 volts, the available voltage from the center tap lead (L0/N) to either adjacent corner of the delta will be 120 volts. Thus in this case, usable 120 volts single phase voltage may be derived while the three phase voltage is 240 volts (Figure 2-42), as opposed to the 139 volt line to neutral voltage available from a three phase, wye connected unit as shown in Table 2-1.

Figure 2-42 also points out one of the aspects of a center tapped delta that is a frequent source of application problems—that of the so called wild leg, or high leg of a delta. In this case voltage is taken from the center tap to the opposite corner of the delta (L3/W). Instead of 120 volts (as would be the case from L0 to either L1 or L2), the voltage from L0 to L3 is 208 volts. When this occurs, the connected load will see 75% high voltage. The reason for this is that the voltage here will be the vector sum of 120 volts plus 240 volts, or the voltage across one phase times the cosine of 30°—0.866. Thus: 240 volts × 0.866 = 208 volts.

b. *Grounding a delta connected alternator.* Any single point in a three lead delta connected machine can be grounded. The ground point on a 240/120 volt four lead delta machine is L0.

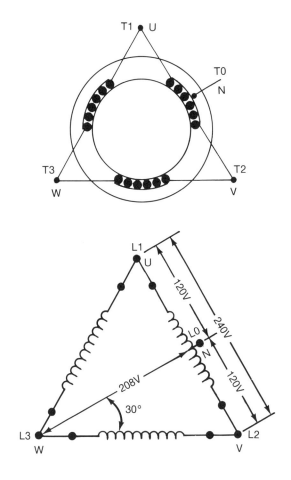

Figure 2-42. Voltages in a Four Wire, Three Phase Delta Connected Alternator

2.3.4 Three phase six and twelve lead reconnectable alternators

a. *Six lead alternators.* Six lead three phase alternators are single circuit machines with all six phase leads (three start leads and three finish leads) brought out for external connection. They may be connected in either a wye (Figure 2-43) or delta (Figure 2-44) connection. Six load lead alternators may also be reconnected for total single phase service. This is covered in para. c.

b. *Twelve lead alternators.* Twelve load lead units, as is the case with six lead units, may be connected either in a wye or delta connection, or for total single phase service. Since these units are constructed with two circuit armatures with all twelve load phase leads (six start leads and six finish leads) brought out for external connection, they may be connected in either low voltage range (parallel connection) as illustrated in Figures 2-45 and 2-46, or high voltage range (series connection) per Figures 2-47 and 2-48.

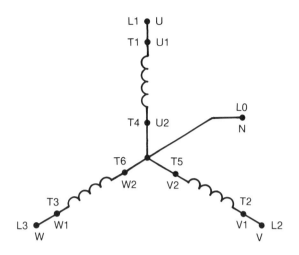

Figure 2-43. Six Lead WYE Connection

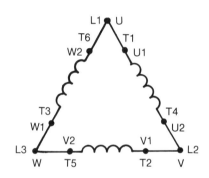

Figure 2-44. Six Lead Delta Connection

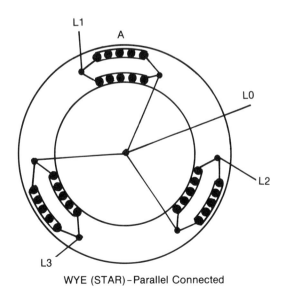

WYE (STAR) - Parallel Connected

Figure 2-46. Low Voltage WYE (Parallel Connected)

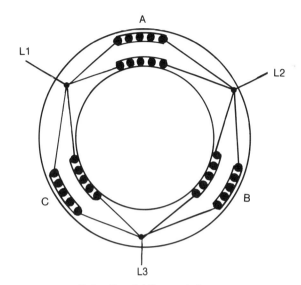

Delta - Parallel Connected

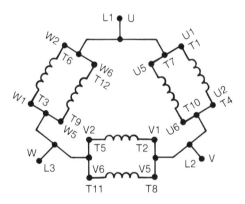

Figure 2-45. Low Voltage Delta (Parallel Connected)

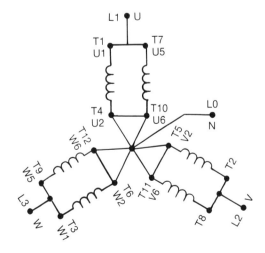

Alternators (Synchronous Generators)

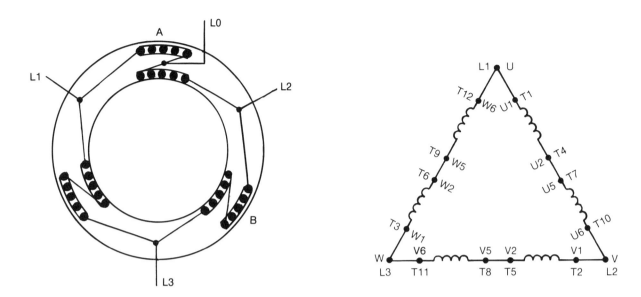

Delta-Series Connected with
"A" Phase Mid-point (L0) Brought Out

**Figure 2-47. High Voltage Delta
(Series Connected)**

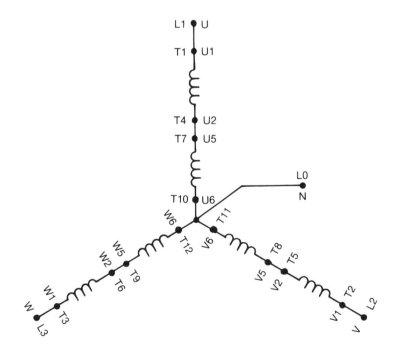

**Figure 2-48. High Voltage WYE
(Series Connected)**

c. *Single phase connection of six and twelve lead alternators.* Three phase armatures may be connected to a total single phase configuration if all of the three phase coil group leads are brought out for external connection. One method of single phase reconnection of the leads is termed a zig-zag connection. Figures 2-49 and 2-50 illustrate this connection as used with both six and twelve lead alternators.

A second method of reconnection for total single phase service is available for twelve lead units only. This method is termed a **double delta connection**. In this connection the outer coil groups are connected as in a six wire delta; the inner coil groups are also connected as in a six wire delta, and then both of these two small deltas are connected in series as shown in Figure 2-51. Note that in the particular double delta illustrated, the two coils of Phase C are connected in series to form the base of the double delta. Regardless of which single phase connection method is used, consult the manufacturer of the alternator for the recommended single phase output rating.

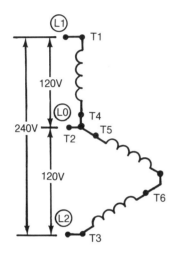

Figure 2-49. Single Circuit, Six Lead Zig-Zag Connection

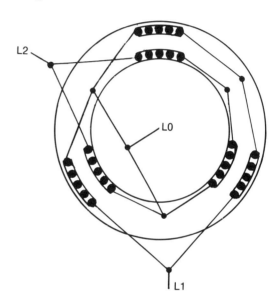

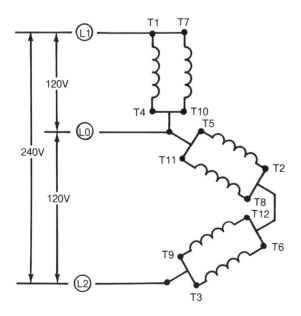

Figure 2-50. Two Circuit, Twelve Lead Zig-Zag Connection

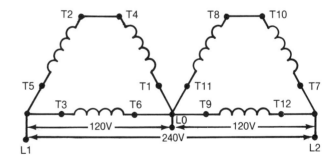

Figure 2-51. Twelve Wire Double Delta Single Phase Connection

2.4 FREQUENCY, FREQUENCY REGULATION, VOLTAGE AND VOLTAGE REGULATION

Frequency and frequency regulation along with voltage and voltage regulation will be covered in two separate sections. We will discuss these subjects from the standpoint of the various aspects of alternator type, design, and construction; the impact of these factors on both voltage and frequency; and the regulation of voltage and frequency. Automatic voltage regulators, engine speed governors and the various accessories associated with these types of devices will be covered in depth in separate sections of this book.

2.4.1 Frequency and frequency regulation

a. *Frequency of alternating current.* The frequency of the alternating current produced by any alternator (ac generator) is dictated by two factors; the number of magnetic poles constituting the field of the unit, and the speed of its shaft rotation (RPM).

b. *Number of field poles.* The number of pairs of poles used to construct the field (always an even number of poles) will determine the number of electrical cycles produced by each complete revolution of the unit's shaft. For example: a two pole field will produce one complete cycle per revolution, a four pole field two complete cycles per revolution, a six pole field three complete cycles per revolution, and so on. Dividing the number of poles comprising the alternator's field by two yields the number of complete cycles produced per revolution of the shaft.

c. *Speed of rotation (RPM).* Once the number of field poles has been established, the alternator has no further control over the frequency of the alternating current it produces. The actual frequency produced will be directly proportional to speed. Since frequency is measured in hertz (cycles per second), and rotative speed in RPM (revolutions per minute), frequency calculations can be made with the following formulas:

$$\text{Frequency} = \frac{\text{RPM} \times \text{Number of Poles}}{120} \quad \text{(Equation 2-4)}$$

$$\text{RPM} = \frac{\text{Frequency} \times 120}{\text{Number of Poles}} \quad \text{(Equation 2-5)}$$

$$\text{Number of Poles} = \frac{\text{Frequency} \times 120}{\text{RPM}} \quad \text{(Equation 2-6)}$$

d. *Standard Alternating Current Frequencies.* There is no single standard frequency for alternating current. A number of frequencies have emerged as standards by commercial power generation companies, and even more for special applications such as 400 hertz for the aircraft industry, 120 hertz for wood working, and 180 hertz used with concrete vibrator equipment. Sixty and 50 hertz have emerged as the two most commonly seen frequencies world-wide for commercial and residential users. Since almost all ac power consuming devices, with the possible exception of purely resistive devices, are frequency dependent to a greater or lesser degree, and the economics of production and distribution of the devices,

and of the electric power itself, have dictated that most geographic areas such as nations, or blocks of nations standardize on either 60 or 50 hertz. For example, 60 hertz has become the standard frequency for most commercial and residential uses in North America, while 50 hertz has become the standard commercially available frequency in most of Europe.

Table 2-2 lists the required input RPM to produce either 60 or 50 hertz for two, four, six, and eight pole alternators.

Table 2-2. Synchronous Speeds for 2, 4, 6, and 8 Pole Alternators

Frequency in Hertz (Hz)	RPM			
	2 Pole	4 Pole	6 Pole	8 Pole
60	3,600	1,800	1,200	900
50	3,000	1,500	1,000	750
RPM Per Hz	60	30	20	15

From the table above it can readily be seen that either a frequency meter or a tachometer may be used to set engine speed, or to determine the frequency being produced. If a frequency meter is used to determine engine speed, or a tachometer is being used to determine frequency, you must know the number of poles in the alternator's field to make the calculation. To calculate RPM from a frequency meter reading, multiply frequency by the RPM/Hz factor from the above table. To determine frequency from a tachometer reading, divide RPM by the RPM/Hz factor.

e. *Frequency regulation.* Regulation of the frequency produced by an alternator is strictly a function of the prime mover speed governor or speed governing system. Governors may range from very simple spring operated devices to complex hydraulic or electronic systems. Frequency regulations may be considered to be the degree of accuracy to which the governor can maintain rotative speed at any level of load relative to the unloaded or no load setting. For most governors, the no load or unloaded speed is usually the highest speed allowed, and then with addition of load, the unit will recover to either the preset speed or to some value lower than the no load level. If the governor maintains speed at a constant value regardless of load level, it is said to be an **isochronous** governor. If the speed droops with addition of load, the governor is said to be a **droop type** governor. Frequency regulation is calculated by the following equation:

$$\% \text{ Freq. Reg.} = \frac{\text{RPM No Load} - \text{RPM Full Load}}{\text{RPM Full Load}} \times 100$$

(Equation 2-7)

f. *Summary.* The frequency of the alternating current produced by an alternator is strictly a function of the speed of rotation of the alternator's shaft. Most alternating current (ac) Engine-generator sets are designed to operate at a specific output frequency. Therefore, we can consider the set's prime mover (engine) and its speed governing system to be a source of constant frequency. How hard the prime mover and its speed governing system must work to maintain this constant frequency is purely a function of the efficiency of the set's alternator in converting rotating mechanical energy into electrical energy, and the nature and magnitude of the load imposed on the alternator. The degree of accuracy to which the prime mover and its speed governing system maintain frequency constant is the frequency regulation of the system.

2.4.2 Voltage and voltage regulation

a. *Alternator output voltage.* In section 1.1.6, electromagnetic induction, we discussed that the value of voltage induced in an alternator's armature windings was dependent upon three basic factors:

1. Strength of the magnetic field, or flux density of the field.

2. Speed with which the flux lines were cut by the conductor coil(s).

3. Number of turns comprising the coil.

Engineers usually combine factors (1) and (2) above and state them together as the number of lines cut per unit time. Therefore, when working with a rotating electrical machine such as an alternator, which is designed to operate at a fixed speed, we can consider that the coils in the armature winding have been designed to produce a specific voltage at a specific speed of rotation, and that the conductors used in the winding have been sized to carry a specific maximum amount of current. While many factors ultimately affect the magnitude of the voltage produced by an alternator, they are all designed to produce a specific voltage under fixed conditions of load, rotative speed and flux density. Varying any one or all of these fixed conditions will directly affect the output voltage of the machine.

An alternator manufacturer will design several different units of the same power output capacity, but with different output voltages, by using identical fields and exciters, merely changing the main armature winding design to produce the desired special voltage.

b. *Voltage regulation.* The EGSA Glossary of Standard Industry Terminology and Definitions, Electrical (EGSA 101E) defines voltage regulation as "The voltage regulation of an engine generator set is the difference between steady state no load and steady state full load output voltage expressed as a percentage of the full load voltage."

c. *Permanent magnet alternators.* Alternators employing permanent magnets for their fields are usually small machines of less than 5 kilowatts capacity. In Section 2.2.3(a) we learned that these units have the advantage of being rather cost effective, but have the disadvantage of having their field strength maintained at a fixed level. In subsequent paragraphs we will discuss the various factors affecting alternator output voltage. However, for our purposes here, we will merely state that machines with a fixed value of flux density do not have the capability of either increasing or decreasing field strength to compensate for these factors. For this reason these machines are usually designed with a no load voltage somewhat higher than rated nameplate value, and allow the factors of speed, temperature and load to reduce this higher voltage to the nameplate range.

d. *Voltage regulated units.* These units employ electromagnetic fields, and have the ability to vary the strength of their field to compensate for the various factors affecting voltage. Voltage regulated alternators fall into two basic classes:

1. *Externally voltage regulated alternators.* These alternators employ an external automatic voltage regulator that senses the unit's output voltage, compares it with an adjustable preset reference voltage, and then corrects the exciter output as necessary to maintain the output voltage at the preset level. These units may be either separately excited by an exciter external to the unit such as a static exciter, or self excited with the exciter incorporated within its frame. See Section 2.2.4, Figures 2-6 and 2-8. The advantage of this type of voltage regulated unit is that the voltage regulator senses voltage and initiates exciter response to maintain a constant output voltage.

2. *Self or inherently voltage regulated alternators.* This type of alternator is usually not only self voltage regulated, but also self excited. This means that the exciter is incorporated in the machine, and not an external component such as a static exciter. These alternators are normally designed with some form of excitation compensation based on load current. Typically the exciter design will incorporate current transformers, and may or may not include some form of no load voltage adjustment. The advantage of these types units is that they do not need an external voltage regulator, eliminating a possibly vulnerable component. In addition, since these units usually employ line current as a source of power, they tend to have very good induction motor starting capability. A disadvantage is that they do not have the ability to compensate for changes in rotative speed, load power factor, or winding temperature rise.

Alternators (Synchronous Generators)

e. *Factors affecting voltage regulation*

1. *Temperature of windings.* The copper wire used in the construction of alternator windings, while being an excellent conductor of electricity, does have some resistance to the flow of electrical current. The resistance of the wire is a function of its cross sectional area and the length of the wire, and is of a given value at a specified temperature. The resistance of copper wire will increase or decrease in direct proportion to any change in the temperature of the wire. When current is flowing in the wire, the resistance of the wire will cause heating (warm up) of the windings in direct proportion to the magnitude of current flowing in the windings. Therefore, when a winding is loaded to its maximum rated current carrying capacity, its increase in temperature (temperature rise) will be at maximum and the resistance of the winding will also be at its maximum. This increased resistance from no load with cold windings to full load will cause the output voltage of the unit to drop gradually as the windings warm up to operating temperature. This drop in voltage is termed **temperature drift**. For non-voltage regulated units, or self voltage regulated units not having some form of temperature compensation, the effects of temperature drift may well have a marked effect on their voltage regulation. For this reason the formula to calculate voltage regulation takes into account temperature drift, as follows:

$$\% \text{ Voltage Regulation} = \frac{\text{VNLC} - \text{VFLW}}{\text{VFLW}} \times 100$$

(Equation 2-8)

Where:

VNLC = No Load Voltage, Cold Windings.
VFLW = Full Load Voltage, Warm Windings.

NOTE: Cold winding temperature is usually considered to be 20°C (68°F), and warm winding is the stabilized temperature at full load. Larger units may require as much as three to four hours of operation at rated current flow to reach a point of constant or stabilized temperature.

Temperature drift due to the warm up of an alternator's windings will not usually create any particular problems for units having an automatic voltage regulator, since the regulator will automatically compensate for this drift. However, the automatic voltage regulator itself may be affected by change in temperature, not only from its own internal warm up, but also any change in the operating ambient temperature. The effects of temperature, both ambient and temperature rise, upon regulator operation will be covered in the voltage regulator section of this manual.

2. *Lagging power factor loads.* The out-of-phase component of current in a reactive load will have a direct impact upon the output voltage of an alternator. In the case of a lagging power factor load (where inductive reactance is the predominant reactance) alternator output voltage is depressed or lowered. This may be thought of as the out-of-phase component of current having the effect of cancelling some of the lines of flux of the alternator's main field. The lower the value of this lagging power factor, the more depressing effect it has upon voltage per unit of load. In non-regulated, or self regulated units, the effect is sometimes referred to as **reactive voltage droop**.

In the case where an alternator has an automatic voltage regulator, the voltage regulator will automatically increase its output, thus increasing the excitation of the main field to compensate for this droop in voltage. The results of this action are:

(a) The automatic voltage regulator's increased power demand may be in excess of its full load current handling capacity, and could result in premature failure of the unit.

(b) The increase in excitation current could cause excess heating in the main field windings, leading to possible premature alternator failure.

(c) The same overloading effects will also apply to the unit's exciter, whether it is static, external rotating, or brushless.

3. *Leading power factor loads.* In contrast to a lagging power factor load which exhibits the effect of depressing system voltage, a leading power factor load (where the predominant value of reactance is capacitive) will cause the alternator's terminal voltage to rise. This can be thought of as the out-of-phase component of current, having the effect of increasing the number of lines of flux of the main field. Thus a leading power factor load will result in a rise in alternator terminal voltage, and the lower the leading power factor, the greater the rise in voltage per unit load.

This rise in voltage can be the source of real voltage regulation problems, especially for alternators equipped with automatic voltage regulators. The rise in voltage is compensated for by the regulator reducing its power output, thus reducing the excitation of the main field. In severe cases, this reduction in the regulator's power output may result in a current flow below the minimum value required for stable operation. This could lead to voltage fluctuation, or even to completely uncontrolled terminal voltage. Once the regulator has shut itself off, the load may well continue to excite the alternator with terminal voltage rising to an uncontrolled value.

4. *Rotative speed.* The natural inclination of any alternator is to vary its voltage directly with the speed of its shaft rotation. This natural tendency can present a problem to inherently regulated or self voltage regulated alternators. If the set's prime mover is equipped with a good governor, the resulting system voltage regulation should be fairly good. However, if the prime mover is equipped with a governor that has fairly broad speed regulation bandwidth, the net voltage regulation for the entire system could be poor.

Here, as with the effects of temperature, voltage regulation of alternators equipped with automatic voltage regulators will provide better voltage regulation since these devices can usually compensate for speed variations ±5% of normal or rated speed.

5. *Nature and magnitude of the connected load.* The effects of the connected load on alternator performance will be covered in later sections. However, for our purposes here we can state that with the exception of a load that is capacitive in nature, the effect of load on an alternator is to depress the voltage. It takes a specific amount of field strength (flux density) to support voltage under any given circumstance of loading, and at no load the least amount of flux is required to maintain output voltage at any desired level. If load is added to the alternator without an increase of the no load level of flux density, the voltage will be depressed.

f. *Regulated voltage bandwidth.* The voltage regulation of an alternator is usually expressed as a percent, plus or minus of rated voltage, within which the unit will maintain its output voltage under any load from no load to full rated load, rated power factor. This value of voltage plus or minus a specific nominal value represents the maximum and minimum operating voltage levels, i.e., the bandwidth within the alternator will maintain its voltage under any condition of loading not exceeding rated values.

g. *Summary.* Three basic factors have a direct impact upon an alternator's output voltage over which the alternator cannot exercise direct control. These three factors are:

1. *Speed of Shaft Rotation.* Rotative speed is strictly a function of the unit's prime mover and its speed regulating system.

2. *Winding temperature.* Winding temperature is a function of the amount of current demanded by the connected load.

3. *Nature and power factor of the connected load.*

Voltage regulation is usually expressed as a percent plus or minus a specific system voltage within which the alternator will maintain its output voltage under any condition of load that is within the nameplated or rated operating parameters of the alternator. The three general categories of alternators that regard voltage regulation are as follows:

1. *Non-regulated.* This type of alternator does not have any form of voltage regulation circuitry or equipment. The unit is designed to produce a given voltage under a given set of operating parameters. Operation at any condition other than designed condition will result in an output voltage other than the rated value. A permanent magnet alternator is an example of a non-voltage regulated alternator.

2. *Self-regulated.* Self voltage regulated alternators have electromagnetic fields, and usually employ some form of current sensing or current responding circuitry. As a rule they do not have any method of sensing and responding to voltage or change in voltage.

3. *Externally-regulated.* Externally voltage regulated alternators are designed with electromagnetic fields and employ externally mounted voltage regulation devices or circuitry. Usually these devices automatically regulate output voltage by sensing the alternator terminal voltage, and initiate correction as necessary to maintain terminal voltage at the preset value.

An engine generator set consists of an engine (prime mover) and an alternator. In the summary of Section 2.4.2 we discussed that the set prime mover and its speed governing system should be considered as a source of constant frequency, and that the amount of work involved in maintaining frequency constant is dictated by the connected load and the efficiency of the alternator. Using this same way of thinking, we can consider the alternator together with its voltage regulator, or regulation system, as a source of constant voltage. How hard the alternator/exciter/voltage regulator or regulation system as a package has to work to maintain output voltage constant is purely a function of the load and load factors. Voltage regulation, then, can be considered as the degree of accuracy to which the equipment maintains output voltage at the preset or designed value. It is important to note that any statement of voltage regulation made by the manufacturer of the alternator, especially for non regulated or self regulated alternators, is usually valid only under conditions of constant rotative speed or frequency.

2.5 TEMPERATURE AND ENVIRONMENTAL CONSIDERATIONS OF ALTERNATORS

In previous sections we discussed some of the effects of heating on alternator windings due to current flowing in

Alternators (Synchronous Generators)

the windings. We also discussed some of the various insulating materials used in the fabrication of the windings. Sections 2.5.1 and 2.5.2 will deal with the grouping of insulation materials into classes according to temperature tolerances, methods of temperature measurement, duty cycles, and the effects of ambient temperature and altitude upon alternator ratings. Section 2.5.3 will discuss the various environmental considerations to be considered in alternator application, and Section 2.5.4 will address the various temperature related devices used with alternators.

2.5.1 Temperature and classes of insulation materials

a. *Classes of Insulating Materials.* The temperature tolerances of the various insulating materials used in the construction of electrical machinery will vary with the composition of the material. These materials are presently grouped into four classes (A, B, F, and H) and each class of material has been assigned a maximum allowable total temperature for continuous duty by various standards writing associations. The standard adhered to by most manufacturers of rotating electrical machinery in the United States is NEMA Standard MG1, Parts 16.40 and 22.40.

b. *Methods of temperature measurement.* Two methods of temperature measurement are currently accepted by NEMA; these are the resistance method and the embedded temperature detector method. A third method, the thermometer method, was used until 1969 when it was dropped by NEMA as an approved method of measurement. The major objection to this method was its poor repeatability of specific measurements. Physical limitations precluded embedding a thermometer directly in the windings; therefore, measurements were taken on the outside of the winding end turns, the best ventilated (coolest) area of the winding. For this reason the temperatures measured by this method were lower than measured by the other two methods.

c. *Resistance method of temperature measurement.* In previous sections we learned that the resistance of a conductor will vary directly with change in the temperature of the conductor. In fact, while the exact change in resistance of a conductor per degree of temperature change is very small, it is quite precise. The resistance method of measurement consists of comparing the resistance of the windings at a known temperature (this reading is taken before the start of the alternator temperature rise test) with the resistance of the windings after the unit stabilizes at a steady state known load. A simple calculation using the start temperature, the amount of change in resistance, and the specific value of temperature versus resistance of the conductor material will give the final temperature of the winding. The resistance method of temperature measurement yields an average temperature of the entire winding. This average temperature will be higher than the value measured by the thermometer method and lower than that given by an embedded temperature detector.

d. *Embedded detector method.* This method of winding temperature measurement consists of insertion of one or more detectors (see para. 2.5.4 of this chapter for specifics of temperature detectors) in the slots of the core during manufacture of the winding. Since these detectors are located right in the core of the winding (which is the hottest part of the assembly), the temperature allowed for this method of measurement will be higher than that allowed for any other method of measurement.

e. *Unit of temperature measurement.* Degrees Celsius (°C) is the standard unit of temperature measurement used in NEMA Standard MG-1. IEC and British standards use degrees Celsius for ambient temperature and the Kelvin scale for other temperatures. Since many specifications state ambient temperatures in degrees Fahrenheit (°F), the following equations may be used to convert °C to °F, and °F to °C:

$$°C = \frac{5(°F - 32)}{9} \quad \text{(Equation 2-9)}$$

$$°F = \frac{9(°C)}{5} + 32 \quad \text{(Equation 2-10)}$$

2.5.2 Temperature considerations of alternator operation

a. *Ambient temperature.* NEMA MG1-22.40 (Ref. 1) defines ambient temperature as the temperature of the cooling air as it enters the ventilating openings of the machine. The distinction here is that if an engine generator set is installed in an enclosure or structure with inadequate ventilation, heat radiating from the set may raise the temperature of the cooling air before it enters the ventilating openings of the alternator. Thus, the windings are assumed to be at the temperature of this cooling air prior to any rise in temperature due to loading. The table of maximum allowable temperature rises contained in NEMA MG1-22.40 is based on a maximum ambient temperature of 40°C (104°F).

b. *Temperature rise.* Temperature rise is the increase in winding temperature above the ambient temperature due to the flow of current in the windings and any internal losses that may occur in the machine during operation. Alternators are designed and rated to have a maximum specific value of temperature rise under operating conditions that do not exceed nameplate rated values.

c. *Total temperature of an alternator's windings.* The total temperature of an alternator in operation is the sum of two temperatures: ambient temperature plus temperature rise. It is important to note that temperature rise is strictly a function of load and load factors. Thus, for any alternator sizing calculations, it must be assumed that the unit being considered for the application will be operating at its full load rated temperature rise. Therefore, it must be assumed that operating the unit in ambient temperatures above its nameplate maximum ambient temperature will cause the unit to become overheated. See Note III of NEMA MG1-22.40 for recommendations dealing with ambient temperatures above the standard 40°C. Since temperature rise is a function of the magnitude of the load, the only way to reduce the temperature rise of the machine is to reduce the load on it. The process of lowering the nameplate capacity of an alternator to compensate for ambient temperatures above the standard 40°C is termed **temperature derating**. This lower or derated capacity should be the nameplate capacity of the unit.

d. *NEMA duty cycle ratings.* NEMA MG1-22.40 recognizes only two classes of operating duty:

1. *Continuous duty.* Continuous duty assumes the capability of operating without interruption under rated conditions of voltage, frequency, full load current and minimum power factor for a protracted period of time. NEMA does not specify the minimum number of hours of operation, so continuous duty must imply 24 hours a day, 7 days a week, etc., for the thermal life of the unit's insulation system. It is generally accepted in the electrical industry that, for operation at the limiting temperature for the class of insulation, thermal life expectancy will be in the area of 20,000 hours or greater.

2. *Standby duty.* NEMA MG1-22.84 (Ref. 1) deals with standby operation as follows:

"*Synchronous generators are at times assigned a standby rating where the application is an emergency back-up power source and is not the prime power supply. Under such conditions, temperature rises up to 25°C above those for continuous operation may occur in accordance with Note IV of MG1-22.40.*

"*Operation at these stand-by temperature rise values causes the generator insulation to age thermally at about four to eight times the rate that occurs at the continuous-duty temperature rise values, ie., operating 1 hour at stand-by temperature rise values is approximately equivalent to operating 4 to 8 hours at continuous-duty values.*"

As with continuous operating duty, NEMA does not specify the number of hours a unit is expected to deliver its rated power, only a gauge to estimate expected life at standby duty temperature rise values. As a worst case approach, consider one hour of operation at a standby rise equal to eight hours at continuous-duty values, with a minimum of 20,000 hours continuous life. We are looking at approximately 2,500 hours life of a standby duty alternator. For this reason, specifications that have such statements as continuous standby, standby continuous, or continuous standby for the duration of an outage, have no real meaning in the context of the NEMA rating. You can expect a given number of total hours of thermal life in the unit, and therefore, as long as the total hours of any given outage plus the total hours of prior outages (at standby temperature rise values) does not exceed this given expected life, the unit should be acceptable.

Table 2-3 lists standby and continuous duty temperature limits for the four classes of insulation materials.

Table 2-3. Table of Insulation Temperature Limits

Measurement in Degrees Celsius (by Resistance)	Class of Insulation							
	A		B		F		H	
	Cont	Stby	Cont	Stby	Cont	Stby	Cont	Stby
Maximum Allowed	100	125	120	145	145	170	165	190
Ambient Temp.	40	40	40	40	40	40	40	40
Maximum Rise	60	85	80	105	105	130	125	150

e. *Overload capability.* When a specification requires a continuous duty alternator to operate at a 10% overload for two hours out of any 24 hour period, multiply the rated load by 100% plus the percent overload (in this case, a total of 110%) to calculate the total amount of load to be imposed on the machine. If this value is equal to, or less than the standby duty rating of the alternator, it meets the specification. If the specification demands this overload not to exceed the continuous duty temperature rise of the class of insulation used in the machine, select a unit large enough to carry the total load (rated plus overload) at the maximum temperature rise for the class of insulation.

Note that it is possible to take the short time overload out of the standby capability of the alternator as long as the total load, rated plus overload, does not exceed the standby rating of the unit. NEMA does not allow any temperature rise in excess of the standby value. Therefore, there is no overload capability in a standby duty rated alternator. If an overload capability is called for in a standby duty rated unit, the alternator selected must have a standby duty rating equal to or greater than the total of the rated load plus the overload.

f. *Altitude considerations.* NEMA bases temperature rises on an altitude of up to 1000 meters (3300 feet). Operation at altitudes in excess of this base altitude will require a temperature rise reduction of 1% for every 100 meters (330 feet) of altitude above the base. This altitude correction is due, basically, to the thinness of the cooling air not being able to cool the machine as efficiently.

For example: To calculate the anticipated increase in temperature rise for a unit to be installed at a site elevation 1500 meters above sea level:

Site Altitude (1500 meters) – Base Altitude (1000 meters) = 500 meters. 500 meters divided by 100 meters = 5% anticipated increase in temperature rise. Therefore, the alternator selected for this application must have a temperature rise of at least 5% below the maximum allowable rise for the class of insulation and duty cycle (standby or continuous duty rise) when operating at sea level to prevent overheating at 1500 meters elevation.

2.5.3 Effects of environment on alternators

a. *Condensation.* Moisture and heat are the two primary enemies of rotating electrical machinery. Condensation usually presents a problem in areas of high humidity where the unit is either in storage, infrequently used, or is an emergency standby machine. During operation, even for units operating at minimal loads, the temperature rise of the machine along with the circulation of cooling air is sufficient to prevent condensation of moisture. For idle machines, strip or space heaters of sufficient wattage to maintain the internal temperature of the unit approximately 5°C above the ambient temperature will usually give adequate protection against condensation. These heaters are usually 120 volt, single phase devices installed between the frame lands in the bottom of the unit.

b. *Fungus growth.* The growth of fungus on alternator windings installed in tropical areas has been a source of trouble in the past. Present day materials used in Class F and Class H are non-nutrient for most known forms of fungi, and therefore tend to be self-defending against the growth of fungus. However, if a specification calls out the application of fungus inhibitors, they may be applied as a final coating in the insulating process during manufacture.

c. *Corrosive atmospheres.* Salt and other corrosive materials will attack the outer coating of winding insulation and may cause premature failure of the alternator. Additional coatings of insulation materials during winding manufacture will protect against this corrosion. Some epoxy compounds used as a final overcoat on the windings will afford an additional degree of protection against corrosive materials.

d. *Abrasive dust and grit.* In very dusty and sandy areas, abrasive dust and sand particles may be drawn through the unit by its cooling fan. These rather high velocity particles tend to grit-blast the windings and wear away the outer layers of insulating material. Filters fitted over the unit's intake air openings (or for units installed in housings or enclosures, filters over the enclosure ventilation openings) may be required to prevent this sort of damage. Units with epoxy type insulating material on their windings are somewhat more susceptible to this type of damage, since these materials tend to be very hard and brittle and chip away under the impact of the particles. Application of a softer, more resilient overcoating helps to defeat the erosion of the underlying epoxy insulating material.

2.5.4 Thermal devices embedded in stator windings.

Overheating of alternator windings can lead to lengthy down time and major repair costs, especially if overheating is allowed to persist for extended periods of time. No system of circuit breakers and/or overcurrent protective relaying can give complete protection against all possible causes of excessive winding temperature. Typical of operating situations that can best be dealt with by temperature related devices embedded in the windings are: loss of ventilation by blockage of intake air openings and passages; buildup of dirt, chemicals, oil, etc. on windings inhibiting heat transfer; extended periods of underspeed (under frequency) operation with attendant reduced cooling air flow; and ambient temperatures higher than anticipated or allowed for during initial system planning.

This section will explain the operation of stator temperature sensors and temperature actuated devices. Each of these devices will be discussed both generally, and specifically as they apply to, and are usually available from most commercial alternator manufacturers.

a. *Applications.* Temperature related devices may be used to either display and/or record actual winding temperatures, or initiate some positive protective action should alternator temperature winding temperatures approach a dangerously high level.

Direct positive action initiated by these devices may:

1. Trigger an audio and/or visual over temperature alarm.

2. Reduce the magnitude of the connected load by activating some form of automated load shedding system.

3. Initiate automated start of additional generator set(s) preparatory to bringing additional electrical generating capacity on line.

4. Dump the entire load by actuating a shunt trip circuit breaker, or protective relaying system.

5. Initiate an orderly shutdown of the entire power generation system.

Passive action by temperature sensing devices may be in the form of temperature readout via a panel meter or long term temperature monitoring via some form of continuous recording device, or simultaneously displaying temperature on a panel meter and recording temperature on a continuous recording device.

b. *Installation.* All of the devices covered in this section are inserted in the slots of the core during the stator winding process. It is not recommended practice to install these devices in an existing machine, since once a winding has been completed and insulated, the only place the device can be inserted is on the outside of the slots, and temperatures in these areas will be considerably lower than those in the slots. In addition, there is a very real risk of damaging the existing winding during the insertion process.

c. *Devices*

1. *Temperature actuated switches (thermostats).* A thermostat is an electromechanical device that will operate a set of electrical contacts when the body temperature of the thermostat is raised above a preset or selected level, and then automatically return the contacts to their former or normal position when the body temperature is lowered below the operating point. A thermostat is a temperature actuated switch, and not a device to be used in actual temperature measurement. One advantage of a thermostat is that its contacts are current carrying, and usually may be inserted directly into an alarm or interrupter circuit without requiring an amplifier.

Thermostats are commercially available with either normally open (NO) or normally closed (NC) contacts; thus, the type of contact (NO or NC) must be specified at the time the alternator is ordered. If the exact type is not known at the time of order placement, suggested practice is to order the unit built with both types of thermostats installed, and allow the customer the option of selecting the desired type at the time of system installation, simply ignoring the unused thermostat.

External devices used in conjunction with thermostats are usually audio or visual alarms, shunt trip circuit breakers, and system shutdown devices and circuitry. Most alternator manufacturers bring the thermostat lead wires out to the connection box, but usually do not provide the external devices or circuitry used with, or activated, by thermostats unless they are also supplying the instrument panels and/or switchgear.

2. *Temperature sensors*

(a) *Thermocouples.* A thermocouple is the remote mountable sensor of a thermocouple thermometer. A thermocouple consists of a pair of conductors of dissimilar metals so joined at two separate points that an electromotive force (EMF) is developed by the thermoelectric effect (Seebeck Effect) when the two functions are at different temperatures. Readout instrumentation accurately measures the developed EMF, and together with the characteristics of the specific thermocouple, translates this EMF data into temperature data.

A variety of thermocouples are available in industry for insertion into stator windings, although iron-constantan thermocouples seem to be the most popular one used by American manufacturers and their customers. Most alternator manufacturers offer these devices as an optional accessory, and are usually installed in sets of three, one per phase. In any case, the exact number and metallic element desired should be specified at the time the alternator is ordered.

(b) *Wire-type resistance temperature detectors (RTDs).* An RTD is the remote temperature sensor of a resistance thermometer (resistance temperature meter). An RTD is a wire wound resistor whose resistance is a known linear function of its temperature. Readout instrumentation accurately measures the resistance of the device, and together with the resistance versus temperature characteristics of the specific RTD, translates this resistance data into temperature data.

As with thermocouples, RTDs are available in industry in a number of standard elements, and are usually installed in sets of three, one per phase. The number of RTDs and the desired metallic element must be stated at the time of order. The following industry standard elements are most frequently specified:

- Copper   — 10 Ohms at 25°C.
- Nickel   — 120 Ohms at 0°C.
- Platinum — 100 Ohms at 0°C.

3. *Temperature detector instrumentation.* Instrumentation used with temperature sensors to either read or record data is designed to be compatible with only one specific type and material of sensor, and will not give accurate data when used with any other sensor. For example, thermocouple thermometers are used with only one type thermocouple, and resistance thermometers are compatible with only one specific element. For this reason, it is vital at the time of order placement not only

Alternators (Synchronous Generators)

to specify the type and number of sensors desired, but also the specific conductor combination (in the case of thermocouples) or metallic element (RTD) that is compatible with existing or prospective readout instrumentation.

It is important to note that the signals provided by temperature must be amplified in order to actuate any of the devices or circuits usually associated with thermostats. Instrumentation and external circuitry associated with temperature sensors is usually not provided by the manufacturer of the alternator unless he also supplies the instrument panels and/or switchgear.

d. *Economic considerations.* At the beginning of this section we discussed overheating problems that could best be addressed by temperature related devices embedded in the stator windings. From an economic standpoint, while thermostats simply provide a GO-NO-GO type of protection, they do this at a minimal cost, and therefore, are most often used in small to medium sized alternators. Temperature sensors and their associated equipment can provide a much more complete protection package, but tend to be quite expensive. For this reason, temperature sensors and their associated external equipment are rarely seen with units of less than 500 kVA, unless the application is one of an extremely critical nature where cost is secondary to reliability.

From a purely economic standpoint, with the possible exception of thermostats and a simple over-temperature alarm system, it is difficult to justify recommending the use of any of these devices and their accessory equipment for standby electric sets, where annual usage is in the order of 50 hours operation (not including exercise time that is rarely at rated full load current), or in the case of smaller sets, less than 150 kW, where the cost of the protective package may well represent a value that is major in proportion to the cost of the alternator it is intended to protect. To put it another way, when an insurance premium approaches a significant percentage of the price of the item to be insured, it tends to be a poor investment.

2.6 ALTERNATOR LOADING CONSIDERATIONS

Alternators may be placed into service to power a single dedicated load such as a single portable light tower, truck or container refrigeration system, etc., or to supply electric power to a central distribution system or bus that may be simultaneously feeding several groups or blocks of loads. Sections 2.3, 2.4, and 2.5 have covered some of the effects of load and environment upon alternator performance. In this and the next section we will be investigating loading of alternators, both steady state and transient conditions, types of loads, and sizing procedures.

2.6.1 Conditions of loading

a. There are essentially two general conditions of loading to be considered when sizing an alternator for a given application. These two conditions are: steady state load conditions and transient conditions.

1. *Steady state load condition.* Steady state loading of an alternator is mainly concerned with the unit maintaining its output voltage within a specified bandwidth, usually expressed as a given percentage plus or minus a nominal value, and limiting the temperature rise of the machine to or below a given level when carrying its full rated load. Also included in steady state loading is any relatively long term level of loading from no load up to, and including, designed service factor overloading. Thus, for our purposes here, we will consider steady state loading means a fixed or constant magnitude of load that is to be maintained for some period of time, and that the alternator is in a stabilized condition.

2. *Transient condition.* A transient condition takes into account temporary voltage excursions from the normal or steady state bandwidth occasioned by a sudden major change in load current demand. During steady state operation, voltage is maintained within specified limits by having its main field's flux held at the density appropriate for that particular level and nature of loading. At the onset of a major change, either load acceptance or rejection, the field density is not at the level necessary to maintain system voltage at the preset value for this new magnitude of load. For this reason, the voltage will either drop to a lower level, or rise to a higher level (depending whether load has been added or removed), until the alternator and its automatic voltage regulator or voltage regulating system can respond to the change and return the voltage to the prescribed steady state bandwidth.

b. *Terms and expressions.* Figure 2-52 is an illustration of a typical light beam oscillograph strip chart showing both steady state and transient conditions of loading. For purposes of clarity, the light beam trace has been removed from the majority of the chart so that the various conditions could be detailed. At the extreme right of the chart there are several cycles of trace as it would appear throughout the entire chart. The following are definitions of the various terms and expressions used in the chart:

1. *Steady state voltage envelope.* This is the bandwidth within which the alternator will maintain its output voltage during conditions of steady state loading. This is usually specified as plus or minus a percent or fraction of a percent of a specified system voltage. Quite frequently this steady state voltage envelope is referred to as **steady state voltage regulation**.

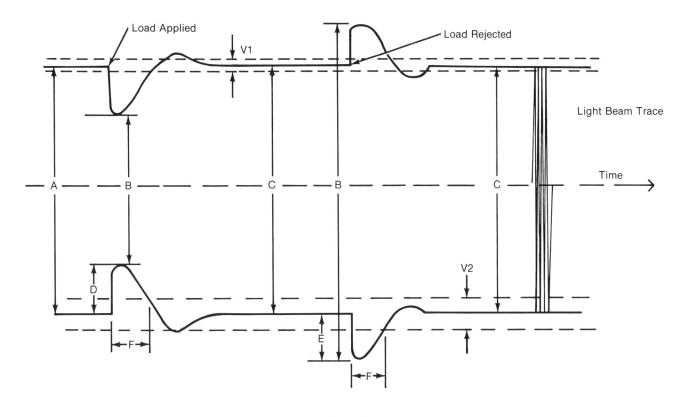

Legend

A. No Load Voltage (100%).
B. Max. Voltage Excursion During Transient.
C. Steady State Load Voltage.
D. Maximum Transient Voltage Dip.
E. Maximum Transient Voltage Overshoot.
F. Recovery Time To ( )% of Steady State Voltage.
V1. Steady State Voltage Envelope.
V2. ± Transient Voltage Envelope.

Figure 2-52. Light Beam Voltage Oscillograph Strip Chart

2. *Transient voltage envelope.* Many specifications will require that following a load change, the alternator voltage shall recover to and stay within a bandwidth somewhat wider than that allowed for steady state in a specified period of time following the onset of the change. This allowable transient envelope will vary from specification to specification, but is usually given between ±2% to ±5%.

3. *Recovery time.* Recovery time is the length of time from the instant of transient to the time the voltage returns to, and stays within, the transient voltage envelope. Some specifications eliminate the transient voltage envelope, and merely specify recovery time as the length of time allowed between start of transient and return of the alternator's voltage to within the steady state envelope. Recovery time may be stated in seconds, tenths of a second, milliseconds, or in some cases, in a given number of cycles.

NOTE: At 60 hertz, one cycle equals 0.0167 sec. (16.7 ms); at 50 hertz one cycle equals 0.020 sec. (20.0 ms).

4. *Transient voltage dip.* Transient dip is the total amount of voltage drop due to acceptance of load, and usually is expressed as a percent of full (100%) steady state voltage. Typically, the allowable transient dip upon application of full rated load on a machine operating at a no load condition will be specified between 15% and 20%. Frequently a specification will spell out allowable transient dips for various incremental load applications.

5. *Transient voltage overshoot.* Voltage overshoot is the total amount of voltage increase upon sudden rejection

of a block of loading, and usually is expressed as a percent of full (100%) steady state voltage. Overshoot, when specified, usually is stated as a maximum allowable value upon rejection of a load equivalent to a given percentage of full rated load.

c. *Load power factor considerations.* Most three phase alternators are rated as having a minimum allowable power factor of 0.8 lagging, and, therefore, may be considered suitable as a power source for any load having a power factor from 0.8 lagging to unity. In Section 1.8.3 we covered the voltage regulation problems associated with leading power factor loading. Fortunately there are few loads that will present leading power factors, but any time a specification mentions the possibility of leading power factor, the alternator manufacturer should be consulted.

d. *Power factor correction.* Low power factor in a circuit is the result of a predominance of either inductive reactance or capacitive reactance in the circuit. In Section 1.8.2 in our discussion of impedance, we saw where the presence of both inductive and capacitive reactances in a circuit have the effect of off-setting each other. Therefore, if an ac circuit has a low lagging power factor, indicating a predominance of inductive reactance, the power factor of the circuit can be raised by adding capacitance to the circuit. This is done frequently in industry, and is called **power factor correction**.

Alternators and distribution systems are rated by their total current carrying (kVA) capacity. Power factor correction of an existing load having a low lagging power factor will increase the load ability of both the alternator and the distribution system. For any given load, the real power demand in kW (the product of voltage times the in-phase component of current) will remain constant. However, raising the power factor of the load will reduce the reactive power kvars (the product of voltage times the out-of-phase component of current). This reduction of kvars will result in a reduction of total current demand (kVA), per unit of kW load, thus freeing up some kVA capacity for additional loading, or possibly reducing the size of the alternator necessary to service the load.

e. Unsuspected leading power factors may occur when an alternator is used to power facilities where power factor correction capacitors have been installed to raise an overall low lagging system power factor. In these instances, sufficient capacitance to correct the entire load power factor may have been installed at the bus or central distribution panel. The problem here is that there may be times when only partial loading may exist, resulting in a situation where too much correction may be present for the existing partial load, and the net effect will be a leading power factor. Obviously, the real solution to this situation is to have power factor correction installed at each load center needing correction, not at the central distribution point. Indeed, this is often done in well thought out installations. The trouble with this solution is cost. It costs more to add correction at many points than it does to add the same total amount of correction at one point. Thus, you may find an installation where cost reduction was the prime objective during construction, and total system power factor correction has been installed at the master distribution station.

These problems of possible overcorrection most often occur where alternators are being supplied as standby for existing facilities. The point to be taken here is that many future problems could be eliminated by careful consultation with the customer as to the exact nature of the load, load power factor, and the possible existence of power factor correction prior to making any recommendations as to sizing of an alternator for a specific situation.

In all instances where a specification makes reference to a power factor lower than 0.8 lagging, the manufacturer of the alternator to be used should be consulted to ensure that this lower-than-standard lagging power factor will not result in injurious heating, either in the main field or in the excitation-voltage regulation equipment.

2.6.2 Electrical loads. Throughout this and subsequent sections, you will frequently see notes to consult with the manufacturer or owner about the specifics of some particular piece of electrical apparatus. This is due to the many variations in design, fabrication, and installation of electric power consuming devices. A piece of equipment may have operating characteristics either that are unique to it, or at least far enough away from what might be considered typical that it would have a real impact upon performance. The intent of typical data given here is to make estimating possible. Take note where typical data is used, and keep in mind that performance of actual equipment may vary from this.

a. *Characteristics of frequently seen electrical loads*

1. *Heating loads.* Heating loads are, for the most part, resistive in nature and operate at unity power factor. However, in industry, heaters known as induction heaters are frequently used. These heaters are wound coils, and as such are inductors, and will produce a lagging power factor. Where induction heaters are being considered, consult with the equipment manufacturer for operating power factor and power demand.

2. *Lighting loads.* Lighting loads in general are considered to be unity power factor in nature; however, caution should be taken when large amounts of fluorescent lighting is involved. While most fluorescent fixtures are

corrected with capacitors to approximately 0.9 power factor, uncompensated fixtures could show power factors as low as 0.5 lagging. Large amounts of fluorescent lighting may cause line voltage disturbances due to high harmonics in the current wave form.

3. *Induction motor loads.* Induction motors vary so widely in design and performance that for our purposes here we will only make some general comments about their characteristics. Induction motors operate at power factors ranging from 0.75 lagging to as high as 0.92 lagging, and their starting current demand could be as high as ten times their running current. As a general rule, motors require about six times their running current in starting current demand. Starting a relatively large electric motor may create transient voltage dips in excess of 40%, with serious effects upon existing loads on the line. For example, during the starting of a motor, line voltage dipping in excess of 40% may cause lighting to dim or be completely extinguished, on-line motors may stop due to insufficient voltage on their holding coils, etc. For these reasons, motor starting will be treated separately in Section 2-7.

b. *Load data.* Load data is the total input power required by a connected electrical load, and as such, is also the total output power required by the alternator to service the load. Data may be expressed in either amperes, kilovolt amperes (kVA), or kilowatts (kW). In some complex loading situations, all three forms may be used to describe the power requirements of various items or blocks of load. Obviously, in order to size an alternator to the load, the data must be converted into a single form or unit to be used throughout all calculations.

In Section 2.5 we saw that kVA represents the total load as seen by an alternator, and therefore, is the form most appropriately used for alternator load calculations. However, when we consider the output power requirements of the alternator's prime mover (engine), that power requirement must be stated in kW, as that represents the total real power consumed by the load. From Equation 1-15 (kW = kVA × PF) we see that in addition to stating load requirements in kVA for alternator calculations, we also need to note the power factor of the load component in order to complete prime mover sizing calculations.

Quite frequently you will see engineering data expressed in terms of per unit (PU). Per unit is the designation of percentage expressed in decimal form, with 100% equaling 1 PU, 200% equaling 2 PU, etc. As an example of per unit use, an engineer may state a requirement for sustaining current during conditions of short circuit of 250% of rated current as "2.5 PU rated current".

c. *Equation symbols.* The following is a list of symbols that will be used in load analysis and alternator sizing equations:

- kVA = Kilovolt Amperes (Apparent Power).
- kW = Kilowatts Output (Real Power).
- PF = Power Factor.
- HPD = Horsepower-Prime Mover (Alternator Driving Power Input).
- EFF = Alternator Efficiency — Expressed decimal form.
- PU = Per Unit.
- $\sqrt{3}$ = Square Root of 3 (1.732).
- E = Voltage in Volts.
- I = Current in Amperes.
- HP = Horsepower, electric motor output power, and all other expressions of horsepower except alternator input power requirement.

The following equations will be used in calculations in this and subsequent chapters. In some cases, they will be formulas used in prior sections. However, for ease of reference, they are repeated here since the formulas for kVA and kW in single phase calculations are different from those for three phase calculations. For this reason single phase and three phase equations are listed separately.

1. *Single phase equations:*

(a) $kVA = \dfrac{E \times I}{1000}$ (Equation 2-11)

or:

$kVA = \dfrac{kW}{PF}$ (Equation 2-12)

(b) $kW = \dfrac{E \times I \times PF}{1000}$ (Equation 2-13)

or:

$kW = kVA \times PF$ (Equation 2-14)

(c) $I = \dfrac{kW \times 1000}{E \times PF}$ (Equation 2-15)

or:

$I = \dfrac{kVA \times 1000}{E}$ (Equation 2-16)

2. *Three phase equations:*

(a) $kVA = \dfrac{1.732 \times E \times I}{1000}$ (Equation 2-17)

or:

$kVA = \dfrac{kW}{PF}$ (Equation 2-18)

(b) $kW = \dfrac{1.732 \times E \times I \times PF}{1000}$ (Equation 2-19)

Alternators (Synchronous Generators)

or:

$$kW = kVA \times PF \quad \text{(Equation 2-20)}$$

(c) $I = \dfrac{kW \times 1000}{1.732 \times E \times PF}$ (Equation 2-21)

or:

$I = \dfrac{kVA \times 1000}{1.732 \times E}$ (Equation 2-22)

3. *Horsepower equations:*

(a) $HPD = \dfrac{kW}{\text{Alternator EFF} \times 0.746}$

(Equation 2-23)

(b) $HP = \dfrac{kW}{0.746}$ (Equation 2-24)

(c) $kW = HPD \times \text{Alternator EFF} \times 0.746$

(Equation 2-25)

(d) $kW = HP \times 0.746$ (Equation 2-26)

4. *SI unit equations.* When the alternator's prime mover (engine) has its power rated in kW, (metric horsepower), a distinction must be made between mechanical kW and electrical kW. Engine manufacturers who state the power capability of their engines in the SI Unit of kW are speaking strictly in terms of mechanical rotating power. An engine generator set converts this mechanical energy to electrical energy. Since losses are inherent in any energy conversion, the electrical power (kW) output of the engine generator set will be less than the mechanical power (kW) input to the set. The difference between these two quantities of power input to power output is the efficiency of the electric generator. For this reason, we must modify the symbols of kW to differentiate between mechanical kW and electrical kW as follows:

kWe = Kilowatts Electrical.
kWm = Kilowatts Mechanical.

The equations associated with calculations of engine generator set input and output power requirements are:

$kWe = kWm \times \text{Alternator EFF}$ (Equation 2-27)

$kWm = \dfrac{kWe}{\text{Alternator EFF}}$ (Equation 2-28)

2.6.3 Steady state (constant) load analysis

a. Steady state or constant loading of an alternator is the total running load at any given time. Steady state, as defined in Section 2.7.1, implies that the unit will be subjected to this load level for a period of time, but does not take into account transient conditions. Load acceptance transients impart a temporary load condition that may represent a greater than full load demand to the alternator. However, the smallest generator set to be considered for any given application is one that has the capacity to carry the largest steady state load. This is the subject we will address in this section. Considerations requiring a larger sized unit to give proper performance during transient conditions is the subject of Section 2.8.

Steady state load analysis, simply stated, consists of listing in tabular form all of the individual loads that comprise the total load, and then adding the kVA requirements of all of these load segments together to determine the total value of the entire load. Table 2-4 is a suggested format for tabulating steady state loads.

1. *Item number.* List each separate load item. An item could contain several identical units of load such as: 4 three phase, 30 amp outlets, or six ventilation fan motors.

2. *Description.* Give brief description of the item. In the example above you would also include the number of the devices in the item.

3. *Running amps.* If load data is given in amperes, list the total running amperes of the item here. If data does not include amps, it is not necessary to calculate their value.

4. *Running kVA.* List the total item running kVA. If data is given in other terms than kVA, calculate kVA using the equations given above. Should the data be given in terms of kW with no stated running power factor, estimate PF as 0.8.

5. *PF.* If given, note PF here. If no PF is stated, estimate the power factor as 0.8 unless the load is known to be purely resistive in nature. In that case, list PF as 1.0.

NOTE: In all cases where data has been estimated or assumed, star (*) the date and note on the bottom of the form: "Assumed value in lieu of specific data".

6. *Running kW.* If data is not given in kW, calculate kW using the equations given above. If load power factor is not given and the item is other than resistive in nature, estimate power factor high (between 0.9 and 0.95). Kilowatts (kW) is used to calculate drive horsepower; therefore, it is best to estimate a greater value and eliminate the possibility of under sizing the driver. Again, star the entry and make the same statement at the bottom of the form as given in the note above.

7. *Transient.* If the item has a greater power requirement for starting than for running, mark (Y) Yes in the column; if not, mark (N) No.

b. *Load balance.* A balanced load demands identical power and current from each of the phases in a three

Alternators (Synchronous Generators)

Table 2-4. Steady State Load List

CUSTOMER/JOB NUMBER _____ ELECTRICAL SYSTEM PARTICULARS
PHASE: ___ VOLTAGE: ___ HERTZ: ___ RPM: ___ AMB.TEMP: ___ °C
TEMP.RISE: ___ °C. SPECIAL FEATURES: _____

ITEM NO.	DESCRIPTION OF ITEM	QTY.	RUN AMP TOTAL	RUN kVA TOTAL	RUN PF AVE.	RUN kW TOTAL	TRANSIENT Y N

phase alternator, or each of the major coil groups of a single phase unit. Thus, when an alternator is supplying power to a balanced load, the load is equally shared by all phases or coil groups.

1. *Three phase unbalanced loads.* In a three phase unbalanced load, the phases of a three phase machine are loaded unequally. This will occur when, in addition to any three phase load, the unit is supplying a single phase load either line-to-line or line-to-neutral, and/or unequal single phase loading line-to-neutral on one or more lines.

2. *Single phase unbalanced loads.* An unbalanced load on single phase machines will occur only on three wire systems, and is caused by uneven loading line-to-neutral.

3. *Steady state unbalanced loads calculations.* When calculating line-to-neutral single phase load current, care must be taken to be sure that the proper voltage is used. In Section 2.3, we learned that for WYE connected alternators, the line-to-neutral voltage is equal to the line-to-line voltage divided by 1.732. Note that the equations for single phase are different from three phase equations. When listing single phase loads on three phase units, calculate the running current of the load using the single phase equations. Enter this value in the amps column. Then calculate kVA using the three phase equations to ensure that no single phase of the unit is overloaded.

4. *Problems associated with unbalanced loads.* Minor load unbalances, usually 5% or less, will not cause any particular problems. However, as the degree of unbalance increases, voltage regulation and voltage unbalance between phases worsens. Every effort should be made to distribute the load evenly between all phases to ensure the best possible performance.

2.6.4 Alternator load bank testing

a. *Load banks.* Alternator testing is usually accomplished by connecting the output leads of the unit to be tested to an artificial load termed a **load bank**. Load banks can be either purely resistive, purely reactive, or a combination of both and as such, may have a power factor anywhere from 1.0 (unity) to near 0 (zero) lagging.

b. *Load on alternator and prime mover.* In our coverage of the power right triangle we saw that kVA (apparent power) is comprised of both kW (real power) and kvar (reactive power), and that power factor is the ratio of the total apparent power delivered to the connected load to the total real power consumed by the load. We also saw that reactive power (kvar) is sometimes referred to as wattless power and consumes no real power. From this we can see that any reactive power component present in the load will be seen as a power demand only by the alternator, and not by the alternator's prime mover.

c. *Alternator production testing.* In general, alternator manufacturers perform production testing of units using zero power factor (lagging) reactive load banks to verify product performance. Reactive load banks are constructed using inductors (wound coils) to provide a lagging power factor. Very low lagging power factor testing will allow the unit tested to be loaded to full rated kVA, but limit the kW (real power) consumed by the test cell driver to a low value, since the greatest portion of the kVA load upon the alternator is wattless power (kvars), not real power (kW). This type of load testing results in substantial energy savings since the unit under test can be subjected to full kVA (rated current) loading while expending a minimum of energy to drive it. To simulate a load at any power factor above zero, resistive elements are added to the load bank to raise its power factor to the desired level or value.

d. *Prime mover production testing.* The manufacturers of prime movers usually employ a dynamometer as a load in production testing to verify performance of their product.

e. *Engine-alternator set testing.* Production testing of a completely assembled generating set is usually conducted at the assembler's facility using unity power factor load banks. These load banks are, for the most part, constructed of multiple resistors, or blocks of resistors, capable of being switched in and out of the circuit as necessary to achieve the desired level of loading. Unity power factor loading of the set, as discussed above, will result in the kVA load on the alternator being equal to the kW load on the engine. Should the alternator be nameplated at 0.8 PF, at full rated kW load on the engine, the alternator will only be loaded to 80% of rated current capacity.

Alternators (Synchronous Generators)

For example: An alternator rated at 100 kW/125 kVA, three phase, 60 hertz, 480 volts will have a full load current rating of 150 amperes, and will be assembled to a prime mover capable of driving a minimum 100 kW load. On a unity power factor load bank, a 100 kW load on the prime mover will represent only a 100 kVA load upon the alternator, and current flow at this load level will be 120 amperes or 80% of rated full load value. Thus, when a generator set is tested using a resistive load bank, in order to pull full rated current from an alternator rated at 0.8 PF, the load demand upon the engine can be as much as 125% of its full load rating.

Reactive adders can be obtained for resistive load banks to lower the power factor, but are not usually found at most assembler's facilities. Commercial engine generator set assemblers load test their product to adjust governor settings, test vibration levels, and to verify performance of the engine/alternator assembly and installed accessories and components. Since the alternator has been load tested to full current rating at the alternator manufacturer's facility, load testing at 80% of full load current at the assembler's facility is considered adequate loading to reveal any problems or damage incurred during shipping or assembly.

2.7 CONSIDERATIONS OF POLYPHASE INDUCTION MOTOR STARTING ON ENGINE-GENERATOR SETS

Electric motors, both squirrel cage induction and synchronous, have several starting characteristics that are very undesirable. During start, the locked rotor or starting current demand can range anywhere from less than three to in excess of 14 times the full load running current of the motor. To compound the problem, the power factor of this transient demand is usually very low, from 0.2 to 0.5 lagging, so that during the motor starting transient period not only does the large current demand depress system voltage, but the low lagging power factor intensifies this condition. This reaction is termed **transient voltage dip**.

While the transient effects of motor starting on the commercial grid may range from slight annoyances to somewhat of a problem, the effects of motor starting on a limited capacity engine-generator set can be devastating if not anticipated. Depending upon the nature of the connected load and the individual motor being started, excessive transient voltage dips imposed upon the power system may result in contactors dropping out, and/or failure of the motor to accelerate to full rated speed.

The purpose of this section is to provide an insight into motor starting, and some basic information to assist in sizing both the alternator and its prime mover to obtain proper performance during these transient conditions. Unfortunately, there are not only large variations in starting requirements and characteristics of individual motors, (and in the inertial content of the loads they must accelerate during start), but also a seemingly infinite number of system variables such as the nature and power factor of any existing load on the unit at the time of start. For these reasons, this section will address the aspects of motor starting on a typical basis.

2.7.1 Three phase induction motor starting

a. *NEMA motor starting code letters.* We discussed some of the adverse effects induction motor starting may have on the power source, and also upon any existing load that may be on the line at the time of motor starting. All of these system problems are directly related to the severity of the transient voltage dip incurred during motor starting. If the magnitude of the starting demand (start kVA) of a motor can be reduced, the overall impact upon the system will be less severe. The obvious solution would be to design motors to have a minimum starting demand. The problem is that motor design is a rather complex science, as motors are designed to have operating characteristics that vary with the intended use of the motor. Thus, a motor design suitable for a given set of operating parameters may require that the motor have a greater or lesser starting power demand than that of another motor designed for some other task. For these reasons, induction motors have a very wide range of starting power requirements. To assist in system planning, the National Electrical Manufacturers Association (NEMA) created a system of motor starting code letters that incorporate a range of starting kVA (SkVA) demand per horsepower of motor output rating. Table 2-5 is a table of the maximum allowable start kVA per horsepower for the various starting code letters extracted from NEMA Standard MG 1-10.37-1987 (Ref 1).

Table 2-5. NEMA Motor Starting Code Letters

NEMA Code Letter	Maximum Locked Rotor kVA per HP	NEMA Code Letter	Maximum Locked Rotor kVA per HP
A	0–3.15	L	9.0–10.0
B	3.15–3.55	M	10.2–11.2
C	3.55–4.00	N	11.2–12.5
D	4.00–4.5	P	12.5–14.0
E	4.5–5.0	R	14.0–16.0
F	5.0–5.6	S	16.0–18.0
G	5.6–6.3	T	18.0–20.0
H	6.3–7.1	U	20.0–22.4
J	7.1–8.0	V	22.4 and over
K	8.0–9.0		

From Table 2-5 we can see it is vital that not only the rated horsepower, but also either the actual starting kVA, or the NEMA starting code letter be given in order to properly evaluate the impact starting any given motor will have upon a specific electric power generating system.

b. *Effects of reduced voltage on induction motor starting.* Any reduction in the voltage imposed on the terminals of an induction motor during its starting sequence from rated voltage to some lower level or value will have two significant effects on its starting performance.

1. If the voltage imposed on the terminals of an induction motor being started is reduced (lowered) from its rated voltage to some given percentage of this full value, the starting power demand (start kVA) of the motor will be reduced to a value equivalent to the full voltage start kVA times the percent of full voltage squared. The reasons for this is seen in a brief review of electric power, Section 1.8 and Ohm's law. Ohm's law states that the amount of current flow (I) in any electric circuit is directly proportional to the voltage, (I = E/R). Further, from Equation 1-9 (Power = I R) we see that power consumed in a circuit will vary directly as the square of the current flow. Therefore, since current flow in a given circuit is directly proportional to the voltage across the circuit, the result of reducing the voltage to some percent of full voltage will be a reduction in power consumed in the circuit equal to the percent of full voltage squared. For example: If the terminal voltage of a motor being started is reduced to 80% of rated voltage (a reduction in its terminal voltage of 20%), the net motor starting kVA demanded will be reduced to 64% (80% squared) of its full (100%) voltage demand.

Under conditions of reduced voltage during start, the amount of starting torque that the motor can produce will also be reduced by a factor equal to the square of the percent of net voltage. Using the example in the above paragraph, a reduction in voltage to 80% of full voltage during start will result in the motor being able to produce only 64% of its full voltage starting torque.

From the two paragraphs above it becomes apparent that reducing motor voltage during a start sequence can be a mixed blessing. On the one hand, you get relief from excessive inrush current demands and excessive transient disturbances. On the other hand, you may arrive at a situation where the motor does not have sufficient torque to accelerate to rated speed, especially if the motor is required to accelerate a connected load with a heavy inertial content.

In some applications where the motor is connected to its load through a mechanical drive train, full voltage starting with its attendant maximum starting torque may provide too hard a start and cause damage to the drive train. In these cases, the soft start afforded with reduced voltage starting may be beneficial.

c. *Effects of full voltage starting.* Full voltage starting is usually considered the best method of induction motor starting. Full voltage starting will provide maximum starting torque; thus, the motor will have a minimum acceleration time. Reduction of acceleration time is particularly advantageous in applications where the motor cycles on and off very frequently. The shorter the time it takes the motor to come up to speed, the less heating effect of the motor windings under the heavy starting current flow, minimizing the effects of heating on the service life of the motor.

Full voltage starting causes the maximum amount of transient disturbance on the electric power supply system. For this reason most public utilities will not allow full voltage (across the line) starting of large motors, and insist that when large motors are brought on the line some form of starting relief be provided. Details of the various methods of obtaining this relief will be covered later in this chapter.

d. *Types of full voltage starting*

1. *Across the line fused disconnect.* This is the simplest of all three phase motor starters. These starters incorporate overcurrent protection in the form of fuses, but they are susceptible to allowing single phasing of the motor.

2. *Across the Line Magnetic Starter.* This type of starter is the most common used in industry. It affords overcurrent protection so that an overcurrent interruption on any one of the three phases will cause the starter to trip the motor off the line.

3. *Part winding start.* For part winding starting, the motor is wound with two or more circuits per phase. During start, only part of each phase's winding is used. When the motor is up to or near rated speed, simple switching closes the circuit in the rest of each phase winding, giving a closed transition start. In some cases the motor is wound with multiple circuits per phase and starting is accomplished in several steps. With part winding starting, full voltage is applied to the motor terminals, but with only part of the winding in the circuit during start, only a part of the full voltage starting kVA is required. Thus, this type of start could be classed as a full voltage, reduced power start.

e. *Types of reduced voltage starting.* A reduced voltage starter incorporates one or more steps or modes during start and then a running mode. The transition from start mode to run mode may be either an open transition or a closed transition.

In an open transition, switching from start mode to run mode involves opening one circuit, and then closing into another circuit. For the most part, when applied to limited capacity engine-generator sets, the switching transients experienced when transferring from one mode to the other create such objectionable voltage spikes and line disturbances that open transition type starters are usually not recommended for use with this type of electric power supply.

In a closed transition, transfer from one mode to the other does not involve an open circuit; thus, many of the objectionable aspects of an open transition are eliminated.

1. *Autotransformer.* Autotransformer starters are available in either a closed or open transition type. For engine-generator power systems, the closed transition starter should be used. Autotransformers are usually supplied with an 80%, 65%, and 50% tap on the secondary winding. During start the transformer steps down line voltage to either 80%, 65%, or 50% of line voltage (depending on which tap is used). When the motor is near rated speed the transformer is removed from the circuit.

2. *Reactor starter.* Series reactor starting is usually used on large motors. This type of starting provides the same starting power reduction as a closed transition autotransformer. Its disadvantage is that it has a lower torque than autotransformer starters to start kVA ration.

3. *Resistor starter.* Resistor starting has similar characteristics to that of closed transition autotransformer starting, but is not limited to single step starting. As with a reactor starter, it has lower starting torque. An important factor is that the series resistor is kW load and often overloads the prime mover.

4. *Star-delta (wye-delta).* Star-delta starting is a relatively simple method of reduced voltage starting, but requires that the motor being started have all six, or twelve leads brought out to the connection box. The motor starts in a wye (star) connection, and then is switched to a delta connection for the running mode. The principal disadvantage of this system is that it usually involves an open transition, and the motor must run in a delta connection, which may be objectionable for some applications. When the motor is started against a low impedance power source such as a public utility, the transient voltage dip and recovery time to full voltage is minimal. However, when wye-delta motor starting is used with a limited capacity engine-generator set, the additional loss in the motor's starting torque due to a significant transient voltage dip usually results in the motor failing to accelerate to near rated speed prior to making transition to running mode. The net effect of this on the power source is almost as if the motor was started directly across the line (full voltage start), with the end result of a no-start situation. For these reasons, alternators being used to power motors equipped with wye-delta starters should be sized as if the motors were being started directly across the line (full voltage starting).

f. *Effects of starter transition timing.* Figure 2-53 is a typical induction motor starting current curve. This curve illustrates inrush current change as the motor accelerates from zero to rated speed for both full voltage starting and starting at 80% of full voltage. Note that there is little current reduction (approximately 10%) until the motor has accelerated to 50% of rated speed (point A), and not until the motor has accelerated to approximately 80% of rated speed (point D) will transition to full voltage result in a start kVA equal to or less than the initial value at the onset of the reduced voltage start. From this it can be seen that if a starter makes the transition from starting mode to running mode before the motor has accelerated to a minimum of 70% of rated speed (point C), the desirable effects of reduced voltage or part winding starting have, for the most part, been negated. For these reasons, when using either reduced voltage starting or part winding starting, transition timing is critical, and often requires a trial and error process at the time of system installation or startup.

g. *Effects of transient voltage dip on reduced voltage starting.* We have seen that the amount of available starting torque when using reduced voltage starters will vary as the square of the percent of full voltage imposed on the motor terminals. Any transient voltage dip experienced during start will further reduce the starting torque by the same factor. For example: if a motor is being started with an autotransformer starter on the 65% tap, the motor will have only 42.25% of its full voltage starting torque available upon start. If the system voltage dips 20% during transient, the voltage at the motor terminals will be only 80% of 65% of rated voltage — or 52% of full voltage. Squaring 52% gives a net value of 27% of full voltage starting torque, not the anticipated 42.25%. This further reduction in starting torque could result in a no start situation.

For this reason extreme care should be exercised when applying reduced voltage starting to engine-generator set power systems. If a given motor will accelerate to full speed with a reduced voltage starter on the 65% tap, but not on the 50% tap when tied to the commercial grid, it could be assumed that the terminal voltage of the motor must be held at or near 65% of full voltage to assure a

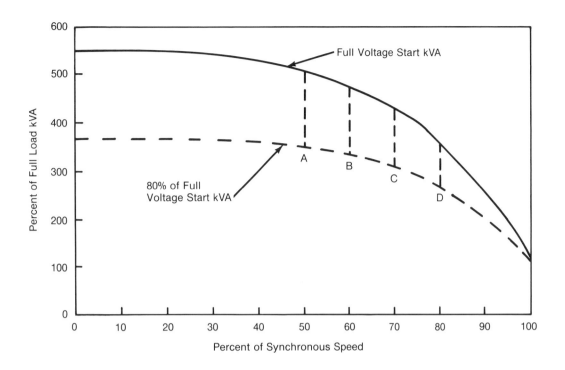

Figure 2-53. Typical Speed/Start kVA Curve for NEMA Design B Squirrel Cage Induction Motor

start. Approximately the same effect could be obtained by having the starter on the 80% tap, and then limiting the transient voltage dip on the engine-generator set to 25%. (100% − 25% = 75%. 75% times 80% equals 60%). Therefore, a 25% voltage dip on a motor with an 80% tap reduced voltage starter would produce about the same effect as starting the unit on the 65% tap (assuming a minimum of 5% transient voltage dip with rapid recovery to full voltage on the commercial bus). If at all possible, the motor manufacturer should be consulted when considering reduced voltage starting to eliminate possible future starting problems.

h. *Summary of motor starting methods*

1. *Full voltage starting.* The advantage of full voltage starting is that the motor being started has the ability to produce maximum starting torque, limiting its acceleration time to a minimum. In addition, the starters are relatively simple in construction, and thus are low in initial cost. The disadvantages of this method of starting are that relatively large amounts of power are required. The high starting torque developed by the motor may, in some cases, cause injurious mechanical shock during the initial starting sequence.

2. *Reduced voltage starting.* Reduced voltage starting has the advantages of requiring relatively low amounts of power during start, and the motors exhibit the trait of a soft start and minimize mechanical shock during the initial start sequence. The disadvantages of this method of starting are that the starters are much more complex and costly, and that starters with an open transition may cause objectionable line voltage surges. In addition, the reduction in starting torque experienced with this method of starting leads to extended acceleration time, and in some cases, results in insufficient torque to allow the motor to accelerate to full speed.

Summary table. Table 2-6 lists the various methods and types of starting, and their principal effects on motor starting.

i. *Wound rotor induction motors.* One of the factors involved in induction motor design is the resistance of the motor rotor. The resistance of the rotor directly affects starting current (SkVA) and starting torque, as well as running performance characteristics. To develop high starting torques at low levels of start kVA requires rotors with a high value of resistance, while good operating performance features such as low full load slip, moderate rotor heating, and high running efficiency require rotors to have low values of resistance.

The wound rotor induction motor, sometimes called a slip ring motor, allows both of these conditions to exist

Alternators (Synchronous Generators)

Table 2-6. Table of Methods and Types of Motor Starting

Method and Type of Starting	PU Line Voltage Applied	PU Rated Start Torque Available	PU Rated Start kVA Required
Full Voltage Across the Line	1.0	1.0	1.0
Full Voltage Part Winding (See Notes 1 & 3)	1.0	.70	.70
Reduced Voltage Autotransformer (See Note 2)	.80 .65 .50	.64 .42 .25	.67 .45 .28
Reduced Voltage Reactor/Resistor (See Notes 1 & 3)	.80 .65	.64 .42	.80 .65
Reduced Voltage Star-Delta	.58	.33	.33

NOTES:
1. Values are typical. Various values of SkVA and torque are available depending upon motor/starter design. Consult the motor/starter manufacturer for specific data.
2. The calculated percentage values of Start kVA have been increased by three percentage points to allow for the magnetizing kVA required for the autotransformer.
3. Values listed are typical. Consult with the customer or the manufacturer of the motor/starter for specific data.
4. In the absence of specific data, indicated typical data may be used for estimating purposes, but should be shown as "assumed values in lieu of specific data". Performance of actual equipment may vary from calculated performance on all calculations and quotations.

as necessary. The rotor is phase wound with as many poles as the stator, and these phases are connected to slip rings. By means of brushes maintaining sliding contact with the slip rings, external resistance (resistors) can be added or removed as necessary to achieve the proper amount of resistance for both starting and running modes. As a result, a wound rotor induction motor has good starting torque while limiting the starting current (SkVA) to approximately 150% or less of full load running current (RkVA). Calculations of starting power requirements of wound rotor induction motors are usually based upon the assumption of starting kVA being 150% of full load running kVA unless otherwise stated in the specification. Starting power factor considerations of these motors are considered similar to that of squirrel cage induction motors.

j. *Considerations of power factor during motor starting.* We have seen in Section 1.8 that the real power (kW) demanded by any electrical load is the product of the apparent power (kVA) delivered to the load times the power factor of the specific load. Therefore, to calculate the impact of motor starting upon the generator set prime mover, we need to know not only the starting kVA requirements of the motor, but also the power factor of the motor during start.

Power factor during the start sequence of a polyphase induction motor rises from a value of near zero lagging at the instant of start, to its running value following a curve that will vary from design to design, and also with the load on the motor. However, since the starting kVA of the motor follows a curve of descending values as the motor is accelerated, and the load power factor is rising at the same time, the power generation industry has generally treated the entire starting transition sequence as though the start kVA remains constant at its maximum value, and the power factor is at some constant intermediate value throughout the entire transition period. Unless otherwise stated in the specification, the starting power factor of a polyphase induction motor is generally assumed to be 0.4 (40%) lagging. If the motor is known to have a rather long acceleration period, this value should be raised to 0.5 (50%) to assure adequate prime mover horsepower to start the load, since the starting PF will probably rise at a much greater rate than the SkVA will decrease.

k. *Impact of transient voltage dip and motor loading upon motor starting.* Experience in industry has generally proven that if the transient voltage dip of an unloaded NEMA design B polyphase induction motor is held to a maximum of 35%, as measured by a light beam oscillograph, magnetic starter contacts will hold in and the motor will have a satisfactory start. From our discussions concerning the reduction in starting torque with reduced voltage at the motor terminals, we can see that if the voltage dip is 35% (a net of 65% of full voltage), the starting torque will be reduced to 42.25% of the motor's full voltage capability. The net effect to the motor will be about the same as if it were being started with an autotransformer on the 65% tap. For most applications, a properly sized alternator will have had its terminal voltage recover to a value greater than 80% of system voltage within a period of approximately 0.5 to 1 second after the onset of the transient condition. Thus, after approximately 1 second the motor should be seeing about 90% of rated voltage, and start torque should have recovered to around 81% (90% squared) of the full voltage value. However, the same caution should be exercised here as with reduced voltage starting. The manufacturer of the motor, or apparatus supplied with the motor, should be consulted to determine the maximum reduction in starting torque the motor can tolerate and still accelerate its connected load to full speed. Further, the generator manufacturer should be consulted for voltage recovery data.

2.7.2 Alternator sizing for induction motor starting

a. *Definition of terms and expressions used in calculating various aspects of three phase induction motor starting and running.* The terms and expressions that will be used in this section may have been defined earlier in this chapter, but are repeated here for ready reference. The equations and equation symbols we will use to perform the various calculations are listed in Section 2.6.2.

1. *kVA*. Kilovolt-amperes (apparent power).

(a) *RkVA*. Running kVA, the apparent power required to run a motor, or operate a given electrical load.

(b) *SkVA*. Start kVA, the apparent power necessary to start a motor with full (100%) line voltage applied to the motor terminals. For part winding starting, this is the kVA that would be required if the complete winding were in the circuit during start.

(c) *NSkVA*. Net start kVA, the apparent power required to start a motor under conditions of reduced voltage, or part winding start. NSkVA = SkVA × SF (start factor).

2. *Start factor (SF)*. The factor by which the SkVA is multiplied to calculate the net starting kVA demanded by a motor being started with some form of reduced voltage starter, or by part winding start.

3. *Power factor (PF)*. The ratio of real power consumed by an electrical load to the apparent power delivered to the load.

(a) *Running power factor (RPF)*. The power factor of a motor or electrical load during conditions of steady state operation. This term is used in motor starting calculations to distinguish operating power factor from the power factor of a motor during starting transient conditions to avoid confusion.

(b) *Starting power factor (SPF)*. The power factor of a motor during start. Unless stated otherwise in a specification, starting power factor should be assumed to be 0.4 (40%) lagging, and noted as an assumed value in all calculations.

4. *Kilowatts (kW), real power.* kW = kVA × PF.

(a) *Running Kilowatts (RkW)*. The real power required to run a motor or operate any given electrical load. RkW = RkVA × RPF.

(b) *Start kW (SkW)*. The real power required to start a motor. SkW = SkVA × SPF.

(c) *Net Start kW (NSkW)*. The real power required to start a motor under conditions of reduced voltage, or part winding start. NSkW = NSkVA × SPF, or: NSkW = SkVA × SF × SPF.

5. *Transient Voltage Dip Factor (DF)*. The per unit resultant voltage squared, when a transient voltage dip occurs.

DF = (1.0 − PU VOLTAGE DIP)   (Equation 2-29)

Dip factor is used to determine the reduction in starting power requirement, and/or starting torque when a transient voltage dip occurs during a motor starting sequence. Use of dip factor is a judgement call since factors such as acceleration time of the motor, length of time for the alternator to recover to full voltage, recovery time of the set prime mover to rated speed, and the type of automatic voltage regulator (volts-per-hertz or constant voltage) all come into play when trying to determine how much effect on the final or net SkW the prime mover will actually see during the entire transient period. Experience has proven that a value of approximately one-half the maximum transient voltage dip used as the percentage voltage dip in calculating dip factor will, in most cases, best describe the net relief afforded to the prime mover during the transient period. For long acceleration times, dip factor is the square of the percent of recovered voltage.

b. *Transient voltage dip.* Most alternator manufacturers will provide transient voltage dip data for their products, whether in curve or tabular form, based upon light beam oscillograph measurements. Figure 2-54 represents typical transient voltage dip data in curve form. Some alternator manufacturers provide similar transient voltage dip data in special slide rules that may also incorporate other pertinent performance information. The method of voltage dip measurement should be stated with the data, since use of instrumentation other than light beam devices may provide dissimilar readings. Any statement by a manufacturer or assembler of start kVA capability of a specific unit without mention of the maximum transient voltage dip anticipated, and the method of measurement of this maximum value, should be considered as incomplete data.

A transient voltage dip of 35% as measured by a light beam oscillograph or oscilloscope (see Figure 2-52) is usually considered as the maximum allowable for unloaded motors started across the line (full voltage). Therefore, in this chapter all calculations for starting induction motors across the line will be predicated upon a maximum transient voltage dip of 35%. All alternator sizing will be based upon selecting the smallest unit capable of carrying the total connected load first, and then starting the largest motor starting requirement with a

Alternators (Synchronous Generators)

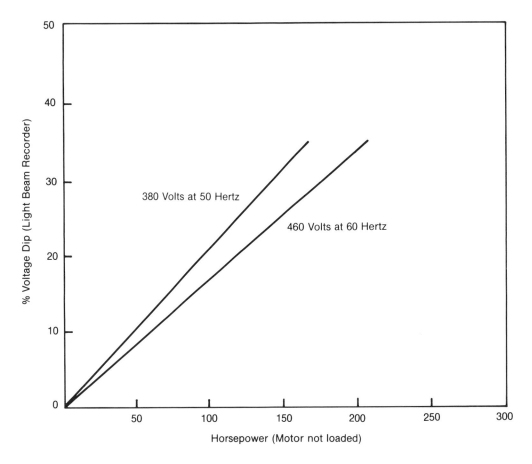

Notes:
A. Code F (5.6 kVA/HP), 230/460V motors starting across the line at 460V, 3 phase, 60 Hz. The voltage on 50 Hz is 380V also with a 3 phase Code F motor.
B. For other code letters or starting methods, use the chart on the following page.
C. See the introduction page of this section for other qualifications.

Figure 2-54. Typical Motor Starting Curve 400 kW Generator

maximum transient voltage dip of 35%, or whatever maximum transient dip is specified as the second consideration.

NOTES:
1. Many alternator manufacturers and engine-generator set manufacturers have specific sizing instructions for motor starting, and applications involving various other types of loads for their equipment. When using equipment for which the manufacturer has provided detailed instructions for sizing, the manufacturer's instructions should be followed, and the data and information provided in this chapter used to amplify as necessary.
2. Some motor starters cannot tolerate transient voltage dips in excess of 20%. For that reason the starter manufacturer should be consulted for maximum allowable transient voltage dip.

c. *Recording data for motor starting analysis.* Figure 2-55 is a suggested format for tabulating all aspects of ac motor starting. Data should be entered as follows:

1. *Sequence*

(a) If a base running load is specified to be on line prior to the start of any motor, enter this load as Item 1, and enter running load per instruction in Section 1.9.3, columns K, L, and M, and skip one line before listing the motor to be started.

(b) If more than one motor is to be started in any sequence, list each motor on a separate line, even if they are identical. After the last motor in a sequence has been listed, enter the word *totals* on the next line, and then skip one line before listing the next sequence.

Alternators (Synchronous Generators)

CUSTOMER: _____ CUST. JOB NO./TITLE: _____ ANALYST: _____ DATE: _____

SYSTEM PARTICULARS: A. MIN. KW: _____ B. PF: _____ C. VOLTAGE: _____ D. NO. PHASES: _____ HZ: _____ E. RPM: _____ DUTY CYCLE: _____

SPECIAL FEATURES: MAX. TEMP RISE: _____ °C, MAX. AMBIENT: _____ °C, OTHER: _____

MOTORS TO START: _____ (Y/N). IF MOTORS TO START, MAX. ALLOWABLE TRANS. VOLTAGE DIP: _____. NOTE: IF NONE SPECIFIED ENTER 35%.

A SEQUENCE OF START	B MOTOR HP	C START CODE	D SkVA/HP	E F.V. SkVA	F TYPE START	G S.F.	H NSkVA	I %TVD	J SPF	K NSkW	L RkVA	M RPF	N RkW	O ALT EFF	P HPD	Q DESCRIPTION OF LOAD
BASE LOADING																

ANALYSIS NOTES: _____

COLUMN F — TYPE START
  ATL — ACROSS THE LINE.
  PWS — PART WINDING START.
  RVS — REDUCED VOLTAGE START.
  WDS — WYE-DELTA START.

COLUMN G — START FACTOR
  ALT — 1.0
  PWS — AS SPECIFIED, OR ASSUME 0.70.
  RVS — AUTO-TRANSFORMER: 80% TAP: SF = .67; 65% TAP: SF = .45; 50% TAP: SF = .28
  WDS — 0.33.

Figure 2-55. Load Analysis Form

Alternators (Synchronous Generators)

(c) If only one motor is to be started in a given sequence, list that motor and skip one line before entering the next sequence.

2. *Motor HP.* Enter the horsepower of the motor.

3. *Start code.* Enter the NEMA motor starting code here. If no code, or actual SkVA is given, enter *G* and star the item. If the actual SkVA (full voltage) is given, draw a line through the start code entry line.

4. *SkVA/HP.* Enter the maximum SkVA/HP of the starting code letter from Table 2-6. If the actual SkVA has been given, draw a line through this entry space.

5. *F.V. SkVA.* Multiply motor HP (Col. B) by SkVA/HP (Col. D), and enter the product here. If the actual full voltage SkVA of the motor is given, enter that value here.

6. *Type Start.* Enter the type starting method as follows:

(a) *ATL* — Across the line.

(b) *PWS* — Part Winding Start.

(c) *RVS* — Reduced Voltage Start.

7. *SF.* Enter the starting factor as follows:

(a) *ATL* — Enter 1.0.

(b) *PWS* — Enter the PU start factor given. If not given, estimate as 0.7 as listed in Table 2-6, and star the item.

(c) *RVS* — Enter the PU start factor given according to the type of reduced voltage start used. If no factor is given, assume 80% (0.8 PU) and star the item.

8. *NSkVA.* Multiply full voltage SkVA (Col. E) by SF (Col. G), and enter the product here.

9. *I TVD.* Using NSkVA from Col. H, consult alternator manufacturer's data for anticipated percent transient voltage dip.

10. *SPF.* If starting power factor is given, enter it here. If none is given, enter 0.4 and star the item.

11. *NSkW.* Multiply NSkVA (Col. H) by SPF (Col. I) and enter the product here.

12. *RkVA.* Enter the given running kVA of the motor. If none is given, use the RkVA value from Table 2-7, Table of 3 Phase Design B Motor Characteristics reprinted below courtesy of The Lima Electric Company, Inc., and star the item.

13. *RPF.* Enter running PF. If none given, enter assumed power factor from Table 2-7 and star the item.

Table 2-7. Table of Three Phase Design B Motor Characteristics

HP	Full Load kVA	Full Load Running kW					
		2 Pole (3600 RPM Synch.)		4 Pole (1800 RPM Synch.)		6 Pole (1200 RPM Synch.)	
		P. F.	Run kW	P. F.	Run kW	P. F.	Run kW
1/2	.8	.70	.6	.65	.5	.60	.5
3/4	1.1	.75	.8	.70	.8	.66	.7
1	1.4	.82	1.2	.76	1.1	.68	1.0
1-1/2	2.1	.84	1.8	.76	1.6	.70	1.5
2	2.7	.87	2.4	.81	2.2	.72	1.9
3	3.8	.89	3.4	.82	3.1	.73	2.8
5	6.1	.89	5.4	.85	5.2	.77	4.7
7-1/2	8.8	.90	7.9	.87	7.7	.80	7.0
10	11.2	.90	10.1	.87	9.7	.82	9.2
15	16.7	.90	15.0	.88	14.7	.85	14.2
20	21.5	.90	19.4	.88	18.9	.85	18.3
25	27.1	.90	24.4	.89	24.1	.87	23.6
30	31.9	.90	28.7	.89	28.4	.87	27.8
40	41.4	.90	37.3	.90	37.3	.88	36.4
50	51.8	.90	46.6	.90	46.6	.88	45.6
60	61.3	.90	55.2	.90	55.2	.88	53.9
75	76.5	.90	68.9	.90	68.9	.88	67.3
100	98.8	.90	88.9	.90	88.9	.88	86.9
125	124.3	.90	111.9	.90	111.9	.89	110.6
150	143.4	.91	130.5	.91	130.5	.90	129.1
200	191.2	.91	174.0	.91	174.0	.90	172.1
250	239.0	.91	217.5	.91	217.5	.90	215.1

NOTE: Data in this table should be used for estimating, and only when manufacturer's data or three (3) phase voltage and current measurements are not available. kVA is based on "National Electric code (NFPA No. 70, 1975)" Data. Power factors were obtained from several meter manufacturer's published data indicated as nominal, but not guaranteed values.

14. *RkW*. Multiply RkVA (Col. K) by RPF (Col. L) and enter the product.

After the sequence of start has been recorded for all motors, and all columns of data have been filled out, add up Columns B, H, K, L, and N for each sequence; enter these totals on the line reserved for each sequence totals. Skip one line in the table after the last sequence total line. Write in Col. A of the last line run totals, and add up the running load totals, columns L and N. The total running kVA listed in column L will be the smallest kVA alternator rating that can be considered for the application. The total running kW of the entire load will be used to determine the minimum size of the engine to be selected to power the set.

d. *Determining transient voltage dip.* The alternator manufacturer's published data must be used as the source for determining the percent voltage dip to be anticipated during transient conditions. This data is usually given for each model produced. However, in some cases a manufacturer may use identical cores for two or three ratings and publish a curve or table of voltage dips versus start kVA for the group. In this case, use the data for the appropriate group of units.

We have covered the fact that each alternator manufacturer may use different design parameters and methods of manufacture when producing their product. For this reason it is not a good idea to substitute data taken from another manufacturer, even though the units have identical ratings, voltages, temperature rises, etc. when data is not available from the manufacturer of the specific unit being used. In the absence of specific voltage dip data, a good approximation of anticipated transient voltage dip may be calculated using the transient reactance (X'd) of the specific alternator. The following equation from "C. Concordia, Synchronous Machines", John Wiley and Sons, Inc., 1951 (Ref 6) has proven to be quite accurate for calculating transient voltage dip:

$$\% \text{ Voltage Dip} = \frac{100 \times X'd}{\frac{RkVA}{SkVA} + X'd} \quad \text{(Equation 2-30)}$$

Where:

$X'd$ = Per unit alternator transient reactance.
RkVA = Rated kVA of the alternator.
SkVA = Start kVA of the load.

NOTE: Whenever using this or any other method of calculating voltage dip in lieu of specific manufacturer's data, specific mention should be made on all calculations that this is a calculated value, and that actual performance may vary from this calculated data.

2.8 APPLICATION CONSIDERATIONS OF SYNCHRONOUS AC GENERATORS TO NON-LINEAR ELECTRICAL LOADS

The application of synchronous ac generators has historically been a rather straightforward process. The primary considerations have been kVA, voltage, frequency, temperature rise, and motor starting/transient performance.

In recent years the introduction of a category of ac electrical loads (termed **non-linear**) has created some very special problems. These problems affect generator design, sizing, and performance.

This section will discuss the nature of non-linear loading and various application problems. It will suggest generator rating guidelines for selection of the generator, its excitation system and accessory components. The references and bibliography at the end of this section contain material for a more in-depth study of subject. Application and performance information should be obtained from the manufacturer of the devices and the manufacturer of the generator set. The rating recommendations given in this section are conservative and may not be precise for a given manufacturer's units or model within his product line. When in doubt, ask; **seek advice from the manufacturer of the product**.

2.8.1 Definitions unique to non-linear loading

a. *Commutation* (rectifier circuits). The transfer of unidirectional current between rectifier circuit elements or thyristor converter circuit elements that conduct in succession.

b. *Harmonics.* Defined as deviations from the fundamental frequency sine wave, expressed as additional sine waves of frequencies that are a multiple of the generated frequency. They are expressed as third, fifth, seventh, etc. harmonics, denoting their frequency as a multiple of the primary wave frequency. See Figure 2-56.

c. *Harmonic content.* A measure of the presence of harmonics in the waveform expressed as a percentage of the fundamental frequency. The total harmonic content is expressed as the square root of the sum of the squares of each of the harmonics amplitudes, expressed as percentage of the fundamental.

d. *Harmonic distortion.* Non-linear distortion of a waveform characterized by the appearance in the output of harmonics other than the fundamental component when the input wave is sinusoidal. Subharmonic distortion may also occur.

e. *Linear load.* AC electrical loads where the voltage and current waveforms are sinusoidal. The current at any time is proportional to voltage.

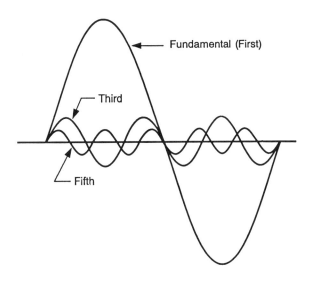

Figure 2-56. Harmonics of a Sine Wave

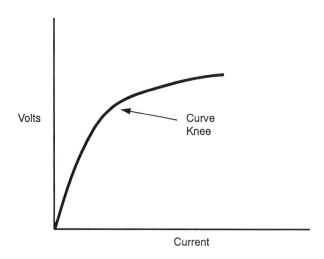

Figure 2-57. Typical Saturation Curve

f. *Non-linear load.* Applied to those ac loads where the current is not proportional to the voltage. Foremost among loads meeting this definition are gas discharge lighting having highly saturated ballast coils and thyristor (SCR) controlled loads. The nature of non-linear loads is to generate harmonics in the current waveform. This distortion of the current waveform leads to distortion of the voltage waveform. Under these conditions, the voltage waveform is no longer proportional to the current.

g. *Saturated coil.* Saturated coils include ballast coils, transformers, squirrel cage induction motors, etc. This type of load is linear in nature until the iron reaches saturation. See Figure 2-57. Once the knee of the saturation curve is reached, non-linearity commences and increases the further it moves towards the flat area of the curve.

h. *Semiconductor static power converter* (thyristor converter). Equipment that changes electrical energy from one form to another employing semiconductor switching devices (rectifiers) such as diodes, thyristors (SCRs), and transistors.

i. *Thyristor/silicon controlled rectifier* (SCR). A semiconductor rectifier controlled by a firing pulse at the gate of the device. The pulse will turn the SCR on and the SCR will stay on as long as current is present at its anode. Since it will block reverse current, it functions like any diode or rectifier with the exception that forward current flow can be controlled — thus the term silicon controlled rectifier. See Figure 2-58.

2.8.2 Gaseous discharge lighting.
Gaseous discharge lighting, the first of our non-linear loads, derives its non-linearity principally from the design and selection of the ballast coil. Cost and size considerations have led manufacturers to operate the ballasts in a highly saturated condition, resulting in distortion of the current wave. See Figure 2-59.

Analysis of the current waveform of this type lighting shows that the harmonic having the greatest amplitude is the third harmonic. In a 4-wire, wye-connected, three phase system, the fundamental currents at any instant will always add up to zero in the neutral. However, the third harmonic of each phase is always in phase with those of the other two phases. As a result, rather than canceling each other (as is the case with the fundamental), they are additive and may well lead to serious neutral loading problems. As an example: a three phase system has 100 amperes load and each phase contains 30% third harmonic. The harmonic current flowing in the neutral will be three times 30% of 100, or 90 amperes at three times the fundamental frequency (180 Hz for 60 Hz systems). Specific studies must be made of facilities having large amounts of this type loading to ensure that the third harmonic current flowing in the neutral is not too large for the conductor size.

High percentages of non-linear loads also cause increased generator heating. The neutral lead of 4- and 10-lead generators is particularly vulnerable to third harmonic currents, because it carries the same load as the wiring system above. Harmonic loads also increase generator winding heating because of increased hysteresis and eddy current losses. This factor is usually only a few percent unless the harmonic content is unusually high. Generator manufacturers' recommendations should be followed.

2.8.3 Adjustable speed motor drives.
Adjustable speed drives (SCR drives) allow infinite speed adjustment

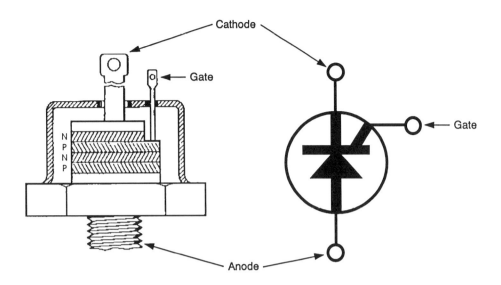

Figure 2-58. Silicon Controlled Rectifier

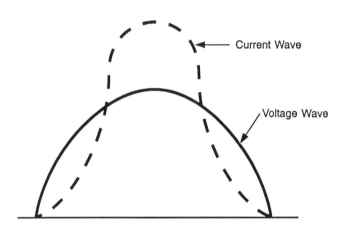

Figure 2-59. Typical Voltage and Current Waveforms Associated with Saturated Coils

of electric motors. SCR drives are used in an ever expanding variety of applications, most common of which are:

a. DC motors used in elevators, cranes, printing presses, etc.

b. AC variable frequency drives (VFD) and ac variable voltage input (VVI) drives used in pumps, fans, process conveyors, machine tools, printing presses and the like.

c. Wound rotor ac induction motors used on pumps, fans, and cranes.

In almost all cases, modern variable frequency drives start at zero frequency and ramp up to the set point, while variable voltage drives start at zero voltage and ramp up to the selected value. In both cases, this ramp up is usually under either current or torque limiting control. The intent here is to avoid large inrush current (start kVA), affording a soft start exhibiting less than 150% of running current. This obviously is an asset when the load is served by an on-site limited capacity engine generator set. In fact, a variation of an SCR adjustable speed drive now becoming quite popular is the soft start starter. This starter accelerates the motor at a controlled rate to running speed as though it were an SCR drive. Once the motor is running at full speed the starter disengages, allowing the motor to operate on electric service voltage and frequency.

2.8.4 Uninterruptible power supplies (UPS systems). A UPS system is defined as a system designed to provide electrical power without delay or transients, during any period when the normal power supply is incapable of performing acceptably. UPS systems usually incorporate battery banks to supply the electrical energy to the load during periods when the primary supply is inoperative. During normal operation, the primary power supply feeds either single or three phase ac to a rectifier battery charger. The charger maintains a float charge on the reserve battery bank and powers a static converter to convert the dc output back into ac to power the load. During a power interruption, power for the system is derived directly from the reserve battery bank without switching. The output to the load may be derived directly from the static power converter to a totally static UPS system (see Figure 2-60) or from a motor generator set powered by a converter as shown in Figure 2-61. In either case, the load as seen by the primary power supply or standby engine generator set is the static power converter/battery charger.

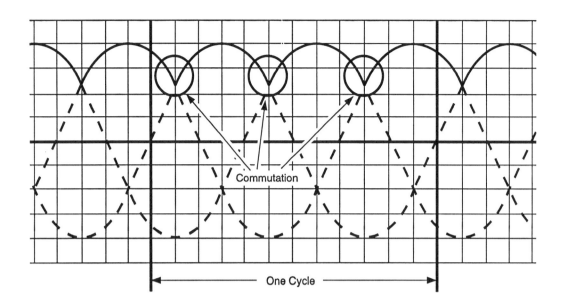

Figure 2-63. DC Output from Three Phase Half Wave Bridge Rectifier Circuit with Full Conduction of Rectifiers

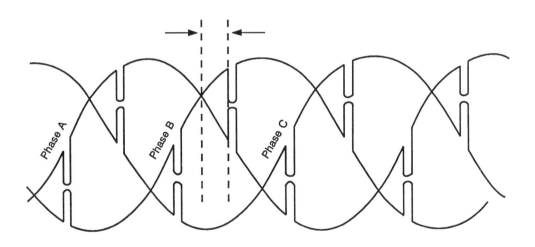

Figure 2-64. SCR Firing Notches on a Three Phase Voltage Waveform

2.8.7 Impact of static power converters upon generator voltage waveform. The notching on the current wave will create harmonic distortion of the generator voltage waveform, which in turn can create problems with the generator, its automatic voltage regulator, the drive system control firing circuits, instrumentation, engine governors, and other loads being serviced by the generator set. When problems are encountered with non-linear loading, a thorough analysis of the entire system and the problem must be made. The generator alone may not be the source of the problem.

Compared to the public utility (infinite bus), an on-site engine generator set is a relatively high impedance source. As discussed above, the reactance of the generator coupled with the reactance of the circuit will further impede commutation of the SCRs, resulting in possible severe notching of the wave form. In addition to this

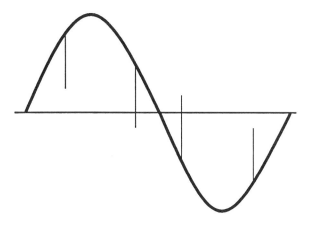

Figure 2-65. Notches on AC Voltage Wave

notching, an oscillation, termed **ringing**, may occur at the circuit's resonant frequency. All of these factors can combine to cause severe distortion of the generator output voltage waveform. See Figure 2-66.

Static power converters, as stated previously, are available in a wide variety of configurations. See references 7 and 8 for a guide to the many facets of these devices.

The theory to which most static power converters conform states that the magnitude of the harmonic current is inversely proportional to the harmonic number:

$$I_h = \frac{I}{h} \qquad \text{(Equation 2-31)}$$

Where:

- h = harmonic order
- I = fundamental current
- I_h = harmonic current per unit of I

Not all harmonics will exist in all configurations. The theoretical harmonics that will be present in a given converter can be calculated by:

$$h = Kq \pm 1 \qquad \text{(Equation 2-32)}$$

Where:

- h = harmonic order
- K = any interger (1, 2, 3, 4, etc.)
- q = pulse number of converter (6, 12, etc.)

Thus a six pulse converter will have 5th, 7th, 11th, 13th, 17th, 19th, etc. harmonics. A twelve pulse will have 11th, 13th, 23rd, 25th, etc.

Static converters are available with three pulse, six pulse, and twelve pulse circuits. Six and twelve pulse circuits are most popular, but the three pulse circuit is used for some battery chargers. The theoretical total harmonic distortion for a three pulse circuit is 66% (0.66 pu), a six pulse circuit is 30% (0.30 pu), and a twelve pulse is 15% (0.15 pu). Circuit inductance reduces the practical value of harmonic currents from that calculated by equation 2-31. Table 2-8 shows practical values of harmonic currents for six and twelve pulse converters, three phase, full

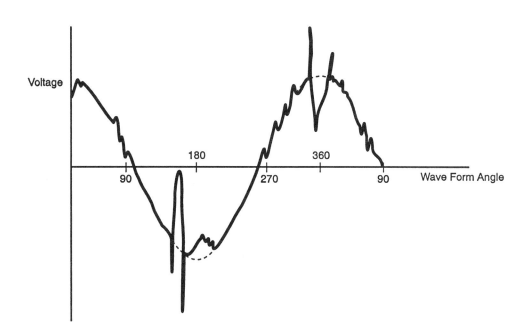

Figure 2-66. Distorted Voltage Waveform with SCR Converter Load

wave bridges. These values may vary widely due to gating phase angle, source impedance, and voltage and phase unbalance.

Table 2-8. Per Unit Practical Harmonic Currents for Six and Twelve Pulse Converters

Harmonic	5	7	11	13	17	19	23	25
6 pulse	0.175	0.111	0.045	0.029	0.015	0.010	0.009	0.008
12 pulse	0.020	0.015	0.045	0.029	0.002	0.001	0.009	0.008

NOTE: Values subject to variation due to circuit and application differences.

The voltage harmonics created by static power converters may cause a variety of problems for the generator. Voltage transients caused by SCR commutation will be very high and have steep wave fronts. Some peak values have been observed to be in excess of twice peak value of the generator voltage. These high transient peaks stress the generator stator insulation and may cause premature insulation failure. If the spacing or surface lengths are marginal, the voltage spikes may cause arcing between bus bars and live parts.

Harmonic currents in the generator stator windings may cause increased heating of the stator windings as well as the main rotor. As discussed above in the section on gas discharge lighting, eddy current losses are increased by the harmonic content; thus, copper losses of the stator winding are increased. Further, the magnetomotive force established by these harmonic currents generates flux in the stator core, rotor core, and air gap. The harmonic flux in the stator core iron will result in additional core iron loss.

The majority of generators used in engine generator sets are salient pole generators with amortisseur (damper) windings. See paragraphs 2.1.1 and 2.1.3. Under steady state conditions, the generator rotor is in synchronism with the stator flux; therefore, no voltages are induced in the rotor to cause eddy current flow. When harmonics are present, the harmonic flux rotates at a speed of n times the fundamental harmonic, but not necessarily in the same direction. Some rotate backwards. These counter-rotating flux waves induce a pulsating flux on the rotor which can induce current in the amortisseur winding. Therefore, the amortisseur winding must be capable of dissipating the losses engendered by the pulsating flux.

2.8.8 Effects of voltage harmonics upon automatic voltage regulators. Voltage harmonics may also create problems for the generator's automatic voltage regulator because the power input side of these devices, which use SCRs, are usually sensitive to voltage waveform. When the notching becomes too severe, the notch may cross the zero reference line as shown in Figure 2-65. When this occurs, the regulator may become unstable and voltage will hunt. Instability of the set output voltage may also affect the operation of the prime mover governor. As the voltage rapidly rises and falls, kW loading on the engine will change rapidly causing the engine to surge or hunt. Another problem associated with the automatic voltage regulator is voltage regulation. Voltage regulators are available with either average or rms sensing and single and/or three phase sensing.

When the total non-linear load represents less than 25% of the total load on the generator, standard voltage regulators and generators will usually give acceptable performance. However, when the total non-linear load represents greater than 25% of the total load, consider the following:

a. Upgrade, if necessary to a regulator with rms sensing. This will afford better voltage regulation with the unit sensing a distorted voltage wave form. Some manufacturers offer a "UPS compatible" regulator that will incorporate rms sensing and some other of the features below.

b. For three phase systems, if at all possible, use a regulator with three phase sensing capability.

c. Isolate the input power from the generator output voltage by either using an isolation transformer, a PMG excitation support system, or a series boost system.

2.8.9 Generator performance data. Two general criteria of generator performance data are usually specified in conjunction with UPS or SCR controlled drive systems. These are (1) maximum subtransient reactance (X''_d), and (2) maximum transient voltage dip. This data is generally available from the generator manufacturer's technical manuals or performance data sheets.

a. *Subtransient reactance, X''_d.* This figure is usually stated for each generator at a specific voltage, frequency, and kVA rating. The X''_d specified for static power converters is usually stated in terms of the kVA of the converter. To calculate the reactances of a unit in terms of the converter voltage, frequency (if other than stated on the generator data sheet), and kVA, use the work sheet in Figure 2-67.

b. *Maximum transient voltage dip.* The power generation industry has not done a good job of establishing state-of-the-art test methods to determine this data point. Methods established by NEMA, IEC, and MIL-STD-705 use peak-to-peak measurement of the waveform on an oscillogram as illustrated in Figure 2-52. These methods are perfectly acceptable for sinusoidal voltage waveforms, but not for distorted waveforms. Wave shape distortion and the timing of the transient in relationship to the voltage wave will cause variations in dip measurements. Standard methodology is to use the average of three duplicate tests to determine dip.

Improved accuracy can be gained by the use of instrumentation employing rms measuring. One such instrument is the Data 6000 Programmable Universal Waveform Analyzer. This test equipment will give repeatable results with non-linear load distorted wave shapes.

2.8.10 Total load considerations. The complete UPS site will normally consist of several of the following:

a. *UPS load.* Determine the UPS input kW/kVA requirement by dividing these values by the efficiency of the UPS.

b. *Battery charging load.* This may be up to 30% of the UPS load, depending upon battery state at the time the generator is brought on line.

c. *Air conditioning/air handling loads.* All of these loads, if present, must be added to establish the total demand. Questions should be asked at the time of inquiry as to the existence of items b and c, and the amount of power demanded by them.

2.8.11 Generator sizing guidelines for static power converter loads. To obtain satisfactory operation of synchronous generators with these loads, the following recommendations are suggested:

a. *Twelve pulse UPS systems and similar standby duty loads*

1. For applications where these loads represent less than 25% of the generator's standby rating, select a standard generator and automatic voltage regulation system.

2. For applications where these loads represent greater than 25% of the generator's standby duty rating, select a generator rated for continuous duty at 105°C rise for Class F and H insulation systems.

3. Using the work sheet in Figure 2-67, calculate the generator subtransient reactance for the actual load determined in 2.8.9 above. Use the calculated load kVA as kVANew to determine X''_d New. Consult with the UPS supplier to determine if this value meets the requirements of his system.

b. *Six pulse UPS systems and similar standby duty loads*

1. Select a generator rated for continuous duty at 80°C temperature rise.

2. Review/upgrade the voltage regulation system.

3. Determine the subtransient reactance per a.3 above. Consult with the system supplier to determine suitability. If no specific information relative to subtransient reactance is available, use a maximum of 12% (0.12 pu) as a UPS base.

4. Various UPS manufacturers have, through experience, determined their own generator set sizing guidelines. As an example: one manufacturer suggests that diesel sets have their standby duty rated at three times the UPS load plus the additional non-UPS load. Natural gas fueled sets should have a standby rating of five times UPS load plus non-UPS load. **It cannot be stressed strongly enough — work closely with the supplier of these systems.**

c. *Variable frequency drive (VFD) and variable voltage input (VVI) drive — standby duty*

1. Select the generator rating that will give a VFD/VVI base subtransient reactance of 15% (0.15 pu) maximum, or lower if recommended by the manufacturer of the drive equipment.

2. Review/upgrade the voltage regulation system.

d. *Continuous duty SCR loads: cranes, induction heaters, VFD/VVI, etc.*

1. Select a generator based upon a maximum of 80°C rise.

2. Review load power factor requirements. If the power factor is lower than 0.80 lagging, consult with the generator manufacturer for sizing recommendations.

3. Review the generator subtransient reactance and compare with system requirements. As a general rule, system requirements of subtransient reactances will fall in the range of 8%–15% (0.08–0.15 pu).

4. Determine maximum transient voltage dip required by the system/load. Typically, these values will be in the area of 8%–10%.

5. Select a main stator insulating system that will afford satisfactory life with continuous voltage transients and spikes.

6. Provide additional mechanical bracing to main stator windings.

7. Review clearances of live parts (bus bars) and increase this spacing if required to prevent tracking during voltage transients.

e. *Automatic voltage regulator review/upgrade*

1. Use a PMG (permanent magnet generator) exciter if available.

2. Use a non-SCR voltage regulator with an isolation transformer if the generator cannot be equipped with a PMG exciter.

Computing generator reactances for values of kVA, voltage, and/or frequency other than those listed in the generator model performance data sheets. For those applications where reactances are desired for rating values of kVA, voltage, and/or frequency other than those listed in the individual generator model performance data sheets as base data, these new or desired values may be computed using the listed reactances as a base as follows:

$$PU\ XNew = PU\ XBase \times \left[\frac{kVANew}{kVABase}\right] \times \left[\frac{VBase}{VNew}\right]^2 \times \left[\frac{HzNew}{HzBase}\right]$$

Where:
- XBase = specific reactance value listed in performance data sheet
- XNew = value of reactance at new parameters of kVA, voltage, and/or frequency
- kVABase = base value of kVA listed on performance data sheet
- kVANew = new value of kVA
- VBase = generator voltage listed in performance data sheet.
- VNew = new voltage
- HzBase = frequency listed in performance data sheet
- HzNew = new frequency

---

kVABase _____   VBase _____   HertzBase _____
kVANew _____   VNew _____   HertzNew _____

$$PU\ XNew = PU\ XBase \times \left[\frac{kVANew}{kVABase}\right] \times \left[\frac{VBase}{VNew}\right]^2 \times \left[\frac{HzNew}{HzBase}\right]$$

$$PU\ XNew = PU\ XBase \times \left[\underline{\qquad}\right] \times \left[\underline{\qquad}\right]^2 \times \left[\underline{\qquad}\right]$$

PU X New = PU X Base × _____ × _____ × _____

PU X New = PU X Base × _____ (multiplier to convert any Xbase to Xnew)

_____ PU X''_d Base × _____ = _____ PU X''_d New
_____ PU X'_d Base × _____ = _____ PU X'_d New
_____ PU X_d Base × _____ = _____ PU X_d New
_____ PU X_2 Base × _____ = _____ PU X_2 New
_____ PU X_0 Base × _____ = _____ PU X_0 New
_____ PU X_q Base × _____ = _____ PU X_q New

Figure 2-67. Generator Performance Data Worksheet

REFERENCES

[1] Motors and Generators, NEMA Standard MG 1-1987, National Electrical Manufacturers Association.

[2] Rotating Electrical Machines, IEC Standard 34, International Electrical Commission.

[3] British Standard BS4999.

[4] National Electrical Code, NFPA 70-1993, National Fire Protection Association.

[5] Glossary of Standard Industrial Terminology and Definitions — Electrical, EGSA 101E, Electrical Generating Systems Association.

[6] C. Concordia, Synchronous Machines.

[7] ANSI/IEEE STD 519-1981, IEEE Guide for Harmonic Control and Reactive Compensation of Static Power Converters.

[8] Update of Harmonic Standard IEEE-519: IEEE Recommended Practices and Requirements for Harmonic Control in Electric Power Systems. IEEE Transactions on Industry Applications, Vol. 25, No. 6, November/December 1989.

BIBLIOGRAPHY

Adjustable Speed Drives, A43, A44, Power Transmission Design, 1991.

Bordeau, S. P., The Fundamentals of Semiconductors, EM Synchronizer, 200.SYN.61, Electric Machinery Company.

Comparing the Three Most Popular Temperature Sensors, Minco. Products, Inc., Electronic Design, January 4, 1965.

Electric Motor and Generator Repair, TM 5-764, Department of the Army.

Fast, Herbert N. Basic Generator Theory, EGSA Paper 78-F-500, Electrical Generating Systems Association.

Fujita, H. and Ailagi, H., Member IEEE, A Practical Approach to Harmonic Compensation in Power Systems, Series Connection of Passive and Active Filters. IEEE Transactions on Industry Applications, Vol. 27, No. 6, November/December, 1991.

Garneau, J., Leroy-Sommer, Canada, Impact of Non-linear Loads on Synchronous Generators. Paper EGSA, June, 1989, Electrical Generating Systems Association.

Glossary of Standard Industry Terminology and Definitions — Mechanical, EGSA 101M, Electrical Generating Systems Association.

Groethe, A. P. Insulation Systems, EGSA Paper 78-F-60, Electrical Generating Systems Association.

Heydt, Dr. G. T., Effects of Electronic Equipment on Plant Power Quality, Plant Engineering, September, 1992 (File 0501).

Jones, C. B., The Effects of Adjustable Frequency Controllers on the AC Line, Square D Co., Columbia, SC.

McPherson, George. An Introduction to Electrical Machines and Transformers, John Wiley and Sons.

Oscarson, G. L. ABC of Motor Starting, E-M Synchronizer 200-SYN-50, Electrical Machinery Co.

Oughton, G., UPS Backup Generator Compatibility, Power Quality Assurance, 1992.

Phillips, G. R., Century Electric Corp., SCR Drives, Performance on Diesel Gensets, Specifying Engineer, March Supplement, 1986.

Sauder, J., Caterpillar Inc., EDS: 703, Understanding Static UPS Systems and Generator Set Application Considerations.

Shaefer, J., Rectifier Circuits, Theory & Design, John Wiley & Sons.

Smeaton, Robert W. Motor Application and Maintenance Handbook, McGraw Hill Co.

Strasser, D. G., Marathon Electric Mfg. Corp., Application of Generators to Non-Linear Loads, EGSA Paper No. 91-FC400, Electrical Generating Systems Association.

Wolfe, F. W., Basler Electric Co., Cooperation Among SCRs, Applying Voltage Regulators. Specifying Engineer, March, 1986 Supplement.

Wyle, P. F., Newage International, Impact of Non-Linear Loads on Stamford A.C. Generators.

Induction Generators

Al Groethe

CHAPTER 3

INTRODUCTION

An induction generator may be similar to, or the same as, an induction motor in its construction. The stator may be of three phase or single phase design and the rotor is of squirrel cage design.

3.1 OPERATING PRINCIPLES

3.1.1 Basic operation. When an induction machine is driven above its synchronous speed, it becomes a generator and produces electric energy. The same machine, when operated as a motor, consumes electric energy to drive a mechanical load at less than synchronous speed.

3.1.2 Synchronous speed. Synchronous speed is determined by the number of poles (p) for which the machine is wound and by the frequency (f) of the connected power source.

$$\text{Synchronous Speed (RPM)} = \frac{120f}{p}$$

A four-pole machine operating on a 60 Hz system will have a synchronous speed of 1800 rpm.

As the load of an induction motor increases, its speed decreases below synchronous speed. The difference between synchronous speed and the motor operating speed is called **slip**.

$$\text{Slip} = \frac{\text{Synchronous RPM} - \text{Actual RPM}}{\text{Synchronous RPM}}$$

If the induction machine is driven above its synchronous speed, its slip becomes negative.

3.1.3 Speed-torque characteristics. Figure 3-1 illustrates the speed torque relationships of an induction machine. Below synchronous speed, the machine puts

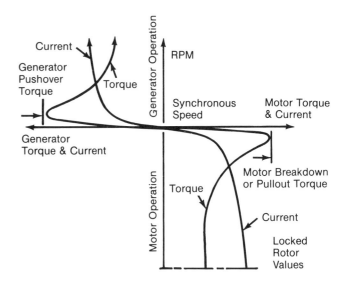

Figure 3-1. Typical Speed-Torque Characteristics of Induction Motor and Generator

out torque (positive) and operating as a motor, absorbs current. Above synchronous speed, the machine absorbs torque (negative) and puts out current.

3.1.4 Performance. Figure 3-2 shows performance curves for a two-pole induction generator through a range of speeds from 3570 to 3650 RPM. The horizontal scale shows motor torque as positive and generator torque absorbed as negative. As kilowatt output increases (lower section) the efficiency and power factor (upper section) also increase.

3.1.5 Construction. Although induction motors can be used as induction generators, significant improvement in performance can be obtained by using an induction machine specially wound for induction generator opera-

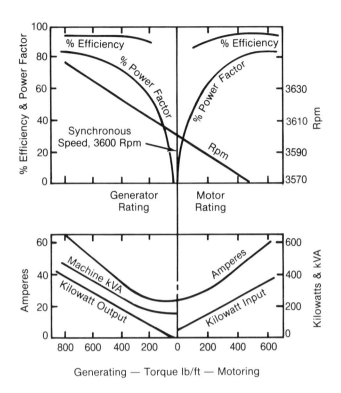

Figure 3-2. Typical Performance Curves for a Two-Pole Induction Generator

tion. The efficiency and power factor will be higher on a true induction generator.

The induction generator may be designed with a lower breakaway torque than an induction motor operating as an induction generator. This will result in lower torque transients on the drive train.

An induction motor for use in a 480 volt system will normally have a 460 volt rating. Since the load current of an induction generator is in the opposite direction, there will be a voltage rise from the system to the machine. This is similar to operating a 460 volt motor at 500 volts. An induction generator is designed for a voltage above the system voltage.

Mechanically, an induction generator can be the same as an induction motor. Induction generators can be supplied with special flanges, couplings, and shaft extensions to facilitate mounting to the driver. As an example, induction generators can be furnished with SAE flywheel housing adapters and drive discs to allow mating to internal combustion engines. Two bearing construction is usually required because of the narrow air gap of induction machines.

Induction generators can be supplied for horizontal or vertical operation. They can be direct-driven, or be belt-driven within limitations.

3.1.6 Excitation. An ac synchronous generator (or alternator) has an exciter to supply direct current power to the field. A voltage regulator is normally used to regulate the output voltage.

An induction generator does not have an exciter or voltage regulator. The induction generator requires an external source of reactive power for its excitation, which is normally obtained from the connected ac power system, usually a utility. The reactive power required for excitation is often 50% of the kW rating and higher than that at part load.

In order to use an induction generator on an isolated, or stand-alone system, it is necessary to use capacitors to supply the excitation. The voltage of such a system is inherently unstable. In order to maintain constant voltage, the required amount of capacitance will vary with each load. Too much capacitance causes the voltage to become dangerously high. Too little capacitance causes the voltage to collapse. The cost of controlling this may exceed the cost of using a synchronous generator.

3.1.7 Frequency and voltage. The frequency and voltage of an induction generator are that of the connected power source, or utility. Changing the speed of an induction generator does not change its frequency or voltage, but does change the amount of output power or load current. Any such speed change is limited to an increase or decrease in the negative slip.

3.2 APPLICATIONS

3.2.1 Prime movers. Prime movers for energy recovery applications include:

- Engines
- Wind Turbines
- Hydraulic Turbines
- Water Wheel Generators
- Expander Turbines

3.2.2 Generator protection. Protective devices used with induction generators include contactors, overload sensors, and circuit breakers. These devices are applied with an induction generator in the same way as with a motor. Additional protection can be provided by the use of meters, overcurrent sensors, over/under voltage controls, unbalanced sensors, and temperature detectors.

If the load of an induction generator exceeds its pushover torque, the generator will offer no resistance to the prime mover and dangerous overspeeds can occur. To prevent an induction generator from reaching its pushover torque, overspeed protection should be included on the prime mover. Reverse power relays are used to sense direction of power flow and to disconnect the induction generator when it begins to operate as a motor. All protective devices that are used must be coordinated throughout the complete system.

3.2.3 Power factor correction.
Induction generators can produce kilowatts only, and are rated in kW. They do not carry a kVA rating like synchronous generators.

The induction generator requires the same amount of reactive magnetizing power regardless of load, so the power factor becomes lower as the output is reduced. Power factor correction capacitors can be added to the output of the induction generator to improve the power factor. The capacitor should not exceed, in kVA rating, the no-load or magnetizing kVA of the generator. It is recommended that the capacitors be connected to the line side of the breaker, or be disconnected from the generator whenever the generator is not on the line.

3.2.4 Starting.
Since induction generators are similar to or the same as induction motors, some three phase applications use this machine as an induction motor for starting purposes. The prime mover speed is then increased until its speed is above its synchronous speed and power flows into the system.

The prime mover can also be used for starting and at about 90% synchronous speed, the induction generator is connected to the line, and the prime mover speed is increased until the induction generator produces power to the system.

3.3 APPLICATION CONSIDERATIONS

Some of the items to be considered in applying induction generators are:

a. Three phase or single phase

b. Voltage

c. Synchronous RPM

d. Maximum kW output at specified temperature rise

e. Prime mover characteristics

f. Ambient temperature

g. Altitude

h. Will the unit ever be operated as a motor?

i. How will excitation be provided?

1. Utility

2. Limited power supply

3. Synchronous generator

j. Will the machine be subjected to adverse environmental conditions?

k. Enclosure (drip proof, totally enclosed, etc.)

l. Special full load speed requirement

m. Special efficiency and/or power factor

n. Special shaft requirements

o. Special mounting flange

p. Direct drive or belted (if belted, what are drive details)

q. Thrust load (which direction)

r. Other electrical or mechanical requirements

3.4 ADVANTAGES AND DISADVANTAGES OF INDUCTION GENERATORS

When compared with a synchronous generator, the induction generator has special advantages.

- No voltage regulator is required. Voltage and frequency are controlled by the utility.
- Excitation is supplied by the utility.
- Simple construction. No brushes, collector rings, or exciter.
- No synchronizing circuit for paralleling with the utility.
- Lower maintenance cost.
- Large power swings do not pull the generator out of synchronism with the utility.

Some disadvantages of the induction generator are:

- Cannot operate by itself; it must be paralleled with utility or other power system (except for special stand-alone systems).
- Power factor varies with load and is lagging.
- Efficiency slightly lower.
- Will not supply sustained short circuit current to a system short circuit.

REFERENCES

[1] Thode and Azbill. Typical Applications of Induction Generators & Control System Consideration, IEEE Trans. Ind. Appl., November 1984 p 1418–1423.

[2] Parsons, Jr., J. R. Cogeneration Application of Induction Generators, IEEE Trans. Ind. Appl. Vol. 1A-20, May/June 1984, p 497–503.

[3] Nailen, R. L. How Induction Generators Work, Electrical Apparatus, June 1980.

[4] Bolin, W. D. Power Cost Reduction Using Small Induction Generators, IEEE Trans. Ind. Appl. Vol. 1A-20, September/October 1984.

[5] Barkle & Ferguson. Induction Generator Theory & Application, AIEE Transactions, Vol. 73, February 1954, p.12–19.

Automatic Voltage Regulation

Fred Wolf

CHAPTER 4

INTRODUCTION

Today's power user requires a reliable source of quality electricity to fulfill his needs. Quality power means, among other things, constant voltage at all load levels. Automatic voltage regulators provide control of generator voltage to satisfy this need.

This chapter discusses the characteristics of automatic voltage regulators and generators that will be important in the selection, application, and servicing of generators using automatic voltage regulators (AVRs). Although much of the discussion may apply to other types of generators, this text will assume that brushless generators are used, since the brushless generator is the most common in use today.

4.1 GENERATOR CHARACTERISTICS

To generate electric power, the prime mover must cause generator conductors to cut through a magnetic field. The synchronous generator develops this magnetic field using the principle of electromagnetism. By flow of dc current through a field winding, a magnetic field is produced in the generator. By increasing the flow of dc current in the field, the magnetic field is made stronger (more lines of magnetic flux). By decreasing the flow of dc current, the magnetic field is made weaker (fewer lines of magnetic flux).

If the generator rotor is made to rotate at constant speed by the prime mover and speed governor, this change in the magnetic field strength causes generator ac output voltage to change. Increasing the dc current will increase magnetic field strength and increase ac output voltage. Decreasing the dc current will result in a decrease in ac output voltage.

For each generator design, this characteristic may be different. Generator manufacturers often plot this characteristic, along with some others, on a graph as shown in Figure 4-1. This characteristic makes possible the control of generator voltage by the adjustment of dc field current.

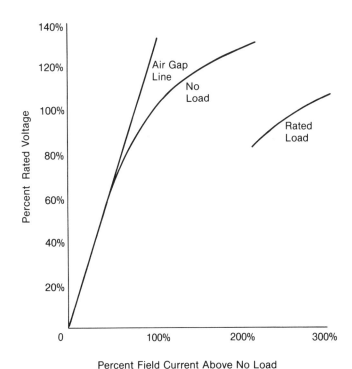

Figure 4-1. Typical Generator Saturation Curve

Automatic Voltage Regulation

A second characteristic of generators that is important to voltage regulators is generator impedance. A simple model of a generator is shown in Figure 4-2. This model consists of a theoretical perfect generator and a generator impedance (resistance and inductance). The perfect generator is defined as a device which will maintain rated voltage at its output regardless of load.

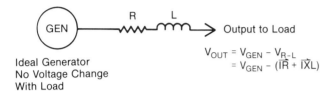

Figure 4-2. Model of AC Generator

If such a generator existed, no voltage regulator would ever be required. All practical generator designs, however, include some amount of the impedance shown in the model, although some designs have more impedance and others less. At zero load current, the output of the model generator is the perfect generator voltage.

By applying load to the model generator and increasing the current, Ohm's Law promises that some voltage drop will occur across the series generator impedance. See Figure 4-3. This voltage drop subtracts from the perfect generator voltage, causing the output voltage of the model generator to drop. This drop in generator voltage with increasing load is the reason for using automatic voltage regulators to adjust generator voltage with changes in load.

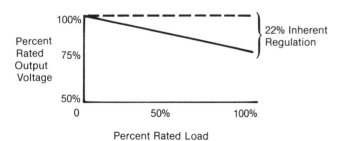

Figure 4-3. Effect of Increasing Load on Model Generator

If the speed of the generator is allowed to change, the generator's third characteristic comes into play. If load and dc field excitation are constant, generator output voltage is proportional to speed. A generator operating at rated voltage with constant excitation will follow the "volts per hertz" curve of Figure 4-4. The curve shows

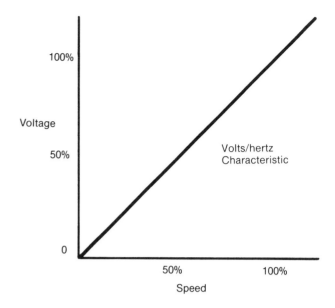

Figure 4-4. Generator Voltage Versus Frequency (Speed)

100% voltage at 100% speed, 50% voltage at 50% speed, and 110% voltage at 110% speed. This characteristic is common to ac generators.

Thus three characteristics of the generator are important to control the voltage of a generator:

a. Generator output voltage can be varied by field current changes.

b. Generator voltage drop is caused by generator impedance.

c. Generator voltage changes with changes in speed of the prime mover.

4.2 MANUAL EXCITATION CONTROL — BATTERY SOURCE

Assume a generator has specifications and characteristics as shown in Table 4-1 and Figure 4-5. To provide excitation, connect the generator (exciter) field to a battery through a variable resistor (VR), as shown in Figure 4-6.

Table 4-1. Generator Specifications

Type: Brushless, synchronous, four-pole
Rating: 100 kW, 0.8 pf, 125 kVA
Voltage: 208–240 or 416–480 volt reconnectable
Frequency: 50/60 Hz
Speed: 1800 RPM at 60 Hz
Excitation Data: (At 60 Hz, 480 volts, series wye)

No Load	20 Vdc	1.0 amps
Full Load	57 Vdc	2.75 amps

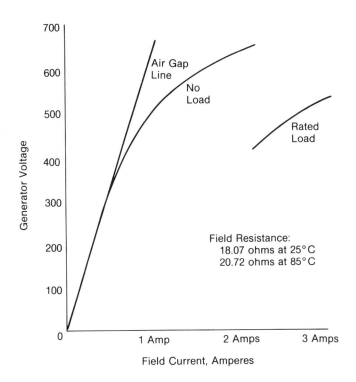

Figure 4-5. Example Generator Saturation Curve

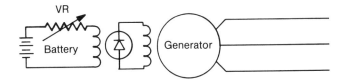

Figure 4-6. Manual Excitation — Battery

If a prime mover/speed governor rotates the generator at rated speed, the VR can be adjusted to change the dc excitation current and adjust generator output voltage. By setting VR to obtain rated generator output voltage and measuring the field voltage and current, the no-load field voltage and current can be recorded for future use, such as in troubleshooting the generator.

By applying load to the generator, the characteristic curves indicate that output voltage will decrease. To restore the voltage to rated, the VR setting can be changed to increase the excitation current and restore generator voltage to rated. Figure 4-7 shows generator voltage and field excitation changes described above. By increasing generator load to its maximum rated value, and adjusting excitation to maintain generator voltage at rated value, excitation voltage and current can be recorded for future use.

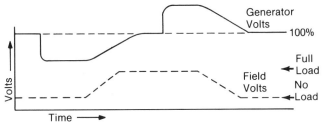

Figure 4-7. Generator Performance Manual Excitation — Battery

Generator manufacturers commonly include excitation voltage and current levels as generator nameplate data, to enable the generator characteristics to be checked in the field. This sort of test will reveal a great deal about the condition of the generator and aid in identifying the source of operating problems to generator or voltage regulator.

A generator set up to operate with a simple battery/VR excitation system in Figure 4-6, shows that suddenly switching on a large load causes a large voltage drop. The generator voltage can be restored by gradually increasing field current while monitoring generator voltage until rated is reached. This method takes a long time to complete.

If faster voltage recovery is needed, a large change in field current could be made. See Figure 4-8. If the change is

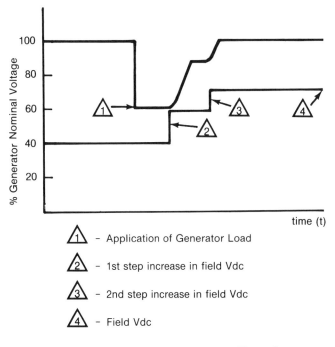

△1 - Application of Generator Load

△2 - 1st step increase in field Vdc

△3 - 2nd step increase in field Vdc

△4 - Field Vdc

Figure 4-8. Generator Performance Manual Excitation Increase by Steps

Automatic Voltage Regulation

too small to restore voltage, a further increase in field current could be made, repeating the process until nominal voltage is restored. Although this process could restore voltage faster than the gradual increase in current, the time to restore is still greater than forcing the restoration.

To force the restoration, reduce the VR to zero resistance and put the maximum available battery voltage across the field. The result (shown in Figure 4-9), will be a rapid increase in generator voltage, but without further action, the voltage will exceed the rated value.

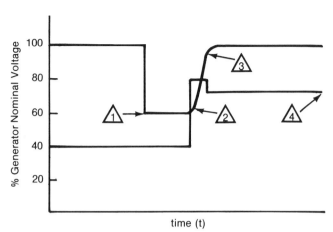

△1 - Application of Generator Load

△2 - Max step increase in field Vdc

△3 - Excitation current manually cut back by operator decreasing field Vdc

△4 - Field Vdc

Figure 4-10. Generator Performance Manual Excitation One Step Increase and Decrease

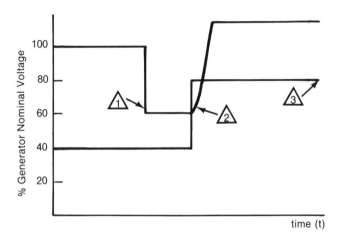

△1 - Application of Generator Load

△2 - Max step increase in field Vdc

△3 - Field Vdc

Figure 4-9. Generator Performance Manual Excitation Increase in One Step

Starting to reduce the field excitation as the generator voltage approaches the rated value prevents this overvoltage condition. With some practice, the performance shown in Figure 4-10 might be achieved. This method of compensating for voltage change can result in the fastest restoration to rated voltage.

4.3 MANUAL EXCITATION CONTROL — SELF EXCITATION

Some generator systems derive power for field excitation from a fixed source such as a battery or a separate excitation generator (permanent magnet generator, for example), but most generators receive excitation power from their own ac output. To understand the operation of a self-excited generator, refer to the manual control scheme shown in Figure 4-11. The generator ac output voltage is converted to dc by the rectifier. The dc output of the rectifier is controlled by a variable resistor (VR) connected in series with the field.

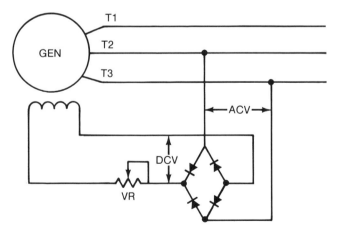

4-11. Manually Controlled Self-excited Generator

To understand the difference of this scheme compared with the battery scheme above, observe the result of applying load to the generator, assuming VR was adjusted for rated voltage. (See Figure 4-12.) When load is applied, the generator voltage will drop because of its impedance. This drop in voltage will proportionally reduce the dc voltage from the rectifier. The drop in dc voltage reduces field excitation, causing the generator voltage to decrease further. This drop results in further drop in excitation. After a time, the generator voltage and excitation will stabilize, with a resultant voltage drop much greater than would be expected with the battery source of excitation power, where the excitation does not decrease with decreasing generator voltage.

affect the excitation as shown in Figure 4-13. Any change in excitation will result in a corresponding change, in the same direction, of the generator voltage. Any change in generator voltage will result in a corresponding change, in the same direction, of the excitation. This type of control is much more difficult to handle than the battery source control with no interaction between generator voltage and excitation voltage.

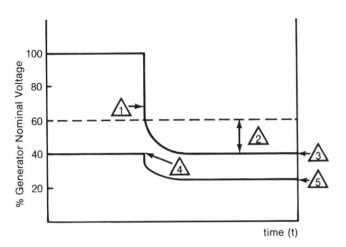

△1 - Application of Generator Load

△2 - Increase in voltage drop caused by decrease voltage input for bridge rectifier

△3 - New steady state voltage

△4 - Rated no load field voltage

△5 - New steady state field voltage

Figure 4-12. Self-excited Generator No Excitation Control

Gradually reducing the VR setting while observing the resultant increase in voltage corrects the low voltage. Although slow to recover to rated voltage, this approach will minimize the effect of increases in generator voltage causing corresponding increases in field current.

The field forcing method of reducing the VR setting to zero resistance and applying all available excitation to the field will result in rapid voltage recovery to rated voltage. With skillful maneuvering of the VR setting during the voltage recovery, voltage overshoot can be avoided, but the skill required will be greater than with the battery scheme, because the changing ac voltage will

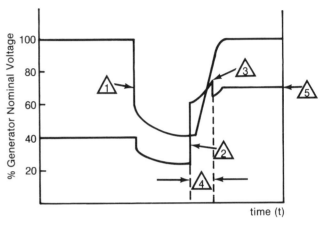

△1 - Application of Generator Load

△2 - 1st step increase in field voltage

△3 - Step decrease in field voltage

△4 - Field reaction to increasing generator voltage

△5 - New steady state field voltage

Figure 4-13. Self-excited Generator Manual Control

4.4 AUTOMATIC EXCITATION CONTROL

Many different devices have been used to replace an operator monitoring a voltmeter and adjusting the VR setting by an automatic adjuster, which can act more quickly and read the voltmeter more accurately. Instead of watching a meter, the automatic device is electrically connected to the generator output voltage for measurement, also called **sensing**. The automatic device controls dc power to the field based on sensed generator voltage.

Figure 4-14 shows the functions of a simple automatic control, referred to from now on as the **automatic voltage regulator** or AVR. The triangle symbol used in this drawing is known as an **operational amplifier**. It has two inputs, one marked with a plus sign (positive) and the other with a minus sign (negative). The third side of the triangle (left side) is the output of the amplifier. The operational amplifier (op amp) is used to symbolize a comparison between two signals. In the AVR, the comparison replaces the operator reading the voltmeter and comparing the reading to the nominal or desired voltage level. In the AVR, the comparison is made between measured or sensed generator voltage and a reference voltage (V_{Ref}) provided by the AVR to represent the nominal or desired voltage level. The adjustable resistor (VAR) is the voltage adjust resistor that allows for changing the nominal or desired voltage level as required to operate the generator. If a positive voltage is applied to the op amp input with the plus sign, the output of the amplifier will be positive. The op amp repeats the input to its positive terminal with the same polarity. If a positive voltage is applied to the op amp minus sign input, the op amp reverses the polarity in its output.

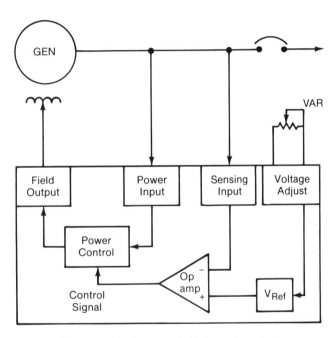

Figure 4-14. Automatic Voltage Regulator

In operation, the generator voltage and the reference voltage are applied to the inputs of the op amp as shown, and the op amp adjusts excitation to the generator until the two inputs are equal. If reference voltage is greater than generator voltage, the output of the op amp will increase, increasing excitation, until the generator voltage equals the reference voltage. If generator voltage exceeds reference voltage, the op amp output decreases, reducing excitation until the generator voltage is equal to V_{Ref}.

Referring again to Figure 4-14, the error signal from the op amp is connected to the power control block. The job of this block is the same as the VR in Figure 4-11. The ac generator voltage used for excitation power is converted to dc before being applied through the power control stage to the exciter field. The ideal power control stage is capable of turning field voltage completely off or applying all available voltage to the field. The power control stage should be able to turn on or off very quickly.

The performance of the AVR is illustrated in Figure 4-15. The top line represents generator voltage changes with time, as load is applied to the generator. The second line is the op amp control signal. The bottom line is the field voltage. The AVR in this example is rather slow to respond, and its regulation accuracy is about 5% from no-load to full-load, as computed by the industry-standard formula:

$$\text{Regulation} = \frac{V_{fl} - V_{nl}}{V_{fl}}$$

These characteristics will be changed to improve the performance of the AVR example.

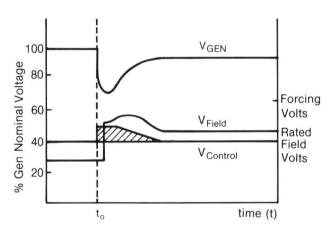

Figure 4-15. Generator Performance with Automatic Voltage Regulator

Time starts with the generator at no-load, speed at nominal (assume speed never changes during the example), and voltage at the desired level (100%). (Refer to Figure 4-15.) The first event to occur is the application of rated load, causing generator voltage to decrease quickly to 80%, then to decrease more slowly to 70%. The voltage regulator responds to the drop in generator sensing voltage by increasing the op amp control signal. The power control takes a little time to respond to the control command, then increases field voltage to the maximum available. Note that the maximum is only 70% of the normally available voltage because power is from the generator output voltage, which has decreased to 70% of nominal.

The field voltage causes exciter field current to increase gradually due to the field's inductance, increasing the brushless exciter output voltage, and gradually increasing generator field current. In time the generator voltage begins to increase toward the nominal value. The op amp senses restoration of voltage and decreases its control signal as the voltage increases. Reduction of the error signal occurs as the excitation voltage available is increasing, and the power control must cut back from full available excitation voltage. By reducing field voltage as the error signal decreases, the exciter field voltage will be reduced to just the amount required to maintain constant voltage at the full load condition.

Why does the voltage return to 95% instead of 100%? If the op amp could detect any error between sensing and reference, no matter how small, the voltage would always return to 100%. With modern technology, sensitivity sufficient to restore rated voltage to within 1% is readily obtainable, and accuracy is available to restore voltage to within 0.25% of nominal.

Because the technology is available, the voltage recovery to nominal should be faster to provide the best gen-set performance. With high sensitivity in the op amp and fast response to the op amp error signal, generator voltage hunting would result due to the generator always lagging behind in voltage response (see Figure 4-16). This performance is totally unacceptable. Something needs to be done to allow the use of more precise and accurate control. This element is needed to compensate for the characteristic slow response of the generator. A method is needed to stabilize the generator response.

4.5 VOLTAGE REGULATOR STABILITY

A review of the manual excitation schemes at the beginning of this chapter will give a clue for solving the need for high sensitivity, fast response, and stable voltage control. The clue is found in the discussion of forcing the field to speed up voltage recovery under load application. The VR operator was to decrease the resistance to zero, watch the generator voltage, anticipate the recovery of generator voltage to 100%, and reduce excitation to prevent overshoot. If the AVR could be made to anticipate restoration of generator voltage and begin to reduce excitation before voltage recovery is complete, perhaps a fast response with rapid, accurate recovery could be achieved.

Let's expand the AVR model (Figure 4-17) to include a stability circuit to accomplish the above task. This circuit must feed information to the error op amp to indicate the progress of the voltage recovery, and allow the power control to anticipate voltage recovery and prevent hunting. Many regulator designs include adjustment of this stability, or feedback circuit, to optimize the voltage transient recovery to match specific generator characteristics.

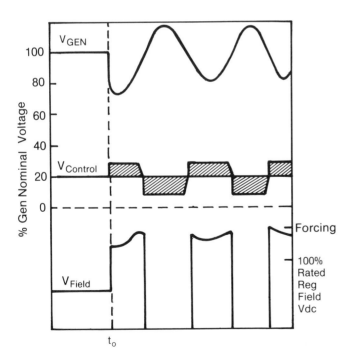

Figure 4-16. Generator Hunting

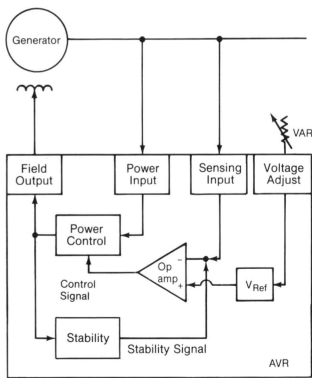

Figure 4-17. AVR Block Diagram

Figure 4-18 illustrates the typical range of adjustment of this control. The first trace illustrates hunting, which is unacceptable performance. The next three traces illustrate settings that may all be considered acceptable, depending on the application. The second trace is marginally stable but offers fastest voltage recovery to nominal voltage. If the overshoot of this trace is not desireable or marginal stability is not allowed, the response to nominal voltage can be slowed to get the performance of traces three or four. Thus any of the lower three traces may be acceptable performance, depending on application needs.

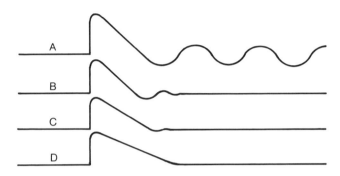

Figure 4-18. Stability Control

4.6 TYPES OF SENSING CIRCUITS

AVRs applied to single phase generators must use a single phase sensing circuit connected to the generator output to measure voltage. The connection of single phase sensing to a low voltage single phase generator is shown in Figure 4-19. The AVR sensing circuit must be rated to accept the nominal generator input voltage and frequency. If direct connection is not possible, a potential transformer (PT) may be used to change the generator voltage to a level acceptable to the AVR sensing circuit. The information that follows, relating the use of single phase sensing and three phase generators, is applicable to the single phase generator.

AVRs applied to three phase generators most commonly use the same sensing circuit as single phase generators, commonly referred to as **single phase sensing**. It would be more correct to identify this circuit as two phase sensing when applied to the three phase generator, because the sensing should always be connected to two phases of the generator and not from one phase to neutral. The line-to-neutral (l-n) connection is not recommended for sensing of generator voltage due to the presence of harmonics in the l-n voltage, which will cause poor line-to-line (l-l) voltage regulation. By connecting from phase-to-phase using two of three phases, the regulator will respond to voltage from both phases instead of just one. The phase-to-phase voltage contains fewer harmonics than the phase-to-neutral voltage to provide more accurate voltage regulation.

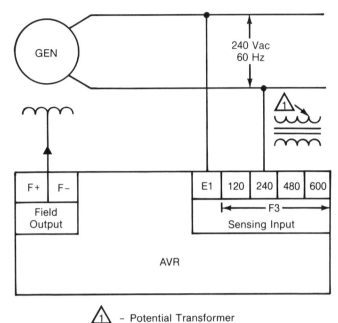

Figure 4-19. Voltage Sensing Single Phase Generator

Three phase generators may also use three phase sensing AVR's connected as shown in Figure 4-20. By monitoring all three phase voltages, the AVR can regulate the average of the three phases. Because the regulator controls only one magnetic field, which influences the voltage of all three phases at the same time, a three phase AVR cannot decrease voltage unbalance, but it will regulate based on the sum of the phase voltages. To measure regulation accuracy with a three-phase sensing AVR, add together the three phase voltages and divide by three to determine generator voltage as sensed by the AVR. Use of a three-phase sensing AVR also provides identical generator system performance when a short circuit is applied from any phase-to-neutral or phase-to-phase.

If the AVR sensing rating will not allow direct connection to the generator output, PTs (potential transformer) may be used to change generator voltage to an acceptable level. PTs selected will affect regulation accuracy of the generator system. Select a PT with metering accuracy, and make certain the PT burden (load) rating is not exceeded to obtain satisfactory voltage regulation. Other devices, such as meters or protective relays, may be connected to a PT used for regulator sensing if total burden (load) of all devices does not exceed PT burden capability or decrease accuracy.

For the three-phase sensing AVR, the open delta PT scheme shown in Figure 4-20 requires only two transformers, and will provide the same regulation accuracy as three transformers at lower cost. If three PTs are used, the wye-wye connection or the delta-delta connection are preferred. Both connections maintain the same phase between primary and secondary, and are preferred over the wye-delta connection.

b. Minimize connecting points or switching devices in the sensing circuit that could open during vibration, shock, or temperature change.

c. If possible, when fuses are necessary for safety, use the same fuses for both sensing and AVR power, so that loss of a fuse will also remove excitation power. See Figure 4-21.

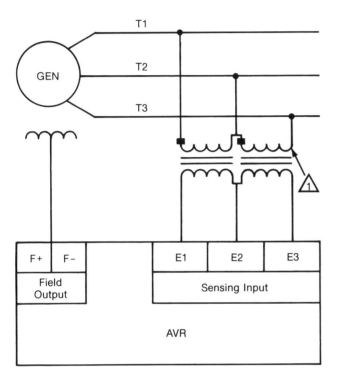

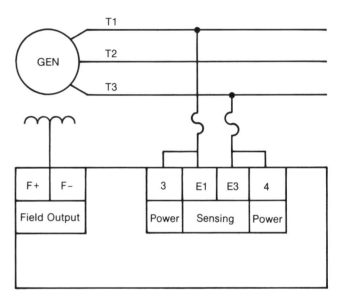

Figure 4-21. Fusing Power and Sensing Circuit

1 - Two Potential Transformers

Figure 4-20. Voltage Sensing Three Phase Generator

When designing a generator system, the importance of the AVR sensing circuit for safe control of generator voltage must be emphasized. Loss of the sensing signal is understood by the AVR as low generator voltage. If excitation power is available, the power control will put full excitation voltage on the exciter field. As the generator voltage increases in response to the full voltage, the power control receives more voltage and applies it to the generator field, further increasing generator output voltage. The only limitation on overvoltage is the generator saturation, commonly 140% to 180% of rated. For this reason, consideration should be given to the following during the design procedure:

a. Use fuses only as required to meet safety requirements. Any unnecessary fuse could cause an overvoltage condition if it opened without cause.

d. If possible, use "failed fuse detection" to shut down excitation. One example is illustrated in Figure 4-22.

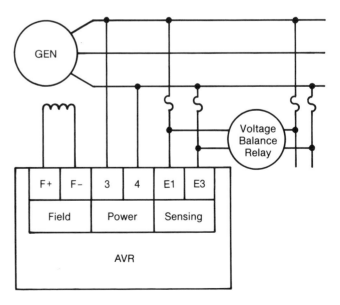

Figure 4-22. Fuse Failure Detection

4.7 POWER INPUT CIRCUIT

To obtain the best performance from an AVR, supply it with the proper input voltage as recommended by the AVR manufacturer. If voltage is too low, the AVR will not have enough output voltage for proper forcing of the exciter field. Figure 4-23 illustrates the performance loss possible with too little voltage supplied. Too much voltage will result in excess stress and decrease in AVR life. Figure 4-24 shows the voltages available from a 12-lead series-wye 480 volt generator. AVR power input may be obtained from phase-to-phase voltage or phase-to-neutral voltage without concern for selecting any specific phase angle. If the regulator power cannot be taken directly from the generator, a transformer can be provided to change the generator voltage to the required level. Fusing of the regulator power input is recommended to protect wiring from damage. Fusing of both input leads should be provided, unless one input lead is connected to the neutral of the generator and codes or safety requirements prohibit neutral lead fusing. See Figure 4-25.

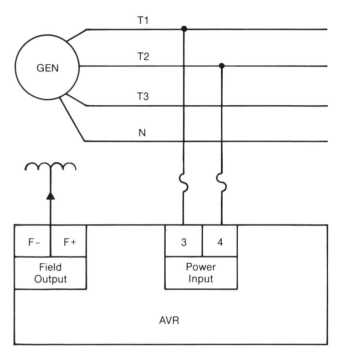

Figure 4-25. Power Input Fusing

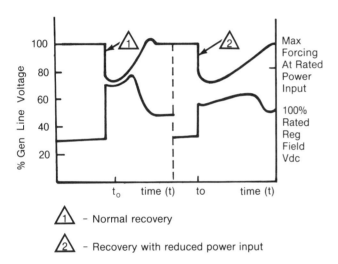

Figure 4-23. Automatic Voltage Regulator Performance, Too Low Input Voltage

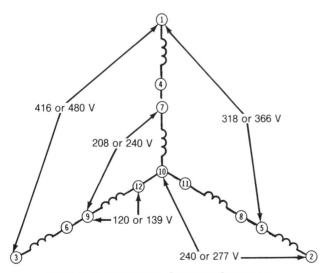

Figure 4-24. Generator Voltages Available

Unless the power circuit includes a power isolating transformer (primary and secondary insulated from one another, as shown in Figure 4-26), the field of the exciter must not be grounded, or extremely high current could flow from the generator output through the AVR to the field ground. Fields are rarely grounded intentionally, but they may become grounded due to insulation breakdown in the field winding, or more commonly, by the use of an engine battery (always grounded) for flashing the field. Using fuses on the power input to the regulator can save wiring but ordinarily will not be fast enough to save the voltage regulator.

To select a power isolation transformer, pick the primary voltage to match a voltage available from the generator, and select the secondary voltage to match the AVR manufacturer's recommendation. The transformer should be a general purpose type, not requiring the instrument accuracy of the sensing input. The transformer power rating should be as recommended by the AVR manufacturer.

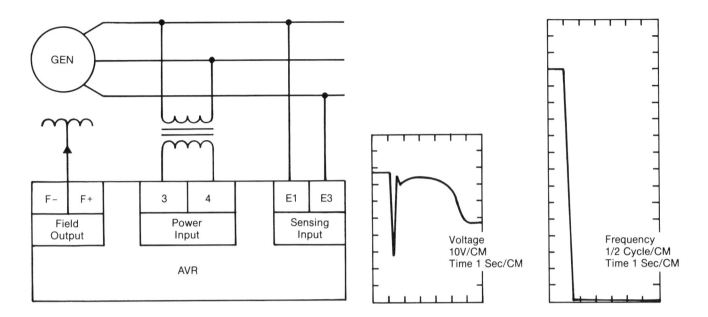

Figure 4-26. Power Input Isolation

Figure 4-27. One Step Load Performance, No Recovery

4.8 FREQUENCY COMPENSATION

Up to now, the effect of frequency (speed) on the generator/AVR has been ignored. Frequency deviation from the nominal value has two effects on generator system performance. The first effect is transient in nature. When a prime mover is required to pick up a large load in a single step, frequency dip and recovery may exceed the allowable amount, as in Figure 4-27 showing frequency unable to recover. If voltage dip is well within specified limits, frequency compensation may be used to increase voltage dip, relieving the initial torque demand on the prime mover and decreasing frequency dip. (See Figure 4-28.)

Circuits developed for this application may be active at all speeds, or they may be blocked from operation until speed has decreased below some threshold. Examples are shown in Figure 4-29.

The second effect of frequency deviation is steady-state, and may be the result of a need to bring the prime mover up to speed slowly or allow for some warmup at idle, or the result of long coastdown or cooldown times. The effect is caused by the generator requiring more field excitation to maintain rated voltage at lower frequency, and by the loss of generator cooling air volume at reduced frequency. The effect is increased generator field and exciter heating. The simple way to prevent the heating is to shut down excitation when operating at reduced frequency. If it is necessary to maintain generator voltage

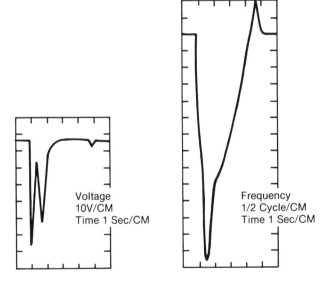

Figure 4-28. One Step Load with Frequency Compensation

at reduced frequency, the generator voltage may be reduced in proportion to speed, which results in constant excitation with changes in speed. The curve active at all speeds is illustrated in Figure 4-29. This characteristic allows motors connected to the generator to continue to operate safely with reduced shaft speed.

Automatic Voltage Regulation

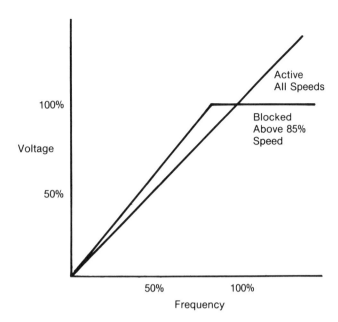

Figure 4-29. Voltage Performance with Frequency Compensation

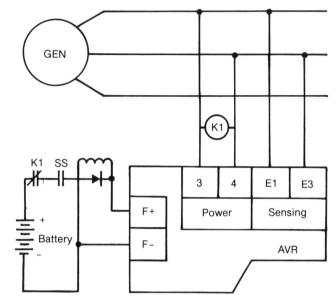

Figure 4-30. Simple Battery Boost

4.9 FAULT CURRENT SUPPORT

Since all excitation power for a brushless self-excited generator comes from the generator output, a short circuit (fault) on the generator output will reduce voltage to or near zero. At such low voltage, no field power will be available to maintain current to the short, and protective devices monitoring current will not be able to use time delay to allow coordination of tripping. For applications not requiring coordination of tripping, this may be the safest design for preservation of equipment and minimizing fire hazard. If fault current is needed for protective devices, one of the following methods of maintaining excitation to the exciter field may be selected:

a. *Battery boost*. Use of a battery to supply excitation power (Figure 4-30). Note the use of a speed switch contact (SS) to prevent boost when the gen-set is idle or shut down and the blocking diode to prevent the regulator from charging the battery. When K1 senses loss of generator voltage, the contact closes to provide field excitation.

b. *Current boost*. Use of generator line current to supply excitation in addition to voltage regulator power (Figure 4-31). By rectifying the line current from the boost current transformer (CT) and connecting the output in series with the AVR output, excitation can be maintained to the generator field.

c. *Voltage/current boost*. Use of generator voltage and line current to supply power to the voltage regulator. By making available to the AVR a constant source of ac

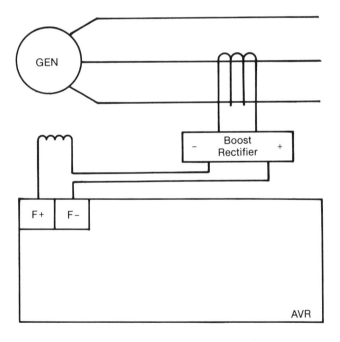

Figure 4-31. Current Boost

input power regardless of generator load as shown in Figure 4-32, the AVR always has the ability to provide the necessary excitation to the generator field.

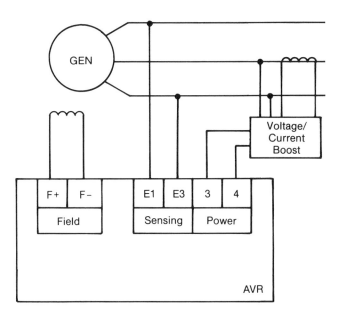

Figure 4-32. Voltage/Current Boost

d. *Permanent magnet generator(PMG)*. By connecting a small generator to the end of the generator shaft that is excited by a rotating permanent magnet for fixed excitation, and connecting the output windings (stationary) to supply excitation power to AVR, the PMG provides another means of supplying a constant source of ac input power to the AVR regardless of load. See Figure 4-33.

e. *Excitation winding*. Use of a separate generator stator winding to supply power to the AVR allows the stator

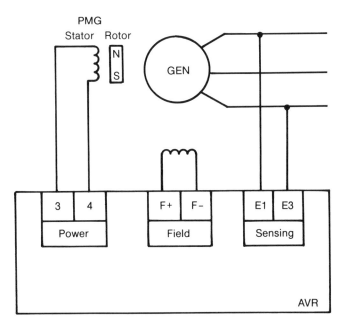

Figure 4-33. PMG Generator

winding to continue to supply power to the AVR even with a short circuit on the generator's normal stator windings.

Many other methods have been devised to provide this capability, all addressing the need to supply excitation power to the field from a source that will not be reduced during a short circuit condition.

4.10 MOTOR STARTING CAPABILITY

The application of a load requiring a large inrush current, such as a motor, can cause generator voltage to dip sufficiently to prevent the motor from starting. The reason for an unsuccessful start may be the dropout of the motor contactor due to low voltage, or the lack of starting torque at the voltage the generator is capable of providing under the heavy load inrush. Although the reason is different, the use of one of the several techniques from para. 4.9 for fault current support will ensure that excitation is available to supply all the starting current the generator system is capable of supplying to roll the motor. Use of these techniques will maintain excitation to keep power to the motor, but the initial voltage dip will only be decreased slightly. If the use of fault current support techniques to improve maximum voltage dip performance are not enough, a larger generator rating may be necessary to reach satisfactory performance. To determine the ability of a boost system to correct a motor-starting problem, connect the AVR power input to a separate source of power, such as the utility power source, and attempt to start the motor. If separate power input is successful, a boost system will likely improve the performance of the gen-set.

4.11 PARALLEL COMPENSATION

If parallel operation of a generator system is required, the AVR must be equipped with paralleling compensation capability. This capability may be provided by an add-on accessory to the AVR or built in to the AVR. The compensation circuit is supplied with a signal from the secondary of a current transformer representing generator current and phase angle, as illustrated in Figure 4-34. The CT signal is compared in the AVR with the sensed voltage to develop a signal proportional to reactive volt amperes (var) load. The circuit causes the generator output voltage to decrease (droop) with increasing reactive (lagging var) load. The circuit is usually adjustable to provide up to 5% voltage decrease (droop) for load increase from zero to rated lagging var. Because this circuit is sensitive to phase angle between voltage and current, the AVR manufacturer's recommendation for installation, wiring, and adjustment should be closely followed to obtain proper parallel load sharing.

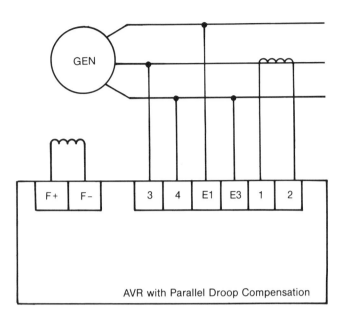

Figure 4-34. Parallel Droop Compensation

4.12 SUMMARY

The automatic voltage regulator functions to control generator excitation, replacing an operator monitoring generator voltage, and providing corrections to the excitation. In order to obtain the best possible performance (much greater than is achievable by manual adjustment) the regulator must include a stability circuit to allow for fast, accurate voltage control. The AVR may sense one phase or three phase voltage. The AVR may include some sensitivity to the speed of the prime mover to assist in block load pickup. For faults or large motor loads, some form of assistance for the self-excited generator may be required. To increase the load capacity of a generator, another unit may be installed, and their outputs paralleled to share the load. If parallel operation is required, the AVR must be equipped with parallel compensation to facilitate the sharing of the reactive load. With the need for fast, accurate voltage control in modern power systems, the necessity to spend time selecting and properly applying the AVR is very important to satisfy the needs of the power consumer.

Circuit Breakers

Ray R. Queen

CHAPTER 5

INTRODUCTION — HISTORY

The need for molded case circuit breakers was created in 1918, when numerous new applications for electrical motors resulted in a demand for a device that would ensure safe operation and, at the same time, protect electrical circuits. During this period, individual motors were used for the first time in industrial plants to operate machine tools, and in private homes to operate appliances such as washing machines and refrigerators. These numerous applications created problems. Plant electricians were constantly changing fuses blown out during motor start-ups because of the lack of properly designed fuses for motor circuit protection. The same problem existed in homes when circuits were overloaded. Inspectors were concerned about fire hazards because plug-fuses were being bridged with pennies, or fuses with higher ampere ratings were being installed. As a result of such dangerous practices, inspection authorities attempted to find a solution to the problem. Meetings with switch manufacturers were initiated in an effort to bring about a solution. Switch manufacturers were asked to develop a switching device that would interrupt a circuit when a prolonged overload condition existed. The device would have to be safe, reliable, and tamperproof. It should also be resettable so as to be reusable after an interrupting operation, without replacing any parts. These early meetings prepared the groundwork for the eventual development of the molded case circuit breaker.

The search for better circuit protection resulted in different approaches to solving the problem. For example, a double-throw switch was designed, which utilized one set of fuses for starting and another set for running. However, the switch was so large and bulky that its use was soon discontinued.

During this period of research and development, The de-Ion arc extinguisher was developed for use in large oil circuit breakers. Although too large in its initial form to be practical for small breakers, the arc extinguisher was eventually modified and coupled with a thermal tripping mechanism. Thus, the first compact, workable breaker was developed in 1923. However, it was another four years until the right combination of materials and design permitted circuit breakers to interrupt available fault currents of 5000 amperes on 120 volts ac or dc.

5.1 DEFINITION

A circuit breaker is defined in the National Electrical Manufacturers Association (NEMA) standards as a device designed to open and close a circuit by non-automatic means, and to open the circuit automatically on a pre-determined overcurrent, without damage to itself when properly applied within its rating. A molded case circuit breaker is further defined in NEMA as one which is assembled as an integral unit in a supporting and enclosing housing of insulating materials.

5.1.1 Function. Molded case circuit breakers are designed to provide circuit protection for low voltage distribution systems. They were originally designed to protect the apparatus connected to the circuit against overloads and/or short circuits. Ground fault protection has been added to many styles of breakers to provide increased protection for the users of the apparatus.

5.1.2 Standards. Molded case breakers are designed, built, and tested in accordance with the NEMA and/or Underwriters' Laboratories, Inc. (UL) standards. In addition, molded case breakers are designed to be applied in accordance with the requirements of the National Electric Code (NEC). With the passing of the federal

Occupational Safety and Health Act, better known as OSHA, compliance with the NEC has been made mandatory.

In addition to the above domestic standards, international standards must be complied with to sell products to various world markets. There are many individual foreign standards on molded case breakers, including British Standards Institute (BSI), German Standards (VDE), Canadian Standards Association (CSA), International Standards (IEC), and others.

5.2 CIRCUIT BREAKER COMPONENTS

Circuit breakers are comprised of five main components: molded case (frame), operating mechanism, arc extinguishers and contacts, trip elements, and terminal connectors. See Figure 5-1.

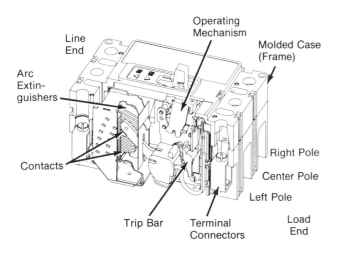

Figure 5-1. Molded Case Circuit Breaker

5.2.1 Molded case (frame). The function of the molded case is to provide an insulated housing to mount all of the circuit breaker components. The cases are molded from moldarta and/or glass polyester material, combining ruggedness and high dielectric strength in a compact design.

Each different type and size of molded case is assigned a frame designation to facilitate identification. The frame identification refers to a number of important characteristics of the breaker, i.e., maximum allowable voltage and current, interrupting capacity, and the physical dimensions of the molded case. Unfortunately, all manufacturers have a different identification system because the breaker's characteristics are different. For example, a 400 ampere, 600 volt breaker supplied by two different manufacturers may have different physical dimensions and interrupting capacity.

5.2.2 Operating mechanism. The function of the operating mechanism is to provide a means of opening and closing the breaker. This toggle mechanism is the quick-make quick-break type, meaning that the speed with which the contacts snap open or closed is independent of how fast the handle is moved. The breaker is also trip free, which means it cannot be prevented from tripping by holding the breaker handle in the ON position. In addition to indicating whether the breaker is ON or OFF, the operating mechanism handle indicates when the breaker is tripped by moving to a position midway between the extremes. To restore service after the breaker trips, the handle first must be moved from the center position to the OFF position to reset the mechanism, and then to the ON position. This distinct trip position is particularly advantageous where breakers are grouped, as in panelboard applications, because it clearly indicates the affected circuit.

5.2.3 Arc extinguishers. The function of the arc extinguisher is to confine, divide, and extinguish the arc drawn between the breaker contacts each time a breaker interrupts current.

In order to interrupt high, short circuit faults, and dissipate the large amount of energy, the de-ion arc extinguisher consists of specially shaped steel grids isolated from each other and supported by an insulating housing. When the contacts are opened, the arc induces a magnetic field in the grids which, in turn, draws the arc from the contacts and into the grids. The arc is thus split into a series of smaller arcs and extinguished very rapidly. See Figure 5-2.

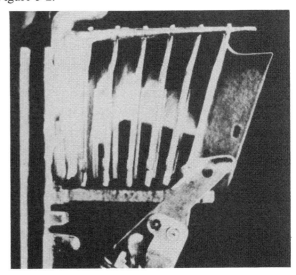

Figure 5-2. De-Ion Arc Quenchers Extinguishing Arc

5.2.4 Trip elements. The function of the trip element is to trip the operating mechanism in the event of a prolonged overload or short circuit current. To accomplish this, an electromechanical or a solid state trip is provided. Today's solid state trips also incorporate optional protection against damaging ground faults.

a. *Electromechanical.* Conventional breakers utilize bimetals (Figure 5-3) and electromagnets to provide overload and short circuit protection. This type of protective action is referred to as thermal magnetic and is the industry standard. To better understand this tripping action, the thermal and magnetic portions will be explained separately and then together.

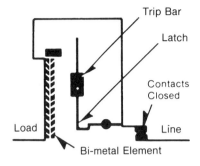

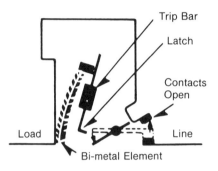

Bi-metal Heats and Bends to Open Contacts on Overload

Figure 5-4. Thermal Action

Figure 5-3. Bi-Metal Type Protection

b. *Thermal overload protection.* Thermal trip action is achieved through the use of a bimetal element heated by the load current. On a sustained overload, the bimetal will deflect, causing the operating mechanism to trip.

A bimetal element consists of two strips of metal bonded together. Each strip has a different thermal rate of heat expansion. Heat due to excessive current will cause the bimetal to bend or deflect. The metal having the greater rate of expansion will be on the outside (longer boundary) of the bend curve. To trip the breaker, the bimetal must deflect far enough to physically push the trip bar and unlatch the contacts. See Figure 5-4.

Deflection is predictable as a function of current and time. See Figure 5-5. This means that a typical 100 amp breaker might trip in 1800 seconds at 135% of rating (Point B). Consequently, bimetals provide a long time delay on light overloads, yet have a fast response on heavier overloads.

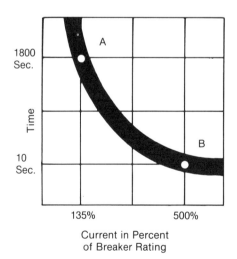

Figure 5-5. Typical Deflection Curve for 100-Amp Thermal Element of a Breaker

Thermal elements are calibrated at the factory and are not field adjustable. A specific thermal element must be supplied for each current rating.

c. *Magnetic short circuit protection.* Magnetic trip action is achieved through the use of an electromagnet in series with the load current. When a short circuit occurs, the fault current passing through the circuit causes the electromagnet in the breaker to attract the armature, initiating an unlatching action, in turn causing the circuit to open. See Figure 5-6. The only delaying factor is the time it takes the contacts to physically open and extinguish the arc.

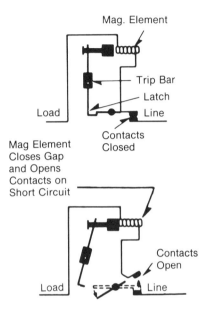

Figure 5-6. Magnetic Action

This action takes place in less than one cycle (0.016 second), and trips the breaker instantaneously. A typical magnetic trip curve is illustrated in Figure 5-7. The breaker will not trip until the fault current reaches or exceeds Point A.

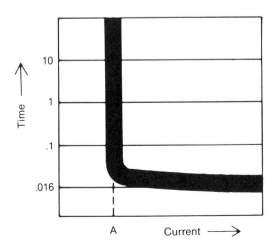

Figure 5-7. Typical Trip Curve for Fixed Magnetic Action

The magnetic trip element may be fixed or adjustable, depending upon the type of breaker and frame size. For example, most thermal magnetic breakers on 250 ampere frames and above have adjustable magnetic trips based on the specific thermal element. Each adjustable magnetic trip is calibrated at the factory for a specific range and is set on the high side. Knobs located on the front of the trip unit can be adjusted to specific requirements. Each of the three adjustment knobs (one per phase) has a high, a low, and a series of intermediate setting positions. See Figure 5-8. The adjustment varies the air gap, which proportionately varies the magnetic trip current rating of the breaker. The magnetic trip is designed so that each point follows a linear scale and each of the intermediate settings has a significant value, within calibration tolerances. For example, Figure 5-9 is a typical set of magnetic trip curve, illustrating how the adjustment knobs

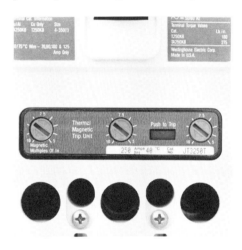

Figure 5-8. Thermal Magnetic Breaker with Adjustable Magnetic Trip

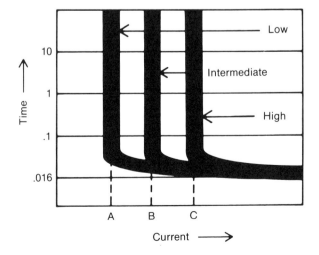

Figure 5-9. Typical Trip Curve for Adjustable Magnetic Action

move the curve from left to right as the magnetic trip is increased. When the adjustment knobs are on the low setting, the breaker will not trip magnetically when the fault current reaches Point A. The fault current must now reach or exceed Point B. However, if the adjustment knobs are moved to the high setting, the breaker will not trip magnetically until the fault current reaches or exceeds Point C. Naturally, there is an infinite number of intermediate settings on the adjustment knobs that enable the curve to be moved to any position between Points A and C.

d. *Thermal magnetic overload and short circuit protection.* This action combines the features of both the thermal and magnetic actions. See Figure 5-10. For example, Points A and B on the trip curve in Figure 5-11 illustrate both the thermal and magnetic action on a typical 100 amp breaker. A 250% overload would take 60 seconds before the bimetal would deflect far enough to hit the trip bar and open the circuit. However, if instead of an overload there was a short circuit that was 4000% (40 times) the breaker rating, the electromagnet in the breaker would attract the armature and trip the breaker in approximately one cycle (0.016 second).

Thermal magnetic trips (identified in the National Electrical Code as time limit) are best suited to the majority of general purpose breaker applications. They are temperature sensitive and automatically re-rate in line with safe cable and equipment loadings which vary with ambient temperatures. They always act to protect the circuit, safeguarding equipment under high ambient conditions and permitting higher safe loadings under low ambient conditions. They do not trip if the overload is not dangerous, but trip instantly on heavy short circuit currents.

e. *Fixed or interchangeable trip unit.* Conventional breakers are available with either a fixed or interchangeable electromechanical trip unit, depending upon the type and frame size. On a fixed trip breaker, the entire breaker must be replaced if a new trip unit is desired. On an interchangeable trip breaker, only the trip unit has to be changed up to the maximum rating of the frame. Figure 5-12 illustrates a 400 ampere interchangeable trip unit. The trip unit can be changed with the breaker already installed, with no modification necessary.

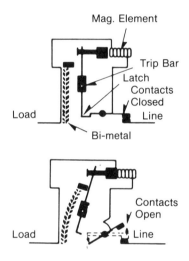

Figure 5-10. Thermal Magnetic Action

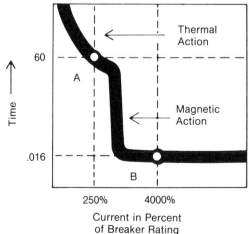

Figure 5-11. Thermal Magnetic Curve

Figure 5-12. Interchangeable Trip Unit

All interchangeable trip breakers have adjustable magnetic elements. The thermal setting is fixed. Trip units, whether loose, or installed in the breaker, generally have the adjustable magnetic setting on the high position when they are shipped from the factory.

f. *Solid state.* Molded case circuit breakers with conventional thermal magnetic trip units are increasingly being replaced by electronic trip units. This is especially true for the larger frame sizes. The result is increased accuracy, repeatability and discrimination. Another major advantage is the option of built-in ground fault protection. Previously, with thermal magnetic breakers, a separate ground fault relay was used to trip the breaker with a shunt trip.

In general, electronic trip systems are composed of three component items as shown in Figure 5-13. First, a current transformer (sensor) in each phase to monitor and reduce the current to the proper level for input into a preinted circuit board. Second, the printed circuit board can be viewed as the brains of the system, since it interprets input and makes a decision based on predetermined parameters. A decision to trip results in the circuit board initiating an output to the third component, a low power flux-transfer shunt trip which trips the breaker. An external source of tripping power is not required.

Because the trip system is electronic, only a low current is needed to verify the electrical performance. This current can be supplied by a simple hand-held test unit. To perform the testing, plug the power cord on the test unit into any 120 volt AC outlet. Connect the two test leads from the test unit into the two receptacles on the front of the breaker as pictured in Figure 5-14. Either the overload calibration, magnetic trip performance, and/or ground fault calibration can be verified. The same test unit is used with all Seltronic breakers.

5.2.5 Terminal connectors. The function of a terminal connector is to connect a circuit breaker to a desired power source and load. There are various methods of connecting the line and load side of circuit breakers: bus bars, panelboard straps, rear connected studs, plug-in adapters, keeper nuts, plug nuts, etc. Whenever conductors (cables) are used on the line and/or load side, terminals are used to connect the conductor to the breaker. Many different types and sizes are available to accommodate single or multiple conductors. They are usually made of copper (for use with copper conductors) or aluminum (for use with copper or aluminum conductors). Whenever aluminum conductors are used, joint compound is recommended to break down oxidation in order to get a better metal-to-metal connection and prevent heating.

5.3 ACCESSORIES AND MODIFICATIONS

5.3.1 Shunt trip. A shunt trip, see Figure 5-15, is used to trip a circuit breaker electrically from a remote location. It consists of a momentary rated solenoid tripping

Figure 5-14. Seltronic Test Kit

device mounted in the breaker. The tripping device must be energized by some control power source with voltage up to 250 volts dc, or up to 600 volts ac. When the solenoid coil is energized from a remote location using some electric pilot device, such as a pushbutton, the plunger moves to activate the trip bar to trip the breaker. At the same time, a cutoff switch opens the circuit to the solenoid coil. This prevents the coil from burning up under a continuous load. Breakers with factory installed shunt trips are available with UL listing (check the factory for UL labeled field-mountable kits). Standard leads extend 18 inches out the side of the breaker. Shunt trips are field-mountable on most industrial breakers.

Note that the shunt trip is merely a means of remotely tripping the breaker, not remotely operating it. To re-

Figure 5-13. Electronic Trip Components

close the breaker, the handle first must be moved to the reset position, and then to the ON position.

5.3.2 Undervoltage release. An undervoltage release, Figure 5-16, basically does as the name implies. It trips the breaker whenever the voltage falls below a predetermined level.

5.3.3 Auxiliary switches (extra contacts). An auxiliary switch consists of normally open and/or normally closed contacts mounted in the breaker that open or close whenever the breaker is opened or closed. See Figure 5-17. To avoid confusion, the contacts are defined as A or B. Contacts A are open when the breaker is open or tripped; contacts B are closed when the breaker is opened or tripped.

Auxiliary switches are connected in control circuits whenever it is desirable to indicate whether the breaker is opened or closed. Here are two examples:

a. Indicating lights are occasionally mounted on switchboards to furnish a visual indication that the circuit is energized. One A contact is required if an indicating light is to be lit whenever a breaker is closed.

b. There are some applications where it is desirable to have an indicating light (red) lit when the breaker is closed and another indicating light lit (green) when the breaker is open. This would require an A and a B contact.

5.3.4 Alarm switches. Alarm switches differ from auxiliary switches in that they function only when the breaker trips automatically, not with the manual operation of the breaker handle. See Figure 5-18. An alarm switch consists of normally open contacts that close when the breaker trips due to an overload condition, short circuit, or operation of a shunt trip. The normally open contacts can be placed in a circuit with an indicating

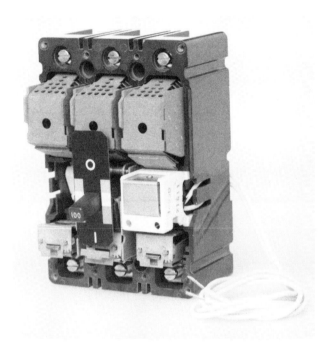

Figure 5-15. Typical Shunt Trip Installation

Figure 5-16. Undervoltage Release

Figure 5-17. Typical Installation of Auxiliary Switch

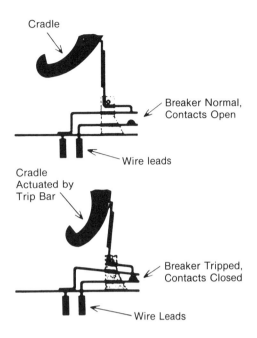

Figure 5-18. Alarm Switch

light, a buzzer, or a bell to provide an audio/visual warning by closing and completing the circuit when the breaker trips.

5.3.5 Motor operators. Where the shunt trip and the undervoltage release can remotely trip breakers, the function of the motor operator is to provide complete remote control by means of a pushbutton or similar pilot device. Positive switching action is accomplished by use of an operating arm engaging the breaker handle. When the motor operator is energized from a remote location, the operating arm moves the breaker handle to either ON or OFF. At the same time, an internal cutoff switch opens the circuit to the motor, preventing damage. In case of a power failure or some other emergency, means are provided for manual operation. Motor operators are designed for 120, 208, or 240 volt ac. If 480 volt power is used, a 480 to 120 volt transformer would be supplied. Some of the larger ratings may be modified for 125 volt dc.

The motor operator is not intended to increase the number of operations of the breaker. It is intended only for infrequent operation in line with UL endurance standards for molded case breakers.

5.3.6 Mechanical interlocks. There are several methods of mechanically interlocking circuit breakers. Each method, although somewhat different, achieves the same result. It interlocks two breakers so that only one may be closed (ON) at a time, yet both may be open (OFF) simultaneously. The three most common types of mechanical interlocks are the walking beam, sliding bar, and key interlock.

The walking beam type interlock operates on a walking beam or teeter-totter principle. An insulated plunger extends inside at the rear of the breaker and blocks the closing mechanism of one breaker when the other breaker is in the closed position. The closed breaker must be open before the open breaker may be closed. Circuit breakers for use with walking beam interlocks require special machining and are always ordered directly from the factory. This walking beam type interlock is standard on most circuit breaker type automatic transfer switches.

The sliding bar-type interlock (Figure 5-19) mounts on a customer's panel in front of two breakers. When the bar is extended toward one breaker (blocking the handle in the open position), the opposite breaker can be closed. The closed breaker must be opened and its handle blocked with the sliding bar before the opposite breaker can be closed. Most breakers do not require alteration for use with this attachment.

The key interlock mounts directly to the breaker cover. Its plunger is extended by turning the key in the cylinder, thereby blocking the breaker in the open position. Several keying arrangements can be supplied: key removable only when plunger extended, or key removable when plunger either extended or retracted. This enables an electrician doing maintenance work to lock a breaker in the OFF position, remove the key and not worry about someone accidentally energizing the breaker. A pair of breakers, remote from each other, can be interlocked so that only one can be closed at a time by using key interlocks operable by the same key, and key removable only when plunger is extended. These interlocks cannot be field mounted.

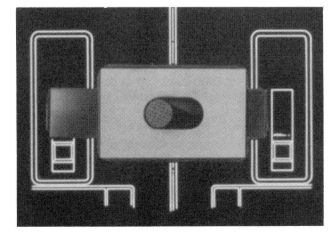

Figure 5-19. Sliding Bar Type Interlock

5.4 HOW TO SELECT A CIRCUIT BREAKER

The proper breaker for a specific application can be selected by determining a few parameters such as voltage, frequency, interrupting capacity, continuous current rating, and unusual operating conditions.

5.4.1 Voltage.
The voltage rating of a circuit breaker is determined by the maximum voltage that can be applied across its terminals, the type of distribution system, and how the breaker is applied in the system.

5.4.2 Frequency.
Most standard molded case circuit breakers up to 600 amps can be applied to frequencies from dc up to 120 hertz without derating. On higher frequency applications, however, the increased effect of eddy currents and iron losses causes greater heating within the thermal trip elements. This requires that the breaker be specially calibrated or derated. The amount of derating depends upon the frame size and ampere rating, as well as the current frequency. In general, the higher the ampere rating in a given frame size, the greater the derating required.

Some 600 ampere breakers and all higher ratings have a transformer-heated bimetal, and are suitable for 60 hertz ac maximum, with special calibration available for 50 hertz ac minimum.

Solid state trip breakers are calibrated for 50/60 Hz applications as standard. Consult the manufacturer for application information at 400 Hz. Special derating and specific cable/or bus size may be required.

5.4.3 Interrupting capacity.
The interrupting rating of the breaker is the maximum amount of fault current it can interrupt without damaging itself. The maximum amount of fault current supplied by a system can be calculated at any point in that system. One rule must be followed for applying the correct circuit breaker. The interrupting capacity of the breaker must be equal to or greater than the amount of fault current that can be delivered at the point in the system where the breaker is applied.

5.4.4 Continuous current rating.
Molded case circuit breakers are rated in amperes at a specific ambient temperature. This ampere rating is the continuous current the breaker will carry in the ambient temperature for which it is calibrated. Most manufacturers calibrate their standard breakers for a 40°C (104°F) ambient.

The selection of a specific ampere rating for a given application depends on the type of load and duty cycle, and is governed by the National Electrical Code. Feeder and lighting circuits generally require a circuit breaker rated in accordance with the conductor current carrying capacity. For example, NEC table 210-25 lists various standard breaker current ratings for different size conductors and the permissible load. The smallest listed rating is 15 amps, because size 14 wire is the smallest approved branch circuit conductor, and a 15 amp circuit breaker is necessary to protect it.

5.4.5 Unusual operating conditions.
Unusual operating conditions mean any external conditions that the breaker may be subjected to other than normal. Here are some examples:

High ambient temperature. If standard thermal magnetic breakers are to be applied in temperatures exceeding 40°C, the breaker must be derated, or specially calibrated for the higher ambient. For some years, all breakers were calibrated for 25°C (77°F). This meant that for applications above this, breakers had to be derated. Since most enclosure temperatures were realistically around 40°C (104°C), a common special breaker was one calibrated 40°C and termed ambient compensated for its enclosure. In the mid-1960s industry standards were changed to make all standard breakers calibrated to 40°C ambient temperature.

Moisture and corrosion. In atmospheres having a high moisture content and/or where fungus growth is prevalent, a special moisture and fungus resisting treatment is recommended for breakers. In high humidity areas, where daily temperature changes cause condensation to form repeatedly, the best solution is inclusion of space heaters in the enclosure.

Breakers should be removed from corrosive environments if possible. If this is not practical, specially treated breakers that are resistant to corrosive environments are available.

Altitude. At altitudes above 6,000 feet, circuit breakers must be progressively derated for voltage, current carrying ability, and interrupting capacity, because the thinner air does not conduct heat away from the current carrying parts as well as does denser air. The thinner air also prevents building dielectric fast enough to withstand the same voltage levels that occur at normal atmospheric pressure.

High shock. When a circuit breaker is applied where there is a possibility of high mechanical shock, a special anti-shock device should be installed. This consists of an inertia weight over the center pole that holds the trip bar latched under shock conditions, but does not prevent thermal or magnetic trip units from functioning on overload and short circuits. The Navy is the largest user of high shock breakers; they are required on all combat ships.

Mounting position. Circuit breakers, in most cases, may be mounted in any position, up or down, horizontally or vertically, without affecting the tripping characteristics or interrupting capacity.

5.4.6 Testing and maintenance (UL testing).

All low voltage molded case circuit breakers which are UL listed are tested in accordance with UL Standard 489, which can be divided into two categories: initial submittal and follow-up.

a. *Factory testing.* Molded case circuit breakers undergo extensive production testing and calibration based on UL Standard 489. Breakers carrying the UL label have factory sealed calibrated elements; an unbroken seal assures that the mechanism has not been subjected to alteration or tampering, and that the breaker may be expected to perform according to UL specifications. A broken seal voids the UL label and jeopardizes the manufacturer's warranty.

Acceptable field maintenance: Molded case circuit breakers have an excellent record of reliability that, to a great extent, is due to the enclosed design, which minimizes tampering and exposure to dirt, dust, and other contaminants. Maintaining reliability depends upon the proper application and installation of the breaker.

To ensure that proper breaker life is attained, it is essential that all terminal connections and trip units be tightened to the proper torque value as recommended by the manufacturer. Poorly cleaned conductors, improper conductors for the terminal used, and loose terminations are all faulty conditions that may cause undue heating and deterioration of the breaker. Any routine maintenance inspection should involve a check for tightness of all terminal connections and trip units.

Circuit breakers that are normally operated manually require no further operations to ensure that their contacts are clean and that the mechanical linkages operate freely. For breakers that are normally not operated on a regular basis, an occasional manual operation will provide the necessary exercise.

b. *Field testing.* Data obtained from field tests of molded case breakers often differs from published information, resulting in confusion on the part of the user as to which is correct. This is caused by the test conditions existing during factory tests, as opposed to test conditions existing in the field. Manufacturers conduct all tests in controlled ambient conditions, using test equipment designed specially for the product tested and, as such, can obtain consistent results.

NEMA publication AB4-1991 is an excellent guide to use if field testing is to be performed. Unless the breaker is equipped for electrical field testing where sufficient test instructions are provided, all testing should be done through a reliable service organization.

Automatic Transfer Switches

Herbert Daugherty

CHAPTER 6

INTRODUCTION

Our nation's growth depends on a steady supply of electricity. However, the supply has not kept pace with the growing demand. Coupled with natural disasters, accidents, breakdowns, malicious mischief, etc., brownouts and blackouts have occurred in many areas. But there are numerous manufacturing, commercial, and institutional facilities that cannot tolerate a loss of electricity.

In hospitals, institutions, and nursing homes, for example, power failures can be fatal to patients who depend on electricity to operate intensive care units, operating rooms, and other life-sustaining equipment. In operating rooms, electricity is vital for light to guide the surgeon's knife, and for the equipment that monitors the patient's vital signs. Electricity is needed to run elevators to get doctors, nurses, patients, and equipment from floor to floor.

In manufacturing plants, power failures can cost more than an entire emergency system. For example, when power fails, idle workers have to be paid. Orders, invoicing, and letter writing stop because electric typewriters, dictating, and accounting machines don't work. Computer damage may occur from loss of voltage. Time clocks stop. Products may be ruined because their production process, such as heat treatment, is interrupted at a critical stage. Additional time is needed for startup. Shipments have to be rescheduled. Customer goodwill is lost.

Because many department stores and shopping malls have no windows, power failures result in darkness, except for limited emergency lighting. Elevators stop (sometimes between floors), thefts and accidents occur, and cash registers can't be operated. See Figure 6-1.

Airports need emergency power to keep radar equipment operating and to maintain communications between

Figure 6-1. Shopping Mall

the tower and pilots, for weather information and other instrumentation vital to airline safety, and to maintain flight reservation information. Power is needed in terminal facilities for passenger comfort, convenience, and safety. See Figure 6-2.

Figure 6-2. Airport

Automatic Transfer Switches

Stronger anti-pollution laws and increasing population mean that waste water treatment plants must have emergency power to keep their facilities in operation during a power failure.

There are numerous other places that use emergency power such as schools, colleges, universities, computer centers, highway toll booths, high rise buildings, police and fire headquarters, subways, theaters, ships in port, radio and television stations, telephone exchanges, banks, museums, stadiums, restaurants, hotels, motels, railroad stations, and bus terminals. Even electric utility companies need emergency power for lights and to cool transformers and maintain oil pressure to turbines that continue to revolve for a long time from their own kinetic energy. In fact, any facility that depends on electricity to preserve life, prevent accidents, theft, panic, or loss of goodwill and revenue needs emergency or standby power. Various codes and standards define the terms emergency and standby. Other words such as alternate, backup, and critical, are often used. In this chapter the words, **emergency power** are used as a general term to cover all cases.

6.1 SUPPLYING EMERGENCY POWER

Emergency power can be supplied by multiple utility services or by on-site power generation.

Ordinarily, normal power is obtained from a utility source. An additional utility service from a separate source is needed to supply emergency power if normal source fails. See Figure 6-3.

However, this requires two (perhaps more) lines coming from different supply locations that are not likely to fail due to lightning or other causes at the same time, although that possibility always exists.

For maximum control and reliability, emergency power is generated on-site using gasoline or diesel engine-generator sets. See Figure 6-4.

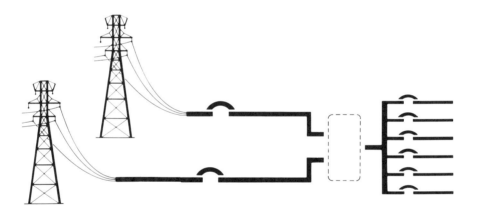

Figure 6-3. Two Utility Sources

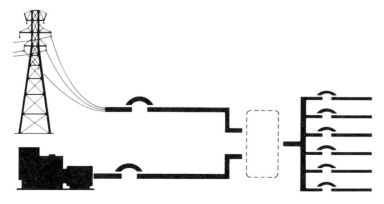

Figure 6-4. On-site Generator Emergency Source

6.2 TRANSFERRING POWER

Whether with a second utility source or on-site generation, a means must be provided to transmit power to the loads from either source, and transfer the loads from one source to the other, such as the transfer switch shown in Figure 6-5.

6.3 MANUAL DEVICES

Transferring can be done with a manually operated device in applications where operating personnel are available and the load is not of an emergency nature requiring immediate automatic restoration of power. Some typical applications are industrial plants, sewage plants, civil defense control centers, farms, residences, telephone buildings, other communication facilities, and certain health care facility equipment system loads. Double-throw knife switches and safety switches have been used for these applications. Many of these devices must be operated under no load conditions; i.e., the operator must be sure the load is disconnected by opening breakers or other load interrupting devices before the switch is operated. Some other double-throw switches are load-break and load-make rated as single-throw switches and are restricted in usage. Devices such as these have not been tested to Underwriters Laboratories (UL) Standard UL 1008 as double-throw transfer switches for all classes of load. Their ability to transfer between two hot unsynchronized power sources, such as occur when retransferring loads to a restored normal source from a hot emergency source, is undetermined. Likewise, they are not short-circuit withstand and close-in rated (discussed in detail later) for use with separate overcurrent protective devices. Because these devices are marginal adaptations, lack a high degree of reliability when adapted as transfer switches, and the restricted operational requirements are easy to abuse, personnel are reluctant to operate them. For facilities that want to transfer power manually, the solution is to use transfer switches designed specifically for manual transfer applications.

Two types of transfer switches are available for manual operation. One arrangement is electrically operated by externally mounted or remote toggle switches or pushbuttons. (See Figure 6-6.) The other type has a quick-make/quick-break operating handle, manually operable without opening the enclosure. (See Figure 6-7.)

Electrically operated units can be arranged for local or remote control station operation. Operation can be from one or more locations and is simple and convenient. The transfer switch can be located close to the load for maximum protection. The control stations can be mounted at a separate location, if required, for maximum operator convenience. Accessibility to the manual electric transfer switch does not cause a problem, or even a consideration, if it is electrically operated. Auxiliary contacts on the transfer switch can be used to signal indicating lamps at the operator's station that the switch has transferred.

The direct manually operated transfer switch is operated through a quick-make/quick-break handle (Figure 6-7) that is accessible from outside the enclosure. The speed of operation is similar to the electrically operated arrangement due to the pre-loading of the main operating springs during the first part of handle travel, thereby permitting identical current make, break, and carry capabilities. All automatic transfer switches should be UL listed, conform to UL 1008 requirements for manual operation, and externally operable without opening the enclosure.

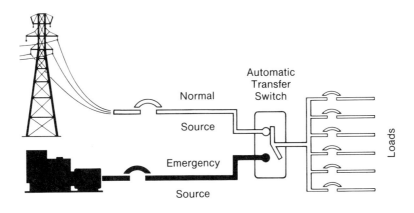

Figure 6-5. Emergency System with Transfer Switch

Figure 6-6. Manual Transfer Switch with Remote Control

Figure 6-7. Manually Operable Transfer Switch

6.4 AUTOMATIC DEVICES

The most convenient and reliable method to transfer power is with an automatic transfer switch. This switch includes controls to detect when a power failure occurs, and triggers controls to start the engine when the emergency power is an engine generator set. When the generator reaches the proper voltage and frequency, the switch transfers load circuits from the normal source to the generator (Figure 6-8).

NOTE: Unless otherwise indicated, the open lines (white with black outline) in all drawings indicate electricity flowing. Solid black lines indicate electricity not flowing.

When the normal source is ready to supply power again, the switch retransfers the load circuits to the normal source and then triggers controls to shut down the engine generator. The complete operation of engine starting, transfer to generator, retransfer to normal, and engine shutdown is handled automatically by the transfer switch every time there is a power failure and restoration.

It should be obvious, that in automatically operated systems, the automatic transfer switch is an important link in the chain of operation of the entire system. When normal power fails, it's the automatic transfer switch that must function to call for emergency power and then put it on the line. If this switch fails, emergency power will not energize the load no matter how much is available. Therefore, it is important to choose the most reliable automatic transfer switch available.

6.5 FACTORS TO CONSIDER WHEN CHOOSING AN AUTOMATIC TRANSFER SWITCH

Automatic transfer switches are usually located in the main or secondary distribution bus that feeds branch circuits. Therefore, the requirements for operating characteristics of an automatic transfer switch are different from those of a branch circuit device. Specifically, four major factors must be given special consideration. An automatic transfer switch must have the ability to:

a. Close against high inrush currents

b. Interrupt current

c. Carry full rated current continuously

d. Withstand fault currents

6.5.1 Close against high inrush currents. When a transfer switch closes on the alternate source, its contacts may be required to handle a substantial inrush, or surge, of current.

a. The amount of current will depend on the load, such as tungsten lamp loads or motor loads.

1. *Tungsten lamp loads.* The resistance of a cold tungsten lamp load is approximately 1/15th to 1/17th of the resistance when it is hot. This means that theoretically the initial current through the lamp will be 15 to 17 times its normal operating current. Such inrush currents may take as long as 14 cycles (0.23 second) to return to a normal or steady-state condition. These characteristics vary depending upon lamp wattage, as shown in Table 6-1.

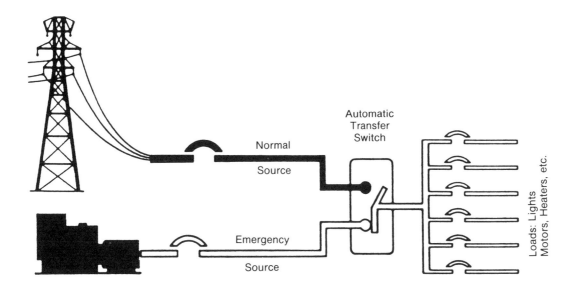

Figure 6-8. Emergency System with Automatic Transfer Switch

Table 6-1. Inrush Current Tungsten Lamps

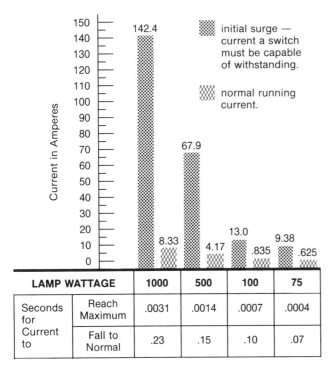

switches. Transfer switches must be capable of transferring a running motor from one source to another, and these two sources are not necessarily synchronized. Therefore, the residual voltage of the motor may be 180° out of phase with the voltage to which the motor is to be transferred.

This means that the motor may momentarily draw as much as 15 times the normal running current. The transfer switch must be capable of closing on such loads. Methods of transferring motor loads will be discussed later.

b. There must be no tendency for the contacts of a transfer switch to weld when closing on loads with high inrush currents. This means that the physical construction of the switch must provide a minimum of contact bounce, and have ample thermal capacity. UL 1008 requires a transfer switch capable of withstanding inrush currents of 20 times full load rating.

Transfer switches are rated for various classes of load, including the following UL 1008 classifications:

1. Total system load consisting of any combination of motors, electric discharge lamps, electric heating (resistive) loads, and tungsten lamp loads, provided the latter do not exceed 30% of the continuous current rating of the transfer switch.

2. Tungsten lamp load consisting entirely of tungsten lamps.

3. Electric discharge lamp load consisting entirely of electric discharge lamps, including fluorescent lamps.

2. *Motor loads.* Motor loads are also a source of high inrush currents. If an inert round rotor motor is connected to a source of power, the inrush currents can be in the range of six times normal running current. A ratio of ten times running current is generally used to compensate for this in designing motor controllers. However, even this may not be a realistic inrush ratio for transfer

4. Resistive load consisting of heating and other noninductive loads.

Characteristics of the above loads vary and are related particularly to the ability to close against inrush currents and to interrupt current.

6.5.2 Interrupt current. When the contacts of a transfer switch go from one source to another, an arc is drawn from the source the contacts are leaving (Figure 6-9).

Figure 6-9. Arc While Interrupting Circuit

The higher the voltage and the lower the power factor, the longer the duration of the arc. This arc must be extinguished before the contacts connect to the other source or there can be a short circuit from one source to the other (Figure 6-10). If the arc is not extinguished, serious damage can occur.

When the switch retransfers from the emergency source to the normal source, it is normally required to interrupt the emergency source current at full voltage.

When the switch transfers to emergency from the normal source, the current interruption can be zero. However, the transfer switch can also be called upon to interrupt full load currents at full voltage under test conditions. Also, and perhaps the most critical time, occurs if a motor is drawing locked rotor current at the instant of transfer.

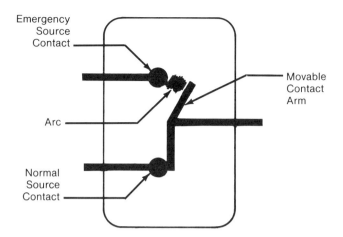

Figure 6-10. Arc in Transfer Switch

Thus, an adequately designed switch must interrupt the arc from both the normal and the emergency sources at all current levels.

To assure that the arc is properly interrupted when the switch transfers from one source to another, wide arc gaps and arc interrupting and quenching means are often utilized.

The **arc gap**, which is the distance from the stationary contact to the movable contact when the movable contact is fully opened (Figure 6-11) must be wide enough to draw out the arc and provide time to extinguish it.

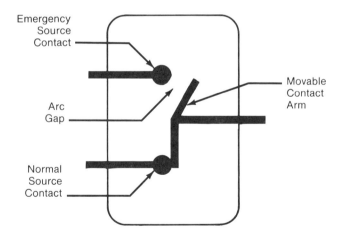

Figure 6-11. Transfer Switch Arc Gap

In addition to wide arc gaps, arc interrupting means such as arc splitters (Figure 6-12) are often used to break up the arc as it is drawn into them by the magnetic field created when the arc forms. This will be discussed in more detail later.

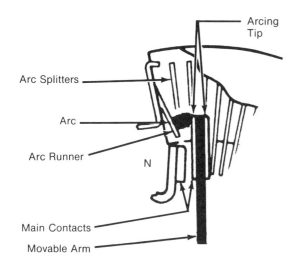

Figure 6-12. Arc Splitters

The effect of a wide arc gap plus arc splitters on extinguishing an arc is shown in Figure 6-13. This series of photos was taken from a film clip shot at 5000 frames per second and made during test runs at ten times rated current.

Notice in Figure 6-13A, just to the left of the center of the photo, the arc begins to form as the movable contact begins to move to the right away from the stationary contact. As the movable contact moves farther to the right in B, C, and D, the arc is drawn into the arc splitters where in E, it is broken up and in F, it is extinguished well before the movable contact reaches the other stationary contact. The action of the movable arm is shown in Figure 6-14.

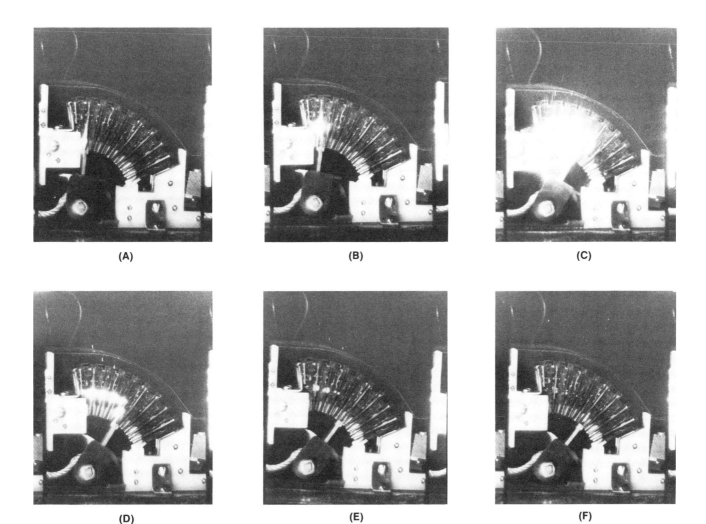

Figure 6-13. Arc Interruption

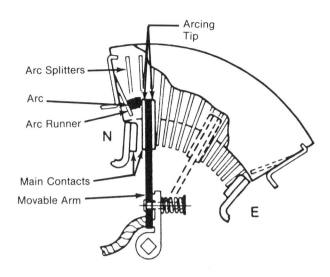

Figure 6-14. Movable Arm Action

6.5.3 Carry full rated current continuously. A transfer switch differs from other emergency equipment in that it must continuously carry current to critical loads. Current flows through the transfer switch during both normal and emergency conditions. On the other hand, the engine generator set is generally required to provide power only during the emergency period.

Furthermore, a transfer switch must provide continuity of power 24 hours a day, seven days a week for an expected minimum life of 20 to 40 years. During this period, fault currents, repetitive switching of all types of loads, and adverse conditions should not cause excessive temperature rise nor detract from reliable operation. Continuous duty operation should be achieved with minimum maintenance.

To meet the requirements for continuous duty, the contact temperature rise of a transfer switch must be well below that established for an eight-hour rated device, and the quality of contact must be sustained through proper contact design. Arcing contacts prevent or minimize arcing at the main contacts that would cause contact erosion and affect the ability of the switch to carry current continuously.

A transfer switch is capital investment equipment and, therefore, the ability to perform without overheating should not be limited to new or unused switches. For this reason most government agencies require a complete temperature test be performed after the endurance test.

When selecting a transfer switch, determine the maximum continuous load current which the transfer switch must carry. Momentary inrushes, such as occur when lighting or motor loads are energized, can be ignored.

Select a transfer switch that is either equal to or greater than the calculated continuous current and rated for the class of connected load. Sometimes the transfer switch size is selected to be the same as the overcurrent device ahead of it on the normal side. Although this may not be necessary, it is a convenience which permits the addition of future loads while remaining within the system's capacity. This will be covered in detail in Section 6.10.

6.5.4 Withstand fault currents. Transfer switches must withstand the magnetic stresses and dissipate the heat energy resulting from high fault currents. Withstand Current Ratings, WCR as they are called, vary depending on the switch size and type. Transfer switches should have withstand and close-in current rating based on the available fault current at its location in the system and the type of overcurrent device.

To properly evaluate application and coordination of transfer switches with protective devices, a system short-circuit calculation should be made to determine the symmetrical fault current magnitude and the X/R ratio at each point of application. The X/R ratio is the **ratio of reactance to the resistance of the circuit**. As the X/R ratio increases, both the fault withstandability of a transfer switch and the fault interrupting capacity of an overcurrent protective device become more critical.

A transfer switch must be capable of withstanding the available fault current at its location in the system until the overcurrent protective device clears the fault. The system designer should determine the available rms symmetrical fault current at the transfer switch location, the X/R ratio, voltage, and types of protective devices (current limiting fuse, molded case breaker, power breaker) before a properly rated transfer switch can be selected. See Table 6-2 for withstand current ratings (WCR) typical for transfer switches.

A number of factors account for high withstand current ratings of transfer switches. These factors include blow-on contact structure, thermal capacity, and operational time delay. These are explained in the following paragraphs.

Transfer switches are rated in terms of the available rms symmetrical fault current with a specific X/R ratio and specific types or classes of overcurrent protective devices. For example, a 600 ampere transfer switch is rated for use with Class L fuses in a system that has an available fault current of 200,000 symmetrical rms amperes. This is based on a maximum size fuse rating of 1200 amps, and an X/R ratio not exceeding 6.6. Most installations normally do not exceed an X/R ratio of 6.6 (i.e., power factor of 0.15). Switches with special requirements such as higher withstand and X/R values are available.

Table 6-2. Typical Transfer Switch Available Fault Current Ratings

Available Symmetrical Amperes RMS Ratings at 480 Volts AC						
Switch Rating Amperes	Long-Time WCR for Use with Any Overcurrent Device		Used with Specific Molded Case CB (Must Indicate on Label)	Used with Current Limiting Fuses		
	Symmetrical Amperes RMS at 480V AC	Time Cycles	Symmetrical Amperes RMS at 480V AC	Symmetrical Amperes RMS at 480V AC	Fuse Size Max.	Fuse Type
30	10,000	1.5	22,000	100,000	60	J
70, 100	10,000	1.5	22,000	200,000	200	J
150	10,000	1.5	22,000	200,000	200	J
260	35,000	3	42,000	200,000	600	J
400	35,000	3	42,000	200,000	600	J
600, 800	50,000	3	65,000	200,000	1200	L
1000, 2000	65,000	3	85,000	200,000	2000	L
1600, 2000	100,000	3	100,000	200,000	3000	L
3000, 4000	100,000	3	100,000	200,000	6000	L

Figures 6-15, 6-16, and 6-17 depict what happens during the first few cycles of current flow in three circuits each with identical available symmetrical rms short-circuit currents but different X/R ratios. Also, in each case, the circuit was closed at a point in the voltage wave to produce the maximum peak current during the first half-cycle of short-circuit current. Note the different values of peak current for each case.

a. *Blow-on contacts.* In many switching devices, the electromagnetic fields that encompass the current-carrying conductor act to force the contacts to separate. Since these electromagnetic forces increase exponentially with current, it can be anticipated that if not properly designed, the contacts may separate during short-circuit currents due to the magnetic effects.

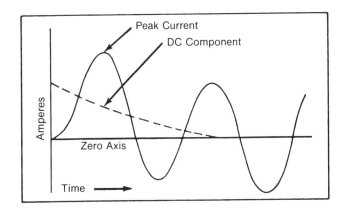

Figure 6-16. Medium X/R Ratio

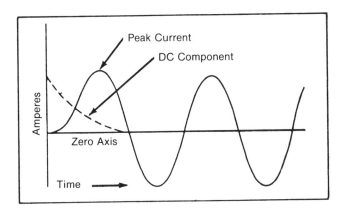

Figure 6-15. Low X/R Ratio

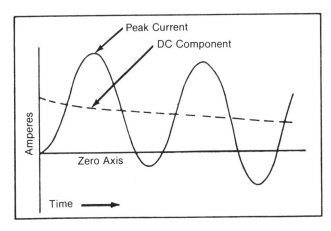

Figure 6-17. High X/R Ratio

Automatic Transfer Switches

b. Thermal capacity. The amount of heat generated at the transfer switch is proportional to the resistance of the parts that carry the current through the switch, multiplied by the square of the current. To provide the thermal capacity needed to cope with this heat, the cross section of the current-carrying parts must be of substantial magnitude to keep the resistance at a reasonable value. It is also necessary to provide adequate radiating surfaces, along with high contact pressures, to keep heating to a minimum.

On large transfer switches, segmented contacts are utilized to provide multiple paths for current flow through the main contacts. This reduces the heat generated and thereby provides longer life without replacement of contacts.

Proper design of the contacts prevent arcing on the main current carrying surfaces, preventing the erosion that leads to heat generation.

c. Time delay. A voltage drop normally accompanies large fault currents. It is important that the voltage drop does not cause an automatically operated switch to transfer while the switch is carrying the fault current.

Two safeguards are provided:

1. A mechanically held mechanism to prevent opening during a fault condition, and a time delay to override momentary dips in voltage. The mechanically held mechanism assures that the switch will not transfer until control voltage is applied to the transfer switch operator. To prevent energizing the operator until the overcurrent protective device clears the fault, a time delay is incorporated into the control circuit.

2. The control current will prevent application of control voltage to the automatic transfer switch operator until a minimum acceptable value for proper transfer has developed.

We've discussed that automatic transfer switches must be able to (1) close against high inrush currents, (2) interrupt current, (3) carry full rated current continuously and, (4) withstand and close against fault currents. In addition, automatic transfer switches should have adequate means to protect the main contacts, and be powered from the live source.

6.5.5 Prevent simultaneous closure of both the normal and emergency sources.
An automatic transfer switch must be designed to prevent the normal source and the emergency source from being connected to the load at the same time. If the two sources were inadvertently connected to the load at the same time, the results could be disastrous.

The contacts should be closed on one source or the other and not left in between, i.e., not closed on either source. If that happened, the load would not be getting power from either source even though power is available. The switch contacts, inherently interlocked, will be in only one of two positions: closed to normal and open to emergency, or closed to emergency and open to normal.

The operating mechanism should be simple. Theoretically, the fewer the operating parts, the higher the reliability. A typical arrangement is shown in Figures 6-18 and 6-19.

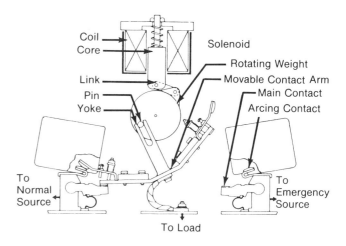

Figure 6-18. Single-solenoid Operator in Normal Position

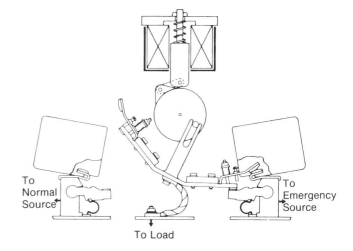

Figure 6-19. Single-solenoid Operator in Emergency Position

In Figure 6-18, the switch contacts are mechanically held closed to the normal source. To move the contact arm from the normal to the emergency source contact, the operator must be energized. The operator drives the contact arm onto the emergency source contact as shown in

Figure 6-19, where it is again mechanically held in place. The same action in reverse power-drives the arm back to the normal source.

6.5.6 Protect main contacts. The arcs that form when the main current-carrying contacts transfer from a live source can erode and eventually destroy the contacts if they are not properly protected. One way to protect the contacts is to lead the arc away from the current-carrying contacts. One method used is an arc runner and arc chute, as shown in Figure 6-20.

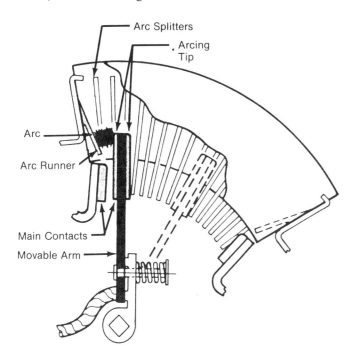

Figure 6-20. Arc Runner

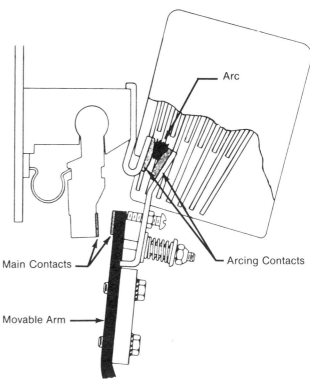

Figure 6-21. Arcing Contacts

As soon as an arc is formed it is led away from the current-carrying area of the main contacts by an arc runner. The punishment from the arc is taken by the arcing tip at the top, and away from the current-carrying area, of the contacts on the movable arm. The photo sequence in Figure 6-13 shows what actually happens with the arc when this switch transfers.

In larger-size switches, separate arcing contacts are frequently used (Figure 6-21). When the switch begins to transfer, the main contacts open first while the arcing contacts remain closed. The arc forms between the arcing contacts and is pulled into the arc splitters away from the main contacts. Then the main contacts open. Thus, the main contacts are protected by the arcing contacts, which are designed to take the punishment from the arcs rather than to continuously carry current, as the main contacts are designed to do.

6.5.7 Powered from the live source. An automatic transfer switch should receive the power to transfer from the source to which it is transferring the loads. That is, if the switch is going to transfer from a failed normal source to a live emergency source, obviously the power to make the switch transfer must come from the emergency source.

6.5.8 Convenient to maintain. Another important factor to consider in an automatic transfer switch is how convenient it is to maintain. Along with many codes and standards, the National Electrical Manufacturers Association (NEMA) recognizes the need for preventive maintenance. Their publication entitled AC Automatic Transfer Switches, NEMA Standard 1CS2-447 (Ref. 1), states, "A maintenance program and schedule should be established for each particular installation to assure minimum down time. The program should include periodic testing, tightening of connections, inspection for evidence of overheating and excessive contact erosion, removal of dust and dirt, and replacement of contacts when required."

To make it convenient to perform this type of preventive maintenance, most components should be accessible from the front of the switch, such as the switch shown in Figure 6-22. Thus, if any parts need to be replaced, it can be done with minimum effort and time.

Automatic Transfer Switches

Figure 6-22. Front Connected Transfer Switch

How to maintain reliability in emergency power systems will be covered in detail in Section 6.11.

6.6 CONTROLLING AUTOMATIC TRANSFER SWITCHES

The controls make a transfer switch automatic. In an emergency power system with an engine generator for the emergency source of power, controls are needed to detect a power failure of the normal source and trigger controls to start the engine. When the engine generator set reaches the proper voltage and frequency, the controls must signal the switch to transfer from the normal source to the generator. When the normal source is ready to supply power again, the controls must detect it and signal the switch to retransfer to the normal source. However, in order to do these tasks properly, time delays and other components, such as those shown on the automatic transfer switch control panel in Figure 6-23, are frequently furnished. A discussion of these requirements follows.

Figure 6-23. Control Panel

6.6.1 Voltage and frequency sensing. To detect a failure, voltage sensing is needed on all phases of the normal source. Voltage sensing on one phase of the emergency source is often provided to determine its availability. Controls are usually provided to sense the frequency on the emergency source. Minimum voltage and frequency are desired.

6.6.2 Time delays. Several time delays are provided; the first time delay overrides any momentary normal source outages that would cause false engine starts and switch transfers. This delay must be short enough so that the emergency source can be connected within code required time periods. The typical range is 0 to 6 seconds.

Another time delay is needed to be sure that when the normal source is restored, it is a bonafide restoration and ready to take the load. This delay is normally adjustable from zero to 30 minutes. The controls should bypass this time delay in returning to the normal source if the emergency source fails and the normal is available.

A time delay is often provided to allow the engine to run unloaded for cooldown before the controls shut it down.

In a power failure, it is usually desirable to have the load transferred to the engine generator set as soon as it reaches proper voltage and frequency. However, there are times when it is desired to sequence various transfer switches onto the generator set. Therefore, for these applications, the controls should include a time delay on transfer to emergency that is adjustable from zero to one minute.

6.6.3 Engine control contacts. The controls must include a contact that signals the engine controls to start when the normal power fails.

6.6.4 Test switch. Since periodic testing is necessary to maintain the emergency power system in good condition, the controls should include a manually operated switch to simulate a normal source failure.

The controls described here are standard and meet most of the application needs for automatic transfer switches. A number of other controls are available to handle specific needs.

6.7 TRANSFERRING MOTOR LOADS WITH AUTOMATIC TRANSFER SWITCHES

Today there is increasing need to supply motors with standby and emergency power. However, two special considerations must be made when motor loads are to be supplied from alternate power systems:

a. How to avoid nuisance breaker tripping and possible damage to the motor and related equipment when the motor is switched between two energized power sources that are not synchronized.

b. How to shed motor loads prior to transfer and delay reconnection to prevent overloading the power source to which the load is being transferred.

Depending on the makeup of the load, either one or both of the above may require consideration for any given dual power arrangement.

6.7.1 Avoiding damage to motors. Motors and related equipment can be damaged when switched between two live power sources. During routine test of the system, or during retransfer from the standby or emergency power source to the normal power source, both power sources are at full voltage. Experience has shown that motors, especially large three-phase motors of 50 HP or more, when transferred from one energized power source to another energized power source can be subjected to abnormal inrush currents, which in turn can lead to damage of motor windings, insulation, couplings, and in some cases, the driven load. The motor overcurrent device may also trip due to abnormal inrush current and require resetting. The abnormal currents are caused by the motor's residual voltage being out of phase with the voltage source to which it is being transferred.

The situation is similar to paralleling two unsynchronized power systems. Various control methods that are being used to overcome this problem are:

a. Inphase transfer

b. Motor load disconnect control circuit

c. Transfer switch with a timed center-off position

d. Overlap transfer to momentarily parallel the power sources

a. *Inphase transfer.* Inphase transfer schemes have been applied to utility company power station auxiliary systems for many years. With thousands of installations such as shown in Figure 6-24, inphase transfer is a popular method of transferring low-slip motors driving high-inertia loads on secondary distribution systems, provided the transfer switch has a fast operating transfer time.

A primary advantage of inphase transfer is that it permits the motor to continue to run with little disturbance to the electrical system and the process that is being controlled

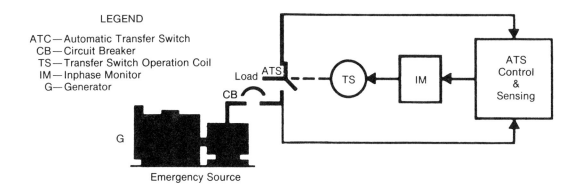

Figure 6-24. Motor Transfer with Inphase Monitor

by the motor. Another advantage is that a standard double throw transfer switch can be used with the simple addition of an inphase monitor. The monitor samples the relative phase angle of the two sources between which the motor is transferred. When the two voltages are within the desired phase angle and approaching zero phase angle, the inphase monitor signals the transfer switch to operate and reconnection takes place within acceptable limits.

Transfer switches that are equipped with inphase monitors, and which operate within ten cycles (166 milliseconds), can safely transfer motors without exceeding normal starting currents.

The operating transfer time of an automatic transfer switch is a primary factor in the proper utilization of an inphase monitor. For example, the advance angle adjustment of the inphase monitor is a direct function of the operating transfer time and the frequency difference between the two sources. (For a given advance angle, the faster an automatic transfer switch can transfer, the wider the allowable frequency difference.)

Figure 6-25 shows the relationship between switch operating transfer time, frequency difference, and phase angle advance required for optimum motor transfer. Note that at any given frequency difference, the required phase angle advance increases as the switch operating transfer time increases.

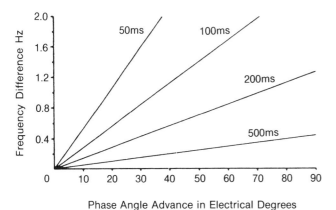

Figure 6-25. Phase Angle Advance Versus Time

b. *Motor load disconnect control circuit.* Motor load disconnect control circuits, such as shown in Figure 6-26, and similar relay schemes are also a common means of transferring motor loads. This arrangement, as well as a following arrangement (transfer switch with a timed center-off position), should not be used if the motors cannot be de-energized momentarily during transfer operations, with resultant disturbances to the electrical system and the process being controlled by the motor.

As Figure 6-26 illustrates, the motor load disconnect control circuit is a pilot contact on the transfer switch that opens to de-energize the contactor coil circuit of the motor controller, for a specific time, prior to automatic transfer. After transfer, the transfer switch pilot contact closes to permit the motor controller to reclose. For these applications, the controller must reset automatically.

The transfer switch motor load disconnect control circuit positively isolates each motor through its own controller, thus preventing possible interaction with other system loads.

A proper motor control disconnect circuit should be arranged to open the pilot contact for approximately three seconds before transfer to the alternate power source is initiated. This pretransfer delay should be non-tamperable since time must be allowed for the motor controller to open and extinguish all arcing before the transfer switch operation. Depending on the motor's time constant, or whether timed reclosing is provided in the motor controller circuit, it may also be necessary to add a second delay in the motor load disconnect control circuit. If the motor controller circuit does not have timed reclosing, and the motor's time constant exceeds three seconds, an additional delay should be included in the motor load disconnect control circuit. A three second delay is usually satisfactory.

If several motors are being transferred and it is desirable not to reconnect them simultaneously to prevent excessive inrush currents if they start simultaneously, the secondary delay can be a sequencer with several timed and sequenced pilot contacts (one contact for connection to each motor controller circuit).

Since all of these arrangements require interconnection of control wires between the transfer switch and motor controller, some consideration should be given to the design and layout of the system to minimize control line runs. To overcome this problem, the transfer switch is frequently located adjacent to, or within, the motor control center.

c. *Transfer switch with a timed center-off position.* Transfer switches with a timed center-off (neutral) position are also used as a means of switching motor loads. Figure 6-27 shows a typical arrangement. This arrangement achieves similar results to the motor load disconnect control circuit.

One advantage, provided that timed sequence reclosing is not required, is that interconnections between the transfer switch and motor controller are not required.

If other loads such as emergency lights are connected to the same transfer switch as the motor, these loads will be without power for the duration of the off period delay.

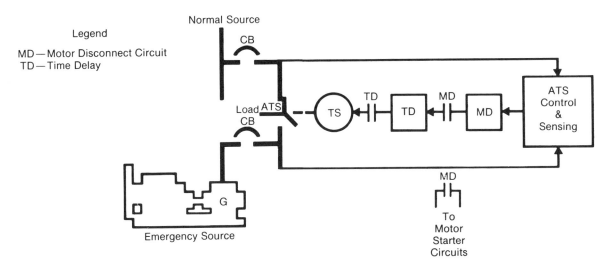

Figure 6-26. Motor Transfer with Motor Disconnect Circuit

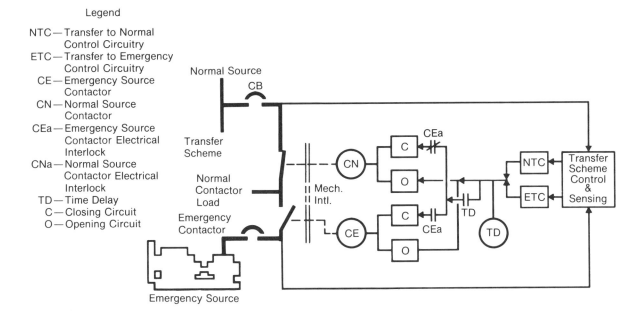

Figure 6-27. Motor Transfer with Timed Center-off Position

Neutral position devices create longer power interruptions on all of the connected loads.

d. *Closed transition transfer to momentarily parallel the power sources, also known as overlap transfer.* This method (Figure 6-28) offers many unique benefits, one of which is that the load is not interrupted during an energized source to energized source transfer. This requires that the sources be properly synchronized by voltage, frequency, and phase angle. In the case of a failed or inadequate source, the controls operate in an open transition mode. For a description of closed transition transfer switches see section 6.10.

Closed transition transfer can effectively be used for testing the emergency system, on retransfer from the standby generator to the utility source, and when transferring to the emergency source in anticipation of a potential outage, for example during electrical storms. The increased use of on-site generators for peak demand reduction and curtailment of interruptible rates are also attracting interest in closed transition transfer, since the loads transferred are not disturbed.

Utility company approval is required for the use of closed transition transfer switches because of the interconnection of the power sources. Most utilities are receptive, however, when assured of the momentary nature of the overlap. Most will grant approval provided proper precautions are taken to assure the sources are properly synchronized before interconnection, and the paralleling mode is not maintained.

6.7.2 Motor load shedding delayed reconnection.

We discussed the possibility of motor damage during transfer between two acceptable sources and the various solutions to prevent it. Here we will discuss the second problem that deals with methods to prevent overloading the emergency source during a normal power outage. On-site generator sets are often used as the emergency source of power. Such generator sets often have limited capability to supply the total inrush and starting currents of the connected load. For economical purposes, generator sets are commonly sized to provide full load current plus a limited motor starting capability. In such cases, it becomes essential to delay reconnection of some of the loads when transferring from the normal power source to the on-site generator.

Article 517 of the National Electrical Code (Ref. 2) requires that the equipment system load (primarily motor load) transfer switches in hospitals be equipped with time delay relays that will delay transfer of the connected load to the generator set. The purpose is to assure that the more important emergency system loads are connected first and established within ten seconds of failure. The transfer switches feeding the motors are then sequentially transferred to the generator set.

Another reason for load shedding is the need to "power down" certain loads, such as those utilizing silicon controlled rectifiers (SCRs), to avoid damage to, or failure of, such components during transfer. Following are various solutions in use today to help solve these problems by adding circuit features to transfer switches.

a. *Transfer switches with individual time delay circuits on transfer to emergency.* This arrangement is in frequent use today. If there are several transfer switches in an installation, the time delay can be adjusted slightly differently on each transfer switch so that the transfer switches close sequentially onto the generator. Consideration should be given to individual motor inrush requirements, the remaining available starting kVA of the generator, and the importance of the respective loads when determining the sequence of transfer.

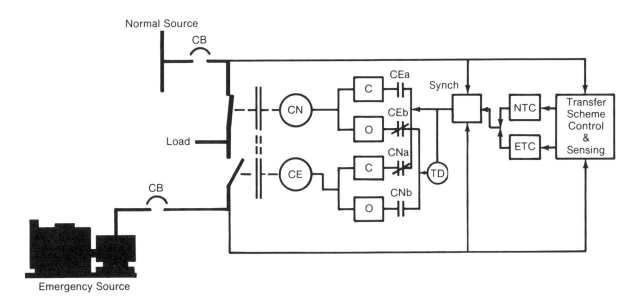

Legend

CEa — Emergency Source Contactor Electrical Interlock
CEb — Emergency Source Contactor Electrical Interlock
CNa — Normal Source Contactor Electrical Interlock
CNb — Normal Source Contactor Electrical Interlock

TD — Time delay to open alternate contactor if proper contactor does not operate to terminate parallel operation.
SYNCH — Synchronizer to synchronize emergency source in voltage frequency and phase angle with normal source.

Figure 6-28. Closed Transition

b. *Transfer switches with signal circuits for definite disconnection of a single load prior to transfer and reconnection after transfer.* This arrangement was described under motor load disconnect control circuit. With the additional delay described in the test, the control circuit not only assures that the motor is disconnected before transfer, but also prevents the motor load from being reconnected until several seconds after the transfer switch has transferred and reconnected any other loads that are fed by the same transfer switch.

c. *Transfer switches as in b., but with multi-signal circuits to sequence several loads onto the generator.* A further refinement of the system described above utilizes several signal circuits when several motors are to be fed by the same transfer switch, and it is desired to reconnect them sequentially rather than all at once. Two to nine circuits are commonly provided. The time delay between reconnection steps is adjustable to allow the starting current to reduce to a safe value before the next motor is signalled to be reconnected. Once the delay is set, it is the same for each step.

6.7.3 Electronic variable frequency drives.

Special consideration should be given to systems that utilize variable frequency drives (VFDs) to control motor operation. These electronic devices often consist of silicon controlled rectifier assemblies to change the voltage and frequency applied to the motor to vary its torque and speed.

With some VFDs, interruption of the input, whether it is caused by transfer switch operation or other momentary power outages, causes the device to see a low voltage input. To maintain motor speed, some control sensors immediately change the conduction angle of the controlling SCRs to compensate for the low line condition. The result is that upon immediate reapplication of power, the controlling SCRs are turned fully on, causing a severe current inrush. This current inrush is of such a magnitude that protective devices such as fuses can blow or worse, the SCRs themselves can be irreparably damaged.

It should be noted that this particular problem is only partly solved by conventional inphase transfer or other load disconnect arrangements, because other extraneous interruptions of power (such as disturbances on the high lines) will also confuse the VFD voltage sensing circuitry.

Fortunately some manufacturers of VFDs have recognized this problem and have included in their systems special high speed voltage sensing circuitry that detects momentary power outages and shuts down the VFD entirely. Upon reapplication of voltage, the SCRs go from an off condition to a soft start current-limiting startup that automatically protects the solid state switching devices.

When other provisions are lacking, the motor load disconnect control circuit previously described (i.e., disconnect before transfer) can be considered for transfer of SCR controlled loads for variable frequency drive motors and other loads such as used by communications companies. The contact, which must be integrated by the VFD system manufacturer, simply signals the SCR control to "power down" before transfer and permits a soft restart after transfer. As indicated above, such controls may protect the SCRs during a power interruption when the transfer switch is operating, but offer no protection when the power interruption is extraneous to the transfer switch.

6.8 GROUND-FAULT PROTECTION

Ground-fault protection (GFP) requires special consideration. This necessitates proper adherence to various Sections of the National Electrical Code (NEC) (Ref. 2).

Section 230-95 of the 1993 National Electrical Code states: "Ground-fault protection of equipment shall be provided for solidly grounded wye electrical services of more than 150 volts to ground, but not exceeding 600 volts phase-to-phase for each service disconnecting means rated 1000 amperes or more."

Section 230-95(b) (FPN No. 6) of the 1993 National Electrical Code recognizes that problems can occur with multiple neutral-to-ground connections, and says: "Where ground-fault protection is provided for the service disconnecting means and interconnection is made with another supply system by a transfer device, means or devices, may be needed to assure proper ground-fault sensing by the ground-fault protection equipment."

Section 445-1 of the 1993 National Electrical Code states: "Generators and their associated wiring and equipment shall comply with applicable provisions of Articles 230, 250, 700, 701 and 702."

Section 250-5(d) of the 1993 National Electrical Code defines "separately derived systems" as follows: "A premises wiring system whose power is derived from generator, transformer, or converter windings and has no direct electrical connection, including a solidly grounded circuit conductor, to supply conductors originating in another supply system..."

Section 250-23(a) of the 1993 National Electrical Code states: "...A grounding connection shall not be made to any grounded circuit conductor on the load side of the service disconnecting means."

"Exception No. 1: A grounding electrode conductor shall be connected to the grounded conductor of a separately derived system in accordance with the provisions of Section 250-26(b)."

Section 250-26 of the 1993 National Electrical Code states: "A separately derived ac system that is required to be grounded by Section 250-5 shall be grounded as specified..."

Section 250-51 of the 1993 National Electrical Code, in defining "effective grounding path" states: "The path to ground from circuits, equipment and conductor enclosures shall: (1) be permanent and continuous; (2) have capacity to conduct safely any fault current likely to be imposed on it; (3) have sufficiently low impedance to limit the voltage to ground and to facilitate the operation of the circuit protecting devices."

These code requirements have a number of bearings on emergency systems as covered by Article 700 and standby systems covered by Articles 701 and 702 of the National Electrical Code. Consider a 480Y/277 volt system with a three pole transfer switch and zero sequence ground-fault sensing as shown in Figure 6-29, with a ground fault as shown.

The ground-fault current has two paths of flow. Path 1 is directly from the equipment ground to the service grounding electrode at the transformer secondary. Path 2 is from the equipment ground to the generator electrode, then through the neutral conductor to the service neutral. The current through Path 2 will not tend to actuate the ground-fault sensor, thereby causing incomplete sensing of the total fault current.

There are several approaches to resolving this problem. These may include the use of isolating transformers, switching the neutral conductor, or deleting the generator's ground. The removal of this ground eliminates the need for a four-pole switch and retains the integrity of the GFP.

When the system designer chooses to switch the neutral, a fourth pole is added to the transfer switch. Some manufacturers use the same pole construction and contact adjustment. Some adjust the contact to break last and make first. Others use a special pole construction with overlap so that the neutral is not interrupted during transfer. An automatic transfer switch with overlapping neutral transfer contacts is shown in Figures 6-30 and 6-31.

This arrangement complies with NEC Section 230-95(b), permits coordinated ground-fault sensing for an alarm circuit on the emergency side, and permits grounding of the generator neutral per Section 250-26. This arrangement also conforms to Section 250-23(a) and provides

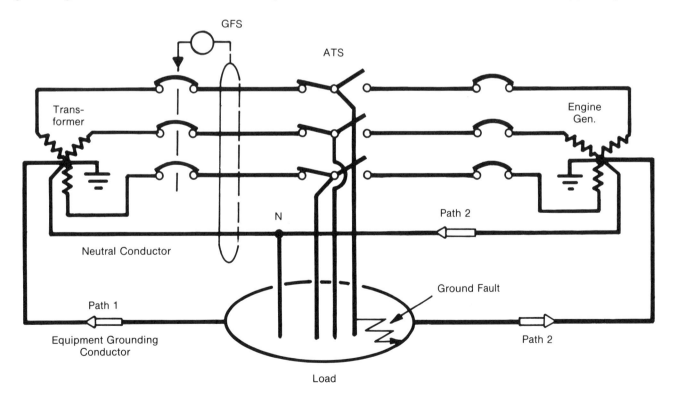

Figure 6-29. Improper Sensing Due to
Multiple Ground Connections

Automatic Transfer Switches

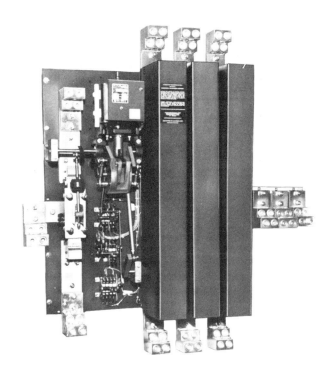

Figure 6-30. Transfer Switch with Overlapping Neutral Contacts

continuity of the grounded neutral conductor, in accordance with Section 250-51, through utilization of overlapping neutral transfer contacts. This feature provides the necessary isolation between neutrals, and at the same time, minimizes abnormal voltages. By means of overlapping contacts, the only time the neutrals of the normal and emergency power sources are connected together is during transfer and retransfer. The overlap duration must be less than the actuating time of the ground-fault sensor, thus avoiding false tripping due to unbalanced current that might exist.

Figure 6-32 shows a typical system using overlapping neutral contacts on a three-pole transfer switch.

6.9 SIZING AUTOMATIC TRANSFER SWITCHES

Serious attention to, and careful selection of, an automatic transfer switch is important to ensure maximum reliability and adequate capability under both normal and emergency conditions. The main points to be considered are:

a. Types of load to be transferred

b. Voltage rating

c. Continuous current rating (carrying capacity)

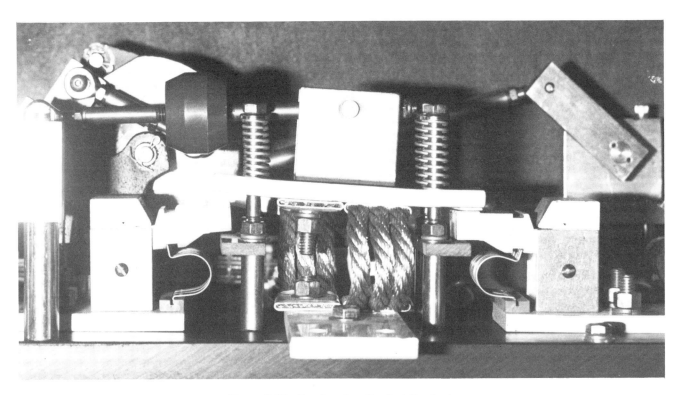

Figure 6-31. Overlapping Neutral Contacts

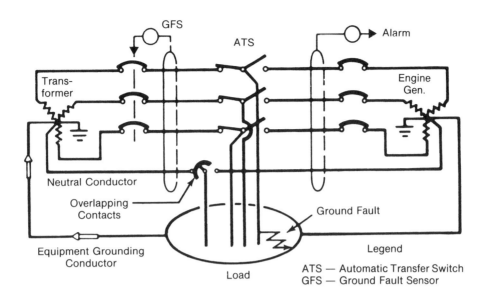

Figure 6-32. System with Overlapping Neutral Contacts

d. Overload and fault current withstand ratings

e. Type of overcurrent protective device ahead of transfer switch

Reliability and economics are the prevailing factors in deciding the best selection for any application.

6.9.1 Types of loads.
Loads, as applied to automatic transfer switches, are classified by UL 1008, as (1) Total System Loads, (2) Motor Load, (3) Electric Discharge Lamp Loads, (4) Resistive Loads, and (5) Incandescent Lamp Loads.

UL requires marking of transfer switches to indicate the type of load they are capable of handling. The marking *Total System Loads* indicates that the transfer switch can be used for any combination of the loads described above under (2) through (5). However, the incandescent load shall not exceed 30% of the total load unless the transfer switch is specifically marked as suitable to transfer a higher percentage of incandescent lamps. Most transfer switches are rated for transfer of *Total System Loads*, though some may be marked *Resistance Only*, *Tungsten Only*, etc., or a combination of these markings. The burden on the system designer is lessened when he chooses switches listed and rated for total system loads.

6.9.2 Voltage ratings.
An automatic transfer switch is unique in the electrical distribution system in that it is one of the few electrical devices that may have two unsynchronized power sources connected to it. This means that the voltages impressed on the insulation may actually be as high as 960 volts on a 480 volt ac system. A properly designed transfer switch will provide sufficient spacings and insulation to meet these increased voltage stresses.

For this reason, the spacings in transfer switches should not be less than those shown in Table 22.1 of UL 1008, regardless of what type of component may be used as part of the transfer switch.

For this discussion, the voltage ratings will be limited to low voltage applications where transfer switches are rated 600 volts or less.

The ac voltage ratings of automatic transfer switches are normally 120, 208, 240, 480, or 600 volts, single or polyphase. Standard frequencies are 50-60 hertz. Automatic transfer switches can also be supplied for other voltages and frequencies when required.

6.9.3 Continuous current rating.
A continuous current (load) is one which is expected to continue at its maximum value for three hours or more. As previously discussed, transfer switches differ from other emergency equipment in that they must continuously carry the current to critical loads, whereas a standby engine generator set generally supplies power only during emergency periods.

Automatic transfer switches are available in continuous current ratings ranging from 30 through 4000 amperes. Typical ratings include 30, 40, 70, 80, 100, 150, 225, 260, 400, 600, 800, 1000, 1200, 1600, 2000, 3000, and 4000 amperes.

Most transfer switches are capable of carrying 100% of rated current at an ambient temperature at 40°C. However, some transfer switches, such as those incorporating integral overcurrent protective devices, may be limited to a continuous load current not to exceed 80% of the switch rating.

For new projects, the system designer may specify a transfer switch that will be able to carry future anticipated loads. In such cases, it is advisable to select a transfer switch with a continuous current rating equal to the total anticipated load.

The transfer switch continuous current rating is found by totaling the amperes required for all loads. Electric heater and tungsten (incandescent) lamp load currents are determined from the total wattage. Fluorescent, mercury vapor, and sodium vapor lamp load currents must be based on the current that each ballast or auto-transformer draws, not on the total watts of the lamps. Motor loads are determined by motor full load currents only. Motor inrush and locked rotor currents need not be considered in sizing a transfer switch UL listed for *Total System Load*. Generally, there is no need to derate a transfer switch for use in ambient temperatures up to 40°C, whether the switch is installed in a switchboard or in a separate enclosure.

Following is an example for finding the ampere rating of the transfer switch that is needed in a given application.

Needed: An automatic transfer switch rated for *Total System Loads*, for a 208Y/120 volt, three-phase, four-wire circuit consisting of the following three-phase balanced load:

115 kW heating load

$$I = \frac{115 kW}{208V \times 3} = 320 \text{ amps}$$

64 kW tungsten lighting load

$$I = \frac{64 kW}{208V \times 3} = 178 \text{ amps}$$

Three 10 HP motors, @ 32 amps each

$$I = \begin{array}{c} 3 \times 32 \\ \text{Total load} \end{array} \quad \begin{array}{c} 96 \text{ amps} \\ 594 \text{ amps} \end{array}$$

Since the tungsten load does not exceed 30% of the total load, select a three-pole transfer switch rated at least 600 amperes. Typical generator line currents versus KW ratings are shown in Table 6-3. For this example a 200 KW generator may be suitable, provided it can handle the inrush currents.

6.9.4 Overload and fault current withstand ratings.

Since transfer switches are often subjected to currents of short duration exceeding the continuous duty rating, the ability of the transfer switch to handle higher currents is measured by its overload and withstand current ratings.

The overload rating refers to the ability of a transfer switch to handle normal inrush currents encountered in switching lighting, transformer, and motor loads. This ability is inherent in transfer switches marked *Total System Loads*. Generally an automatic transfer switch should have a minimum overload rating of 15 times the continuous duty rating for 0.5 second.

As was mentioned earlier in para. 6.5.4, the withstand current rating pertains to the ability of an automatic transfer switch to withstand the magnetic and thermal stresses of high fault currents until the fault is cleared by an overcurrent protective device. The overcurrent protective device is usually located external to the transfer switch although there are transfer switches that do include integral overcurrent protection. To differentiate between the two, one recognized standard defines the following type designations:

Transfer switch, Type A means an **automatic transfer switch that does not employ integral overcurrent devices**.

Transfer switch, Type B means an **automatic transfer switch that employs integral overcurrent devices**.

In a Type A transfer switch, the main contacts stay closed while the system's coordinated protective devices clear the fault, whereas a Type B transfer switch depends upon trip elements to open the contacts of the transfer switch during a short circuit fault. The latter arrangement depends upon "interruption rating," which should not be confused with "withstand rating." This discussion addresses itself to the conventional Type A transfer switch.

To evaluate and coordinate transfer switching and overcurrent protection, a short circuit study should be made on the system to determine the available fault current magnitude, and the X/R ratio of the circuit at each point in the distribution system where a transfer switch is to be located. This study is usually made by the system designer.

Underwriters' Laboratories UL 1008, Standard for Safety, Automatic Transfer Switches (Ref 3) includes minimum requirements for withstand current ratings. See Table 6-4.

Available fault currents often exceed the minimum UL requirements. Transfer switch size with appropriate withstand rating and overcurrent protective device continuous and interrupting ratings are determined by the

Table 6-3. Generator Output Current Ratings

Amps. Single Phase		Kilowatts	Amps. Three Phase			
@ 120 Volts	@ 240 Volts		@ 208 Volts	@ 240 Volts	@ 480 Volts	@ 600 Volts
104	52	10	35	30	15	12
156	78	15	52	45	23	18
208	104	20	69	60	30	24
260	130	25	87	75	38	30
312	156	30	104	90	45	36
416	208	40	139	120	60	48
521	260	50	173	150	75	60
624	312	60	208	180	90	72
780	390	75	260	226	113	90
1041	521	100	347	300	150	120
		125	433	376	188	150
		150	520	451	226	180
		175	607	526	263	211
		200	694	601	300	241
		250	866	751	376	301
		300	1040	902	451	361
		350	1213	1052	526	421
		400	1388	1202	601	481
		500	1732	1502	751	601
		600	2080	1803	902	722
		700	2426	2103	1052	842
		750	2602	2256	1128	902
		900	3123	2706	1353	1083
		1000	3470	3007	1504	1203

Table 6-4. UL 1008 Minimum Withstand Current Requirements

Continuous Current Rating*	Withstand* Current
100 or less	5,000
101 to 400	10,000
401 and above	20 times continuous but not less than 10,000

*RMS symmetrical amperes

Table 6-5. UL 1008 Preferred Withstand Current Ratings

RMS Symmetrical Amperes	
5,000	42,000
7,500	50,000
14,000	65,000
18,000	85,000
22,000	100,000
25,000	125,000
30,000	150,000
35,000	200,000

system requirements, thus introducing the need to coordinate the transfer switches with overcurrent protective device ratings. For purposes of standardization and coordination with overcurrent protective devices, UL 1008 has established a list of preferred optional fault current ratings. See Table 6-5. The currents listed coincide with published interrupting ratings of overcurrent protective devices.

The symmetrical rms current withstand rating, in itself, does not provide sufficient data to establish the complete withstand rating of the transfer switch. Invariably, a short circuit is asymmetrical during the first few cycles. This is referred to as the **transient current period** before the current subsides to its symmetrical value. The degree of asymmetry, which in turn determines the maximum peak instantaneous current and the I^2t thermal energy that the transfer switch must withstand during the first few cycles, is governed by the circuit power factor and the point on the voltage wave at which the short circuit is initiated.

The testing procedure detailed in UL 1008 requires specific ranges of test power factors for specific short circuit currents to determine transient or asymmetrical withstand ratings. See Table 6-6. It should be pointed out that the transfer switch manufacturer is not limited to the withstand current and power factor ranges given in Table 6-6. Transfer switches are often tested to higher currents and lower power factors.

Table 6-6. UL 1008 Test Power Factor Ranges

Withstand Test Available Current (Symmetrical Amperes)	Test Power Factor Percent	Equivalent X/R Ratio
10,000 or less	40–50	1.73 to 2.29
10,000 to 20,000	25–30	3.18 to 3.87
20,001 or more	20 max.	4.9 min.

For the system designer, a table similar to Table 6-7 is useful, and brings out a number of important points regarding short circuit characteristics. Columns 1 and 2 indicate an inverse relationship between power factor and the X/R ratio. As the X/R ratio increases, the power factor decreases. Columns 2 and 3 indicate the direct relationship between the instantaneous current and the X/R ratio. The factors shown in Column 3 are used to multiply the calculated rms symmetrical short circuit current, to obtain the maximum available instantaneous peak current in the first loop of the asymmetrical current in any phase of the faulted circuit. For example: a 400 ampere continuous duty transfer switch rated to withstand 20,000 amperes symmetrical at an X/R ratio of 3.18 can withstand the magnetic stresses of 20,000 × 1.978 = 39,500 peak amperes (from Table 6-7, at X/R = 3.18, the multiplying factor is 1.978).

Thus, the continuous current and symmetrical withstand current ratings should not be the only criteria considered when selecting a transfer switch. Caution must be exercised to be certain that the peak current withstand of the transfer switch is not exceeded.

In addition to the symmetrical withstand rating, the transfer switch X/R rating and the circuit X/R ratio at the transfer switch location must be taken into consideration. Failure to do this can lead to transfer switch misapplication. The system X/R ratio should never exceed the transfer switch test X/R rating or that of any other device in the system. For example: the 400 ampere transfer switch in the previous example was tested to withstand 20,000 symmetrical amperes at X/R = 3.18 and withstood 39,560 instantaneous peak amperes safely. If this switch is installed in a circuit where the circuit X/R = 5.46, the transfer switch would be subjected to 20,000 × 2.231 = 44,620 instantaneous peak amperes. Since the

Table 6-7. Power Factor Versus X/R Ratio

Power Factor (Percent)	X/R Ratio	Max. Peak Current Factor
0	∞	2.828
1	100.000	2.785
2	49.993	2.743
3	33.322	2.702
4	24.979	2.663
5	19.974	2.625
6	16.623	2.589
7	14.251	2.554
8	12.460	2.520
9	11.066	2.487
10	9.950	2.455
11	9.035	2.424
12	8.273	2.394
13	7.627	2.364
14	7.072	2.336
15	6.591	2.309
16	6.170	2.282
17	5.795	2.256
18	5.465	2.231
19	5.167	2.207
20	4.899	2.183
21	4.656	2.160
22	4.434	2.138
23	4.231	2.110
24	4.045	2.095
25	3.873	2.074
30	3.180	1.978
35	2.676	1.894
40	2.291	1.819
45	1.985	1.753
50	1.732	1.694
55	1.519	1.641
60	1.333	1.594
65	1.169	1.553
70	1.020	1.517
75	0.882	1.486
85	0.620	1.439
100	0.000	1.414

magnetic stresses are proportional to the square of the instantaneous currents, the transfer switch may fail if subjected to that maximum short circuit current. To satisfy the increased peak instantaneous current requirement, it may be necessary to select another 400 ampere or larger transfer switch with an X/R rating greater than 5.46, or include provisions to limit the available current.

Referring to Table 6-6, for transfer switch withstand rating greater than 20,000 amperes, the test power factor is 20% or less. However, for design purposes a power factor of 15% corresponding to an X/R of 6.6 is more appropriate. This is to coordinate the transfer switch X/R rating with overcurrent protective devices. Refer to Tables 6-8 and 6-9.

Automatic transfer switches, tested at an X/R ratio of 6.6, may be safely installed at any point in a low-voltage power system where the circuit X/R ratio is 6.6 or less at the service entrance equipment. It is only necessary to satisfy the continuous current rating, available electrical fault current, and voltage rating.

Both the I^2t thermal withstand and the instantaneous peak current withstand values can be determined only by actual short circuit current performance tests. If the transfer switch contacts are not solidly mated, or if the transfer switch contacts tend to part due to magnetic effects during the test, it is possible that the I^2t portions of the contacts to form small globules of molten metal, which upon cooling, will weld the contacts together.

6.9.5 Protective device ahead of transfer switch.

The type and size of overcurrent device ahead of the transfer switch, as indicated in Table 6-10, plays a significant role in transfer switch application. Transfer switches can have multiple available fault current ratings, depending on the type of overcurrent protective device.

The available fault current rating of a transfer switch protected by current limiting fuses can be as high as 200,000 amperes symmetrical. When a short circuit occurs, the current-limiting fuse will pass or let through only a small portion of the available short circuit current. The current-limiting fuse instantaneous peak current and I^2t current, when the fuse is sized correctly, will be less than the transfer switch peak current and I^2t withstand ratings. For example, a 600 amp transfer switch can be used with 1200 ampere Class L fuses (twice the rating of the transfer switch) in a system with an available fault current of 200,000 amperes symmetrical. See Table 6-10.

Table 6-8. Circuit Breakers Versus Test Power Factor

Interrupting Rating (Symmetrical)		Test Power Factor (Percent)	Equivalent X/R Ratio
Molded-case Breakers	Power Breakers		
10,000 & less	—	45–50	1.73–1.98
10,001	—	25–30	3.18–3.87
20,001 & more	—	15–20	4.9–6.6
—	All ratings	15 max.	6.6 min.

Table 6-9. Fuses Versus Test Power Factor

Fuse Class	Interrupting Test Current (Symmetrical)	Test Power Factor (Percent)	Equivalent X/R Ratio
H*	10,000	40–50	1.73–1.98
K	50,000	20 max.	4.9 min.
	100,000		
	200,000		
J	200,000	20 max.	4.9 min.
L	200,000	20 max.	4.9 min.
R	200,000	20 max.	4.9 min.
T	200,000	20 max.	4.9 min.

*When rated above 100 amperes

Table 6-10. Transfer Switch Rating Versus Available Symmetrical Short Circuit Amperes

Switch Rating (Amps)	Available Symmetrical Amperes RMS Rating at 480 Volts AC and X/R Ratio of 6.6 or Less			
	When Used with Class J & L Current Limiting Fuses		Used with Specific Molded Case Breakers (Must State Manufacturer's Specific Breaker on Label)	Used with Any Overcurrent Device
	Symmetrical Amperes	Max Fuse Size (Amps)	Symmetrical Amperes	Symmetrical Amperes
100	200,000	200	22,000	10,000
260	200,000	600	42,000	35,000
400	200,000	600	42,000	35,000
600	200,000	1200	65,000	50,000
800	200,000	1200	65,000	50,000
1000	200,000	2000	85,000	65,000
1200	200,000	2000	85,000	65,000
1600	200,000	3000	100,000	100,000
2000	200,000	3000	100,000	100,000

Also, when a circuit breaker is used to protect a transfer switch, the transfer switch withstand current rating must be at least equal to, and preferably greater than, the available short circuit current at the transfer switch.

It may be necessary at times, in order to provide a transfer switch with the required withstandability, to select a transfer switch with a higher continuous current rating. For example, if a short circuit study indicates that an 800 amp continuous current rated transfer switch must have a withstand of 50,000 symmetrical amperes when used with a molded case circuit breaker, it would be necessary, according to Table 6-10, to select a 1000 ampere rated transfer switch with a 55,000 ampere withstand current rating in order to obtain the required withstand rating.

Sometimes the feeder overcurrent protective device may be rated considerably higher than the continuous current, such as when supplying motor loads as permitted by Table 430-152 of the National Electrical Code. The system designer should be sure that the transfer switch is UL listed for use with the higher rated overcurrent device. Some transfer switches may be suitable for use with overcurrent devices rated at 125% of the transfer switch rating, while other transfer switches may be rated for use with overcurrent devices ranging up to 400% or more of the automatic transfer switch rating. Larger breakers are slower opening and have lower impedances, both of which contribute to the severity of fault current that the automatic transfer switch will see until the overcurrent device clears the fault.

6.9.6 UL long-time fault current, withstand, and closing ratings. Underwriters Laboratories (UL) made several revisions to the UL-1008 standard effective April 13, 1989. A major revision permits optional long-time withstand and closing ratings which allow the switch to be marked for use with *any* manufacturer's circuit breaker within its rating. Such umbrella ratings give the application engineer more flexibility when specifying and coordinating the transfer switch. If a transfer switch manufacturer does not test for the long-time rating option, the switch must be marked to show the *specific* manufacturer's circuit breaker with which the switch was tested. The specific marking can limit the product's application and acceptance by the inspecting authority. See Table 6-10.

6.9.7 Special considerations. NFPA 20, Standard for Centrifugal Fire Pumps-1990 (Ref. 3), includes requirements for transfer switches and overcurrent protective devices. In recognition of the rather special considerations given to this application, paragraph 7-8.1 of NFPA-20 requires the transfer switch to have a rating of not less than 115% of motor full load current, as well as being able to interrupt the locked rotor motor current. For example a 75 HP, 208 volt motor used to drive a fire-pump requires 190 amperes at full load. The applicable transfer switch must be rated for 219 amperes minimum.

Transfer switch sizes are sometimes selected on the basis of the size of the engine generator set that was chosen to accommodate a specific load. However, this can be misleading where the load is primarily inductive, such as motor load. Generator sets are often sized to deliver the needed starting kVA which is significantly greater than the continuous running kVA. For example, a load consisting of one 100 HP, three-phase motor may require a 175 kW engine generator set to provide the starting kVA. This equates to 607 amperes at 208 volts. The running current of a three-phase 100 HP motor is in the range of 250 amperes. Thus, a transfer switch rated for total system loads need only be rated 250 amperes.

6.9.8 Selection summary. Proper sizing of transfer switches requires consideration of all the factors outlined in this section and are summarized below:

a. System voltage at the point of application

b. Continuous load current

c. Maximum available fault current at the point of application

1. Symmetrical amperes

2. System X/R rating at point of application

d. Type and size of overcurrent protective device ahead of transfer

e. Special considerations

6.10 CLOSED TRANSITION TRANSFER SWITCHES (CTTS)

In a typical emergency power system, there is an inherent momentary interruption of power to the load when it is transferred from one available source to another. In most cases this outage is inconsequential, particularly if it is less than 1/6 of a second.

There are some loads, however, that are affected by even the slightest loss of power. There are also operational conditions where it may be desirable to transfer loads with zero interruption of power when conditions permit. For these applications, closed transition transfer switches can be provided.

When transferring loads in this manner during a test or when re-transferring to normal after power is restored, the switch will operate in a make-before-break mode provided both sources are acceptable and in synchronism.

If either source is not present, or not acceptable, such as when normal fails, the switch must operate in a break-before-make mode (standard open transition operation).

6.10.1 Applications. Typical load switching applications for which closed transition transfer is desirable include data processing and electronic loads, certain motor and transformer loads, load curtailment systems, or anywhere load interruptions of even the shortest duration are objectionable. It should be understood that a system with a CTTS is not a substitute for a UPS (uninterruptible power supply). In addition to providing line conditioning, a UPS has a built-in stored energy that provides power for a prescribed period of time in the event of a power failure. A CTTS by itself simply assures there will be no momentary loss of power when the load is transferred from one live power source to another.

6.10.2 Utility approval. With closed transition transfer, the on-site engine generator set is momentarily connected in parallel with the utility source. This usually necessitates getting approval from the utility company.

6.11 MAINTAINING EMERGENCY POWER TRANSFER SYSTEMS

Blackouts such as the one that plunged the northeast United States and parts of Canada into darkness on November 9, 1965, and the one in New York on July 13, 1977, emphasized a very important aspect of emergency power systems: emergency systems must be tested, inspected, and maintained.

Facilities that followed regular programs of preventive maintenance had little, if any, difficulty when those blackouts occurred, while many of those who did not had major problems and some complete failures.

The National Electrical Manufacturers Association (NEMA) recognizes the need for preventive maintenance in its Publication ICS2 (Ref 1) which states: "a maintenance program and schedule should be established for each particular installation to assure minimum down time. The program should include periodic testing, tightening of connections, inspection for evidence of overheating and excessive contact erosion, removal of dust and dirt, and replacement of contacts when required."

Furthermore, the National Fire Protection Association NFPA 70B (Ref. 5) defines electrical preventive maintenance as "the practice of conducting routine inspections, test, and the servicing of electrical equipment so that impending troubles can be detected and reduced, or eliminated." They also state, "the purpose of this recommended practice is to reduce hazard to life and property that can result from failure or malfunction of industrial-type electrical systems and equipment." NFPA 99-1990 (Ref. 6) requires that an emergency power transfer system in a health care facility be tested for at least 30 minutes at least every 30 days. NFPA 110-1988 (Ref. 7) also shows detailed maintenance procedures.

Because of the proven need for preventive maintenance of emergency systems, this section will discuss what should be done to properly maintain automatic transfer switches.

A preventive maintenance program should be comprehensive enough to minimize the probability of experiencing a major malfunction in the emergency power system. The procedure should be such that it can be performed in the shortest period of time. The following suggestions will help to achieve these goals.

6.11.1 Installation precautions. Preventive maintenance begins with the installer when the automatic transfer switch arrives at the job site. Be sure the shipping carton or crate is not damaged. Look for handling instructions on the outside of the carton or crate so that it can be moved and unpacked without damage to the contents. For example, there may be a warning that the carton is top heavy.

The equipment should be placed in a safe, dry area and partially unpacked for inspection and removal of the manufacturer's handling and installation instructions. If the equipment is to be stored for an extended period of time, it should be protected from physical damage, dirt, and water. Ambient temperature in the storage area should be above $-30°C$ but not over $65°C$. Heat should be used to prevent condensation inside the equipment.

If a fork lift is to be used after the equipment is unpacked, the equipment should be in an upright position and care should be taken not to damage any handles, knobs, or other protrusions. If the equipment is to be lifted, lifting plates should be secured to the frame with hardware as described in the mounting instructions. Lifting cables or chains should be long enough so that the distance between the top of the equipment and the hoist is greater than one-half the longest horizontal dimension of the equipment.

The installer should closely follow the manufacturer's instructions. If there is any question about any part of the unpacking or installation procedure, contact the manufacturer's representative or the manufacturer.

After the equipment is installed, it should be thoroughly cleaned and vacuumed, taking special care to remove any loose metal particles that may have collected during installation. Before energizing the circuits, the switch should be manually operated in accordance with the manufacturer's instructions. Next, the phase rotation between the normal source and the emergency source should be checked. Then the switch can be energized and given a functional test in accordance with the manufacturer's instructions.

6.11.2 Drawings and manuals. Before attempting to test, inspect, or maintain an emergency power transfer system, at least one set of the manufacturer's drawings and manuals should be on hand. These should be kept in a readily accessible location. For convenience, additional copies should be readily available. These can usually be obtained easily from the manufacturer. Drawings and manuals are an indispensable aid in understanding the sequence of operation of the equipment, establishing test and inspection procedures, determining the type and frequency of test, troubleshooting, stocking spare parts, and conforming to code requirements for periodic preventive maintenance.

6.11.3 Testing. Regular testing of an emergency power system will uncover problems that could cause a malfunction during an actual power failure. Furthermore, regular testing gives maintenance personnel an opportunity to become familiar with the equipment and observe the sequence of operation. By being familiar with the equipment, personnel on duty during an actual power failure can respond more quickly if a malfunction occurs.

Field surveys have proven that an automatic emergency power system should be tested under conditions simulating a normal utility source failure. Experience shows that the rate of discovery of potential problems, when a system is tested automatically with the emergency generating equipment under load for at least one-half hour, is almost twice the rate when a system is tested by manually starting the generating equipment and letting it run unloaded.

To simulate a normal power failure, a test switch is recommended on the transfer switch. Some manufacturers supply this as standard equipment. The advantage of a test switch is that it does not cause a loss of power to connected loads as happens if the main disconnect is opened to simulate a failure. A test switch opens one of the sensing circuits and the controls react as they would during a power failure. In the meantime, the loads stay connected to the normal source. Power is interrupted to the loads only when the load is transferred from the normal to the emergency source and vice versa. This interruption is typically a few cycles.

On the other hand, if the main disconnect is opened to simulate loss of normal power, the power to the load could be interrupted for up to ten seconds. The interruption could be much longer if a malfunction occurs during start-up of the emergency generator. When a test switch is used, fewer maintenance personnel are needed because the potential of experiencing an extended power outage during the test is significantly reduced. While it is not always recommended, the control circuits of an emergency power system can be designed so that a clock will automatically initiate a test of the system at a predetermined time. This type of test can be restricted to running the engine generator set unloaded, or it can include an actual transfer of loads to the generator output. Using a clock to initiate testing under load or no-load conditions is not recommended for the following reasons:

a. With no load interruption, the engine generator set is tested unloaded and engine performance can be adversely affected. Also, the transfer switch is not tested, and the interconnecting control wires are not verified as being intact. The transfer switch is a mechanical device that typically is lubricated for life. However, to keep the lubrication adequate, the switch should be operated periodically.

b. There may be installations where automatic tests are not easily integrated into the operation of a building and it may be difficult to stop the clock.

c. A clock malfunction or improper setting could result in a test at a most inopportune time.

d. Clocks are usually set to initiate a test on weekends or during early morning hours. If a malfunction occurs, skeleton crews are usually not equipped to handle it.

e. Maintenance personnel should be present to observe the operation of the system to ensure that all components operate properly. Various codes and a good preventive maintenance program require that logs be kept and specific parameters recorded. Since clock testing is automatic, the parameters are not recorded. It is easy to assume that if everything is back to normal at the conclusion of the test, everything must have operated properly.

When a test switch is used to test an emergency power transfer system in the automatic mode, all relays, timing, and sensing circuits operate in accordance with the engineered sequence of operation. If the test procedure is written in accordance with this predictable sequence of operation, precious steps and time can be saved. It is possible for one person to initiate a test and then move to another location to initiate another action, or to monitor and record certain test results. The automatic sequence of operation should not vary greatly in time so that maintenance personnel movements can be correlated to that sequence. Furthermore, test procedures can include emergency procedures to cover any malfunction. Key people should be trained to react instinctively when certain malfunctions occur. For example, in a multiple engine generator system with automatic load-dump capabilities, an overspeed condition on one set may cause certain loads to be shed by retransferring them from the emergency source of power to the normal. The test procedure should simulate such a condition and include steps to confirm that the loads were retransferred, steps to reset the failed generator and put it back on the line, and steps to confirm that the load is again transferred to the emergency source of power.

6.11.4 Check list. A minimum check list for an emergency power system should include at least the following items:

a. Inspect for loose control wires and cables

b. The time from when the test switch is activated until the engine starts to crank

c. The time required for the engine generator to achieve nominal voltage and frequency after the test switch is activated

d. The time that the transfer switch remains on the emergency generator after the test switch is released

e. The time that the engine generator set runs unloaded after the load circuits have been retransferred to the normal source of power

f. Proper shutdown of the engine generator set and engine auxiliary devices

6.11.5 Record keeping. Accurate and comprehensive test records should be kept for several reasons. They may be required by national, state, and local codes. Also, the records will serve as the basis for determining whether corrective maintenance or replacement of certain elements in the system is necessary or desirable. They can help chart the wear of the system over its service life, and can be used as reference data for recommendations to modify existing equipment.

6.11.6 Inspection. In addition to putting the system through a test run, visual inspection is necessary to guard against system malfunction.

Before starting a scheduled operational test, the equipment should be checked for loose control wires and any discoloration of power connections caused by excessive heat due to loose connections. Also inspect for excessive dust and dirt particles and "craters" left in the dust by excessive moisture. With proper care, this type of inspection can be done without de-energizing the equipment.

If the transfer switch is subjected to a fault current resulting from a short in the distribution system, the switch should be checked immediately to determine if it was damaged. Such damage is a potential source of fire, can jeopardize the safety of personnel, or adversely affect the operation of the switch. To check the switch for such damage, first de-energize the switch by opening both the normal and the emergency breakers, and then look for any of the following:

a. Damage to insulating parts

b. Evidence of arcing between live parts and to ground

c. Welded or significantly eroded main and/or arcing contacts

While the switch is de-energized, a dielectric test should be conducted to determine if the switch will withstand twice the rated voltage plus 1000 volts rms. The manufacturer should be consulted prior to performing this test, because improper testing could cause damage to voltage sensitive parts. The length of downtime for this inspection is kept to a minimum, if the design of the transfer switch allows inspection of the live parts without major disassembly or removal of switch subassemblies, such as a portion of the drive mechanism, etc.

After the inspection and any necessary repairs are completed, re-energize the transfer switch and test it.

6.11.7 Detailed inspection suggestions. An in-depth inspection should be performed at least annually. This must be done with the equipment de-energized. However, this is frequently difficult to schedule when critical loads are involved. Thus, such inspections are usually performed after regular working hours or on weekends. At those times, labor rates are at a premium. For that reason, the downtime must be kept to a minimum. (Bypass-isolation switches are helpful for making in-depth inspections without downtime. These are discussed in Section 6.11.9.)

To save time, careful preparation with regard to the inspection procedure and the availability of tools and replacement parts is helpful. However, the design of the equipment and the skill of the maintenance personnel are the prime factors in saving time when an inspection of this type is performed. For complex systems, it is often advantageous to factory-train maintenance personnel. Some manufacturers will provide a factory technician to conduct in-house inspection and maintenance training classes for personnel at start-up time. However, the instructions included in equipment manuals will usually be adequate.

Equipment should be designed so that the inspection can be performed without removing equipment from the enclosure. All connections should be accessible from the front and no special tools should be needed to install or remove covers and parts.

The main operating mechanism of an automatic transfer switch should be rugged and simple. The parts that are lubricated should be checked at this time. If, due to environmental conditions, the mechanism requires lubrication, the manufacturer's recommendations should be followed. The mechanism should also be checked for loose or deformed parts and slowly operated to ensure that the mechanism is performing properly.

UL Standard 1008 requires that an automatic transfer switch be capable of from 2000 to 6000 cycles of operation under load, depending on amperage size, without parts replacement. However, the main current carrying contacts should be checked during this inspection, as required under NEMA ICS2-447.85. To do this easily, the pole covers and arc chutes should be removable without any disassembly of the operating mechanism. Expensive time is consumed if the operating mechanism must first be partially disassembled, or if the operating mechanism must be lubricated and adjusted after it is reassembled. The best rule to follow is to choose equipment that does not require major disassembly for inspection and preventive maintenance. Contacts on transfer switches that employ integral overcurrent protection cannot be conveniently inspected. Most large transfer switches are furnished with main current-carrying contacts and arc-interruption contacts. Main contacts should not show arc damage. Arc damage or wear on the arcing contacts is a result of the magnitude of the current voltage, power factor, and, most important, on the number of interruptions. If the arcing contacts are damaged or worn, they should be replaced.

To properly check the arcing contacts, a manual operating means is necessary.

Arcing contacts on the main power poles should make and break simultaneously. If one arcing contact breaks before the other, that contact would probably show greater wear and should be readjusted so that future wear is reduced. Since the arcing contacts make before the main contacts and break after them, there is always a main contact gap when the arc contacts just begin to touch. See Figure 6-33.

A decrease in this gap may be attributable to excessive arc contact wear. By using the manual operating handle, the switch can be stopped when the arc contacts touch. The main contact gap is measured at this point. Check the manufacturer's specifications to determine the need for adjustments.

Main operator control contacts mounted on the transfer switch should also be inspected for proper adjustment. The operation can be checked by using the manual handle and slowly operating the switch.

6.11.8 Preventive maintenance and repair. Dirt, dust, and moisture cause most malfunctions in automatic transfer switches. Eliminating these is the best preventive maintenance that can be performed. Dust should be removed by a vacuum during the annual inspection. Do not use compressed air, because flying dust and other particles can settle on uncovered control relay contacts and armature surfaces, and adhere to lubricated surfaces. If condensation moisture is a problem, a small space heater can be used.

There are no consumable spare parts in an emergency power transfer system. However, replacement parts should be kept in stock in case a malfunction occurs. The

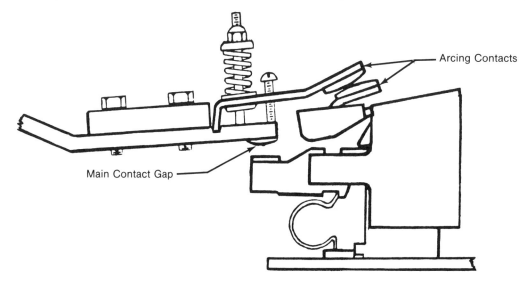

Figure 6-33. Main and Arcing Contact Operation

quantity of replacement parts is determined by the following factors:

a. *Manufacturer's recommendations.*

b. *Replacement parts common to all installed equipment.* If a majority of the manufacturer's recommended replacement parts are common to all transfer switches on the premises regardless of type, system voltage, or amperage size, a minimum quantity of replacement parts is required. For example, if all control relays are identical and are powered from the secondary of a control power transformer, the same coil can be used on all control relays.

c. *Availability of replacement parts from a local source.* If loads are critical, parts should be purchased and kept on hand in stock.

The contents of each replacement kit should be checked to make sure all items are correct. The installation instructions should be studied so that maintenance personnel know how to change the parts. No special tools should be required to install the parts. An inventory list of the spare parts kits should be kept.

6.11.9 Automatic transfer and bypass-isolation switches for testing and service.
In any installation — hospitals, computer centers, etc. — where interruption in power needs to be minimized when the automatic transfer switch is tested or serviced, an automatic transfer and bypass-isolation switch meets the need. As the name implies, this switch consists of an automatic transfer switch and a bypass-isolation switch. A typical automatic transfer and bypass-isolation switch is shown in Figure 6-34.

The automatic transfer switch is the same as discussed earlier. It is located in the lower half of the open cabinet in Figure 6-34.

The bypass-isolation switch, in the upper half of the cabinet in Figure 6-34, is connected electrically and mechanically to the automatic transfer switch. The bypass-isolation switch must meet the same requirements as automatic transfer switches, because it must handle the same loads.

The automatic transfer and bypass-isolation switch unit is in the same location in the electrical system as the

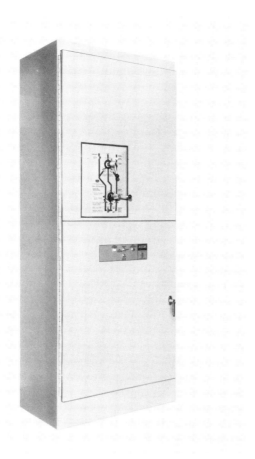
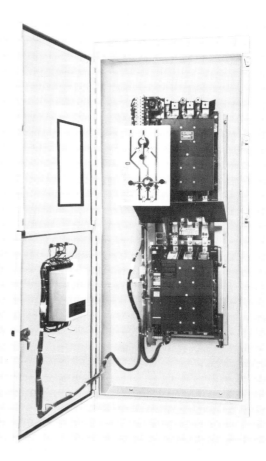

Figure 6-34. Automatic Transfer and Bypass-Isolation Switch

automatic transfer switch (Figure 6-35). Two types of bypass-isolation switches are available. One type momentarily interrupts load during bypass. Another type bypasses without load interruption. The following discussion refers to the second type.

With the automatic transfer and bypass-isolation switch set for automatic operation as it is in Figure 6-35, power comes from the utility, through the transfer switch and to the loads. The arrow configurations in the drawing indicate drawout type contacts. They are considered closed when shown close together as they are here. In this automatic position, the switch functions the same as an automatic transfer switch, as we've discussed. That is, if the normal power fails, the switch signals the engine generator to start and then automatically transfers the loads to it (Figure 6-36). When normal power is restored, the switch automatically retransfers the loads to the utility, as shown in Figure 6-35.

To test or service the automatic transfer switch, the controls that are accessible from the front of the enclosure with the door closed (Figure 6-35) are manually moved to Bypass to Normal and Test. This is done without interrupting the load. In this position, the loads are fed from the normal source, but through the bypass-isolation switch, not through the transfer switch. See Figure 6-37. Now the emergency power system, including the transfer switch, can be electrically tested and operated without interrupting the load.

The transfer switch can be bypassed to the normal source or to the emergency source (Figure 6-38). (Thus, the bypass switch also functions as a manual transfer switch.)

In either the Bypass to Normal, or Bypass to Emergency position, the automatic transfer switch can be electrically tested and operated without interfering in any way with the loads.

If the controls on the front of the enclosure are now moved to the open position, the automatic transfer switch contacts, etc., can be inspected. In this position, there is no power to the automatic transfer switch. Thus, the switch can be physically removed from the enclosure, if necessary (Figure 6-39) — still without interfering with the load in any way.

While in the open position, the loads are fed through the bypass switch. This can go on as long as necessary because the bypass switch has the same ratings as the transfer switch. Furthermore, if the loads are being fed from the normal source and it fails, you can simply start the engine generator and manually switch the loads to the generator with the bypass switch, and then switch them back when normal power is restored. Thus, the automatic transfer and bypass-isolation switch makes it easy for any facility to test and service its automatic transfer switches on a periodic basis without interrupting power to the loads.

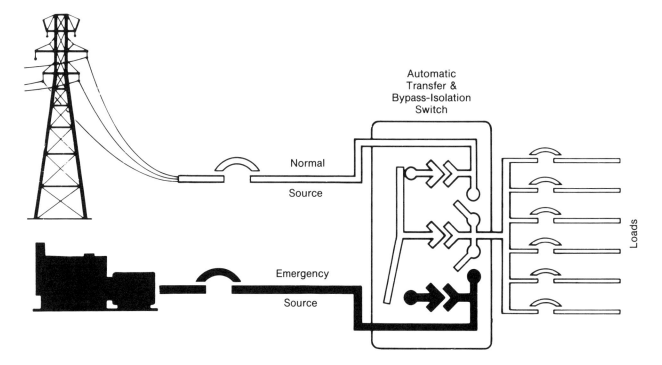

Figure 6-35. Schematic of Automatic Transfer and Bypass-Isolation Switch, Normal Position

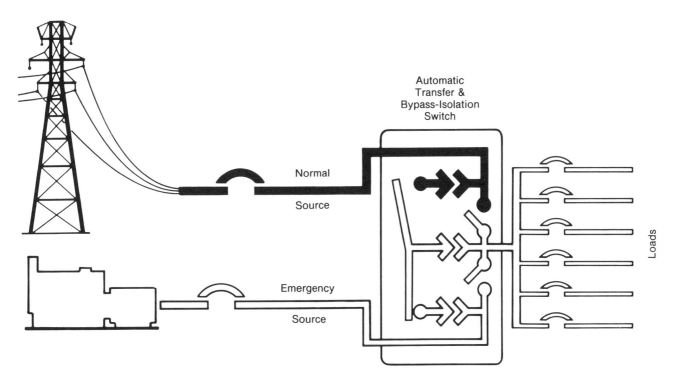

Figure 6-36. Schematic of Automatic Transfer and Bypass-Isolation Switch, Emergency Position

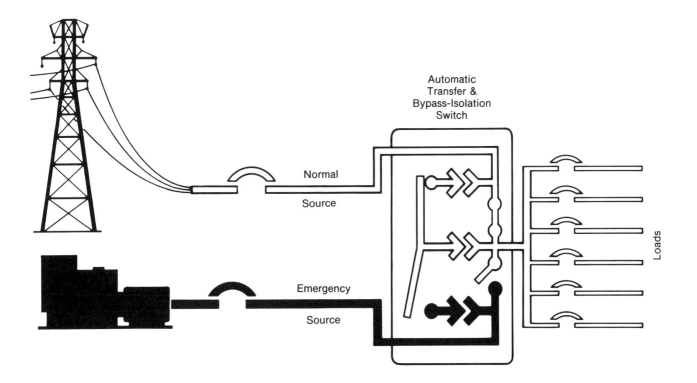

Figure 6-37. Schematic of Automatic Transfer and Bypass-Isolation Switch, Bypass to Normal Position

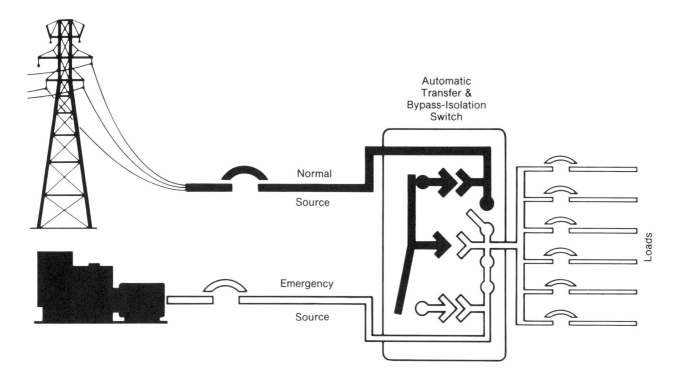

Figure 6-38. Schematic of Automatic Transfer and Bypass-Isolation Switch, Bypass to Emergency Position

REFERENCES

[1] AC Automatic Transfer Switches, ICS 2-447-1988, National Electrical Manufacturers Association.

[2] National Electrical Code, NFPA 70-1993, National Fire Protection Association.

[3] Standard for Automatic Transfer Switches, UL 1008 Fourth Edition, Underwriters Laboratories Inc.

[4] Centrifugal Fire Pumps, NFPA 20-1990, National Fire Protection Association.

[5] Electrical Equipment Maintenance, NFPA 70B, National Fire Protection Association.

[6] Health Care Facilities, NFPA 99-1993, National Fire Protection Association.

[7] Emergency and Standby Power Systems, NFPA 110-1993, National Fire Protection Association.

Figure 6-39. Removing Automatic Transfer Switch

Generator Switchgear

Herbert Daugherty

CHAPTER 7

INTRODUCTION

Generator switchgear is the assembled equipment that controls and protects the electrical output of generating systems. It can include switching devices, interrupting means, control logic circuitry, instrumentation, metering, and protective and regulating devices, together with their supporting structures, enclosures, conductors, electrical interconnections, and accessories.

Switchgear can be designed for indoor or outdoor use, and depending upon the application, is constructed according to various codes and standards. Some standard and code-making organizations involved with switchgear construction are:

a. National Electrical Manufacturers Association (NEMA)

b. National Fire Protection Association (NFPA) — this organization sponsors the NEC (National Electrical Code)

c. Institute of Electrical and Electronic Engineers (IEEE)

d. Underwriters Laboratory (UL)

e. American National Standards Institute (ANSI)

In addition, installations outside of the US generally come under the jurisdiction of the International Electro-Technical Commission (IEC) as well as national agencies within the country involved.

7.1 VOLTAGE CLASSIFICATIONS

Generator systems are generally classified by voltage. Low voltage systems are those with nominal ratings up to and including 600 volts. Typical voltages are 480Y/277, 208Y/120, or 480, 240, or 208 volt, three-wire. Medium voltage systems are generally divided into 5kV (4160 or 2400 volt) or 15kV (13,800 volt) systems.

7.2 SWITCHGEAR TYPES

Selection of the proper switchgear for a particular application depends on many factors including system voltage, function of the systems, protection and coordination considerations, budgetary restraints, and whether the equipment is to be located indoors or outdoors.

7.2.1 Low voltage metal-enclosed switchgear. See Figure 7-1. Metal-enclosed switchgear is enclosed on all sides and top with sheet metal (except for ventilating openings or inspection windows). It contains the power switching or interrupting devices with buses and connections, control and auxiliary devices. Access to the interior of the enclosure is provided by doors or removable covers, or both. Typically included in the switchgear, as required, are:

a. Low voltage power circuit breakers (fused or unfused) in accordance with ANSI/IEEE C37.13

b. Bare bus bars and connections

c. Instrument and control power transformers

d. Instruments, meters, and relays

e. Control wiring and accessory devices

f. Feeder circuit breakers and connections

The low voltage power circuit breakers are contained in individual metal compartments and controlled either remotely or by the system controls. The circuit breakers may be stationary or removable (drawout) type. When drawout circuit breakers are used, mechanical interlocks are provided for proper operating sequence.

Figure 7-1. Metal-Enclosed Switchgear

Outdoor switchgear is essentially the same as indoor switchgear, with a structure built around it for weatherproofing. A walk-in aisle can be provided in front of the circuit breakers to protect workers and equipment from weather during maintenance and system operation.

Bus bars are constructed from either copper or aluminum. Copper has a higher conductivity than aluminum and is more easily plated. Copper joints must be silver-plated. Aluminum joints are either silver-plated or tin-plated.

Low voltage circuit breakers are available in molded-case, insulated case, and metal frame power type. When specifying circuit breakers, the required frame size and desired trip rating must be determined. The choice must be made between drawout or stationary, and manually or electrically operated.

7.2.2 Metal-clad switchgear. Metal-clad switchgear (Figure 7-2) construction differs from low voltage switchgear in several respects. The main switching and interrupting device must be of the removable (drawout) type, arranged with a mechanism for moving it physically between connected and disconnected positions, and equipped with self-aligning and self-coupling primary disconnecting devices and disconnectable control wiring connects.

The major components of the primary circuit, that is, the circuit switching or interrupting devices, buses, voltage transformers, and control power transformers, are completely enclosed by grounded metal barriers that have no intentional openings between compartments. Specifically included is a metal barrier in front of, or a part of, the circuit interrupting device to ensure that, when in the connected position, no primary circuit components are exposed by the opening of a door. All live parts are enclosed within grounded metal compartments. Automatic shutters that cover primary circuit elements when the removable element is in the disconnected, test, or removed position are provided. Primary bus conductors are covered with insulating material throughout. Instruments, meters, relays, secondary control devices and their wiring are isolated from all primary circuit elements by grounded metal barriers, with the exception of short lengths of wire such as at instrument transformer terminals. The door through which the circuit interrupting device is inserted into the housing may serve as an instrument or relay panel and may also provide access to the secondary or control compartment within the housing.

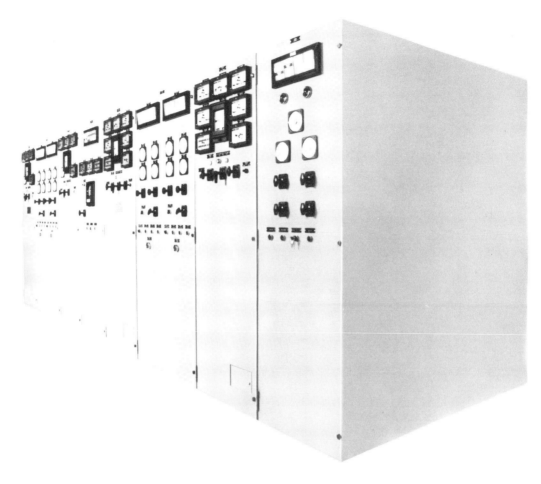

Figure 7-2. Metal-Clad Switchgear

Medium voltage circuit breakers are available in either 5kV or 15kV class. Although air-type circuit breakers have been in use for many years, vacuum breakers, that is, circuit breakers having arc interruption in a vacuum, have come into increasing use. Since medium voltage circuit breakers are not equipped with integral overcurrent trip units, they are utilized in conjunction with separate overcurrent relays, rated for the appropriate trip level of the circuit or device being protected.

7.2.3 Metal-enclosed interrupter switchgear. Metal-enclosed switchgear is designed to house medium voltage interrupter switches, power fuses (current limiting or non-current limiting), instruments and instrument transformers, control wiring and accessory devices. The interrupter switches and power fuses may be stationary or removable (drawout) type. When removable types are used, automatic shutters that cover primary circuit components when the removable element is in the disconnected, test, or removed position, and mechanical interlocks are to be provided for proper operating sequence.

Bus bars in metal-enclosed switchgear need not be insulated, nor do connections, unless there is insufficient clearance between phases or from phase-to-ground.

7.3 APPLICATIONS

Applications for generator switchgear vary from the simple, single generator control panel, to the multi-generator, automatic synchronizing and paralleling systems, and to the computer controlled parallel with utility systems used for cogeneration.

7.3.1 Generator control panel. Engine/generator controls are available as standard packages in enclosures suitable for engine, wall, or floor mounting (Figure 7-3) with or without generator circuit breakers. The generator control panel will typically consist of:

a. *Starting controls.* Means for manually and/or automatically cranking the engine and responding to engine safety devices to shut down the engine. See Figure 7-4.

b. *Visual and/or audible alarms.* For abnormal conditions such as low oil pressure, high water temperature, overspeed, and overcrank.

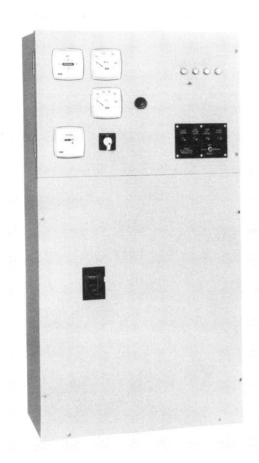

Figure 7-3. Generator Control Panel

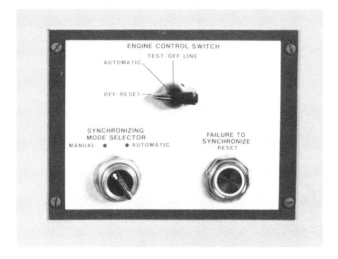

Figure 7-4. Starting Control

c. *Generator circuit breaker.* These are typically molded-case type, stationary mounted, and include lugs for generator and load cable connections.

d. *Generator instrumentation.* Typical instrument configuration consists of ammeter, voltmeter, and frequency meter, along with phase selector switch to read amps and volts for each phase.

7.3.2 Automatic transfer systems. When the engine/generator is utilized with an automatic transfer switch to provide standby power, the generator controls, over current protection, and the automatic transfer switch are often combined in a common switchgear section, interwired and intercabled to provide a complete, coordinated system. Systems of this type include two-source systems, three-source systems, and three-source priority load systems.

The basic two-source system (Figure 7-5) typically consists of a normal source (usually the electric utility) and an emergency source (usually an on-site generator). The system includes a transfer switch, as described in Chapter 6, engine generator controls, and emergency and/or normal circuit breakers. The system provides all necessary controls to start the engine and transfer loads to the emergency source. When the utility source has returned, the system retransfers the loads to normal and shuts the engine down after allowing it to run unloaded for a cool-down period.

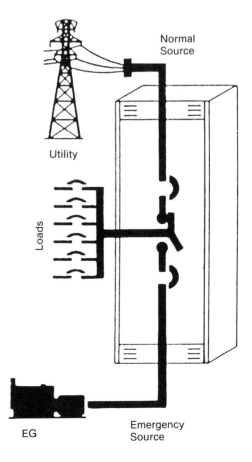

Figure 7-5. Two Source Transfer System

Three-source systems (Figure 7-6) allow a second standby source to be connected, to provide a redundant backup power source. The system includes a second transfer switch, as well as a second set of generator controls, and a generator circuit breaker. In this system, when normal power fails, both engines are signalled to start. The first engine generator to provide acceptable voltage and frequency is connected to the emergency load. The other generator will then be shut down after a time delay, and kept in readiness, to be started automatically upon failure of the running generator set. As in the two-source system, when the normal source is determined to be adequate, the load will be retransferred, the running generator allowed to cool down, and then be shut down. All operations are handled automatically by the system controls.

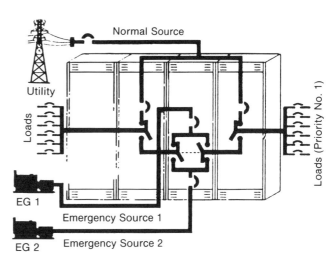

Figure 7-7. Three Source Priority Load System

7.3.3 Paralleling switchgear. There are many situations that can be handled best by operating two or more generator sets in parallel on a common bus. One reason for paralleling generator sets is for economy. For example, an existing distribution system may not lend itself to being split into several sections and handled by separate non-parallel units. Also, when the loads are expected to expand substantially, the initial investment is minimized by installing one smaller generator set, and then adding more sets in parallel as the loads increase.

The most important reason for paralleling in a standby system, however, is increased reliability. When a part of the emergency load is deemed very critical, it may be desirable to have more than one generator capable of being connected to that load. When there is a normal source outage, all generators in the system are started. The probability of having a generator start and achieve nominal voltage and frequency is increased according to the number of sets available. The first set ready to handle the essential load does so. As the other generators are running and connected to the bus, the remaining loads are connected in declining order of priority.

Basically, there are two types of paralleling systems: sequential paralleling and random paralleling. In sequential paralleling, the engine/generator sets are connected to the bus in a predetermined order. The lead engine is connected to the bus first. When the engine/generator selected as number 2 is ready to be connected, a synchronizer is connected between the output terminals of generator 2 and the bus. When the generator is in synchronism, its paralleling circuit breaker is closed, connecting it to the bus. Usually, a restriction is imposed to limit the time the controls will consume in attempting to synchronize and parallel a set to the bus before reconnecting the controls to the next set in sequence.

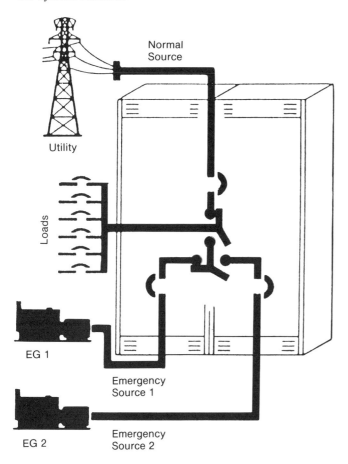

Figure 7-6. Three Source Transfer System

The three-source priority load system (Figure 7-7) is similar to the three-source system described above, except that it allows the second generator to be connected to another, less critical, load. The second generator remains as a redundant backup for the more critical load. The system adds a six-pole priority selector switch to allow for switching of the priority and non-priority loads.

In sequential systems (Figure 7-8), the usual approach to system design will incorporate only one automatic synchronizer. This may produce questionable operation in two respects. First, any malfunction in the synchronizer or its switching circuitry will result in complete loss of automatic paralleling capability. Second, the time required to put all sets on line is substantially increased as the number of generators increases. For example, it may take as much as two minutes to synchronize and parallel five generators in a sequential system. If a unit fails to synchronize in the first sequence, the time delay must time out before the next set in sequence can be put on line. Sequential paralleling is generally utilized in prime power applications, where these limitations are acceptable. In such installations, time is not as critical, since generators are, as a rule, connected and disconnected one at a time. In addition, these installations are generally attended by trained personnel capable of manually paralleling units in the event of system malfunction. In unattended emergency power applications, however, where duration of outage is of prime concern, sequential paralleling is questionable.

Random paralleling (Figure 7-9), or more accurately random access, to the bus employs a synchronizing device for each generator set in the system. Thus, the reliability of the system is substantially increased due to the multiplicity of parallel logic paths.

Random access permits simultaneous synchronizing of each set to the bus and therefore achieves parallel operation of sets in a much shorter period of time than the sequential method. With such a system, from inception of outage to all sets on-line, the time frame is seconds as compared to minutes. Obviously, where duration of outage is a consideration, the random access method is preferred. In fact, codes mandating emergency loads to be reconnected within ten seconds would require this mode of operation. With diesel or natural gas driven engine/generator sets, it is reasonable to expect that the emergency bus will be established within the ten second limit in a random access system, because any one of the generators can be first on-line. The only time that the ten second on-line requirement would be violated is if all sets failed to start in this time period. This, of course, is highly unlikely.

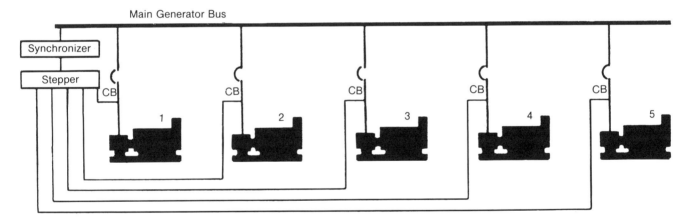

Figure 7-8. Sequential System

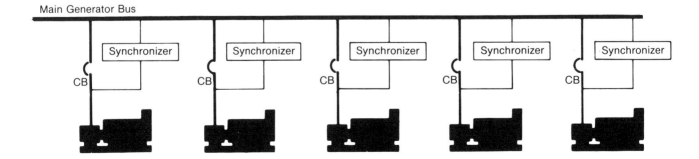

Figure 7-9. Random Access System

Whenever engine/generators are paralleled, the loads should be divided and controlled so that the system will not be overloaded. Overloading an emergency system will cause voltage and frequency deviations and possibly cause the failure of the complete system. The loads should therefore be grouped into blocks consistent with the prime mover size. See Figure 7-10. This means that load prioritization is necessary. The system can then control the connection of load to the bus in a prioritized sequence as generators are placed on line. Similarly, the system must disconnect, or shed, loads in reverse order of priority, to ensure maximum continuity of power to the highest priority loads if bus capacity reduces due to loss of generating units.

Having established the basis for load connection and shedding, it is necessary to consider the means to achieve this switching. There are several ways to switch the loads. In an emergency power system, one convenient means is to utilize the automatic transfer switches for load connect and load dump operation. See Figure 7-11. Another method involves the use of remote control switches, or contactors to open and close, adding and shedding the loads. Downstream circuit breakers can also be tripped to shed load. However, if shunt tripped molded case circuit breakers is the method used, consideration should be given to the fact that these breakers must be manually reset to reconnect the load. In short, there are many approaches to load switching. The preferred approach for any application is determined by the requirements of the application.

When power sources are connected in parallel on a common bus, the voltage and frequency of each unit are the same. Therefore sensing voltage or frequency will not detect a malfunctioning engine/generator set. To determine proper operation on the bus, measure the power output of each generator set. When a set is delivering power to the bus, it is operating properly. When it is receiving power from the bus or "motorizing", the possibility of a malfunction exists. It is not unusual to have small amounts of power flow to the set for short periods of time when the bus is lightly loaded. See Figure 7-12.

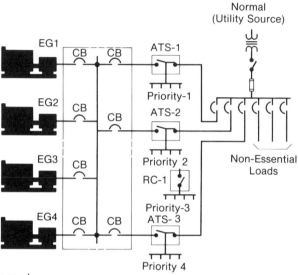

Legend
ATS — Automatic Transfer Switch
CB — Power Circuit Breaker
EG — Engine Generator
RC — Remote Control Switch

Figure 7-10. Prioritized Loads

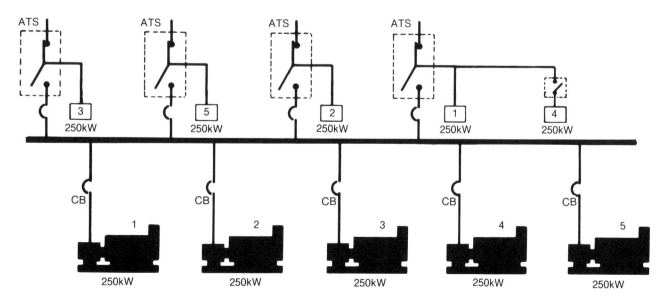

Figure 7-11. Load Control with Automatic Transfer Switches

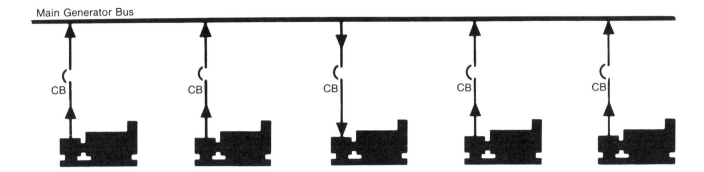

Figure 7-12. Reverse Power

At these very low load levels, the exchange of synchronizing currents is necessary to maintain the sets in synchronism. The device monitoring reverse power, therefore, must be set high enough to ignore this harmless condition, but low enough to detect a malfunction. In order for the device to be properly set for the specific engine being utilized, it must have an adjustable trip power range. Since small amounts of power in the reverse direction are acceptable for a short period of time, it should also have an integral time delay, to ignore transient conditions caused by light loading and switching large blocks of load.

Whatever method of paralleling is used (sequential, random, or even manual paralleling), the sets must first be synchronized. Synchronism is always determined in the same way; that is, the sine waves must be equal. The sources to be paralleled must be equal in (1) phase angle, (2) frequency, (3) voltage, and (4) phase rotation. All four conditions must be satisfied before connecting the sources in parallel. See Figure 7-13.

Paralleling sources that are not synchronized can cause substantial damage to the engine, the generator, and/or the system. Due to tolerances in the equipment, there will always be some allowance for differences in parameters. However, when these differences become too great, the result is out-of-synchronism paralleling.

To produce minimal disturbances, the differences should be minimized. The following maximum allowable differences are typical:

    Voltage 5%

    Frequency . . . . 0.5 Hz

    Phase angle . . . . 5°

The synchronizer senses the difference in these parameters and takes corrective action to reduce any differences to within the acceptable limits.

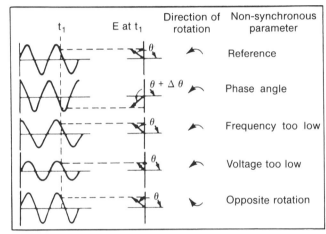

AC sine waves representing differences in phase angle, frequency, voltage and rotation.

Figure 7-13. Paralleling Conditions

Generator output is normally monitored by instrumentation furnished on the paralleling switchgear. This instrumentation facilitates periodic inspection and testing, preventive maintenance, and calibration essential to proper system operation and longevity.

Since the controls of an automatic synchronizing and paralleling system, as well as the electronic isochronous governors and electronic voltage regulators are highly accurate, the instrumentation that measures the performance of these controls should also be accurate. Switchboard-type 1% accuracy instruments are normally used for this application.

The minimum complement of instruments includes an ammeter, with means of switching to measure current in each line; a voltmeter, with switching means to read each line-to-line voltage; a frequency meter to measure engine speed; and a wattmeter to indicate engine performance. See Figure 7-14.

Figure 7-14. Instrumentation

The ammeter furnishes data on generator loading and voltage regulator performance and adjustment. The voltmeter permits adjustment of the voltage regulator for proper operating voltage. The frequency meter permits adjustment of the governor for proper operating frequency. The wattmeter allows adjustment of the governor for proper load division between engine generator sets operating in parallel.

The generator circuit breaker serves two functions. First, it provides protection against overload and short circuits. Second, it serves as a switching device connecting the set in parallel with the common bus. The breaker used must be fast-closing and capable of repetitive operations.

For low voltage systems, an air power circuit breaker is normally used. These breakers fall into two categories:

a. *Insulated case circuit breakers.* These breakers may be used when the switchgear is built to UL 891. They have high interrupting ratings without fuses — up to 150,000 amps — and are capable of high switching endurance — up to 8000 operations.

b. *Metal frame power circuit breakers.* Used when the switchgear is built to either UL 891 or UL 1558. These breakers have high withstand or short time rating and have high interrupting capability with fuses — up to 200,000 amps.

Both types of breakers are high speed closing, with maximum five-cycle closing time in either the manual or electrically operated modes. Both types can be furnished with solid-state trip elements with wide adjustment bands, including adjustable long delay, short delay, instantaneous, and ground fault settings. To provide fully automatic operation, the generator breakers must be electrically operated. To accommodate inspection and maintenance, drawout breakers are usually preferred to fixed breakers.

7.3.4 Paralleling with the utility. Generators are often operated in parallel with the utility source to reduce energy costs by reducing the peak kilowatt demand of a facility or in a cogeneration system. Either application requires the approval of the local electric utility, who must review and approve the protective relaying provided for the interconnect before parallel operation will be allowed.

In a peak demand reduction application, on-site generator sets can be used to reduce utility peaks, either by transferring loads or by connecting the generator in parallel with the utility. The transfer method is often used where generator(s) and transfer switches are already in place in a standby application and can be readily utilized for this additional purpose. Peak shaving by transfer is a simple, economical means for reducing electric costs. See Figure 7-15.

Optimum savings can be achieved by operating in parallel with the utility. This, however, requires careful design consideration to ensure safety to both equipment and to personnel. A circuit breaker is necessary for interconnection with the normal building distribution. See Figure 7-16. Upon detection of any malfunction or abnormality when the generator and utility are connected in parallel, this interconnect device must be opened immediately, isolating the two sources. This assures that the generator will not backfeed the utility grid when the utility power has failed. Voltage and frequency protective devices and an automatic synchronizer are necessary to ensure that the sources can be safely reconnected before the interconnect device can be closed.

Adequate protective relaying must be provided to cause tripping of the interconnect under abnormal conditions. This relaying will vary as a function of the application, the location of the interconnect point, and the requirements of the local utility company. A partial listing of

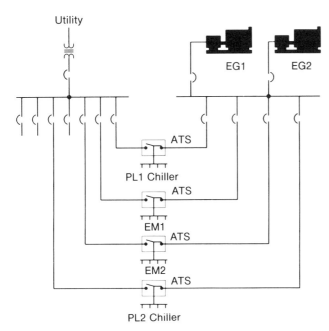

Legend

ATS – Automatic Transfer Switch
EM – Emergency Load
EG – Emergency or Standby Engine Generator Set
PL – Peak Shaving Loads

Figure 7-15. Peak Demand Reduction System

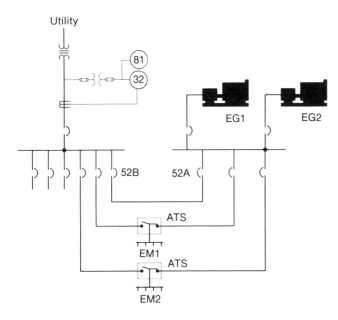

Legend

32 – Directional Power Relay
52 – Circuit Breaker
81 – Frequency Relay

Figure 7-16. Parallel with Utility

protective devices commonly utilized follows. The device numbers indicated are as defined by ANSI/IEEE C37.2-1979, Electric Power System Device Function Numbers.

Device 25. Synchronizing or synchronism-check device. Operates when two ac circuits are within the desired limits of frequency, phase angle, and voltage to permit or to cause the paralleling of these two sources.

Device 27. Undervoltage relay. Operates when its input voltage is less than a predetermined value.

Device 32. Directional power relay. Operates on a predetermined value of power flow in a given direction, or upon reverse power flow such as that resulting from the motoring of a generator upon loss of its prime mover.

Device 40. Field relay. Functions on a given or abnormally low value or failure of machine field current, or on an excessive value of the reactive component of armature current in an ac machine, indicating abnormally low field excitation.

Device 46. Reverse-phase or phase-balance current relay. Functions when the phase sequence is reversed or when the polyphase currents are unbalanced or contain negative phase-sequence components above a given amount.

Device 47. Phase-sequence voltage relay. Functions upon a predetermined value of polyphase voltage in the desired phase sequence.

Device 50. Instantaneous over current or rate-of-rise relay. Functions instantaneously on an excessive value of current or on an excessive rate of current rise.

Device 51. AC time overcurrent relay. Operates when its ac input current exceeds a predetermined value and in which the input current and operating time are inversely related.

Device 52. AC circuit breaker. Used to close and interrupt an ac power circuit under normal conditions, or to interrupt this circuit under fault or emergency conditions.

Device 59. Overvoltage relay. Operates when its input voltage is more than a predetermined value.

Device 67. AC directional overcurrent relay. Functions on a desired value of ac overcurrent flowing in a predetermined direction.

Device 81. Frequency relay. Responds to the frequency of an electrical quantity, operating when the frequency or rate of change of frequency exceeds or is less than a predetermined value.

Device 86. Locking-out relay. Electrically operated hand or electrically reset relay or device that functions to shut down or hold equipment out of service, or both, upon the occurrence of abnormal conditions.

Device 87. Differential protective relay. Functions on a percentage of phase angle or other quantitative difference of two currents or of some other electrical quantities.

In low voltage applications, the overcurrent sensing required to protect the system and feeders is incorporated in the circuit breaker itself. In medium-voltage applications, these functions are provided by a combination of separate protective relays such as Device 51V, three-phase overcurrent protection with voltage restraint, and Device 87, differential relay.

In cogeneration, two forms of energy, generally electricity and heat are produced by a single process. An on-site generating system with heat recovery fits this definition. See Figure 7-17. In a conventional engine driven generator set, considerable energy is wasted in the form of heat loss to exhaust or jacket water. In a cogeneration system, this otherwise wasted heat is recovered and converted into useable energy for process systems requiring steam or heat. A cogeneration system can operate in an isolated mode, but generally it is preferable to operate in parallel with the utility. When the electricity produced in such a system is less than the demand of the facility, additional power is imported from the utility. If thermal requirements dictate that more power is produced than can be used in the facility, the excess power can be exported into the utility grid (with proper precautions).

When operating in parallel with the utility, a generator loading control or import-export control is necessary to control the kW output of the generator set. This device operates in conjunction with the engine governor to control the fuel input to the engine, increasing or decreasing its kilowatt (or true power) output as desired. Another accessory required is a var/power controller. This device biases the voltage regulator to cause the generator to maintain a constant power factor (or constant VA) or reactive power output. Generator voltage can therefore track normal utility voltage excursions without kVA overloading.

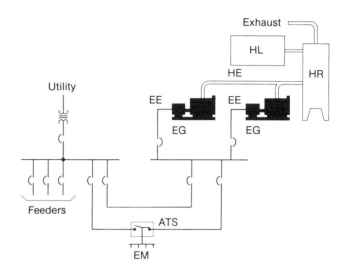

Legend

ATS – Automatic Transfer Switch
EE – Electric Energy
EG – Engine Generator
EM – Emergency Load
HE – Heat Energy
HL – Heat Load
HR – Heat Recovery System
(– Main, Tie or Feeder CB or Switch

Figure 7-17. Cogeneration System

Liquid-Cooled Spark-Ignited Engines

Robert Holtgreive

CHAPTER 8

INTRODUCTION

The successful use of internal combustion engines began with the development of the Otto cycle engine in 1876. Ever since the invention of gunpowder, men had dreamed of doing work with an engine using combustion to move a piston in a cylinder (Ref. 1). Otto's patent of 1876 described the four stroke cycle that we know as the Otto cycle, Figure 8-1. The idea of wasting three strokes to get one power stroke was highly ridiculed but quickly proved to be far superior to other engines of the time. With Otto's invention, the work horse was doomed and man's drudgery was relieved. An advertisement of the Otto Gas Engine in the American Gas Journal, October 16, 1885 listed "special engines for electric light work." Thus generator sets became one of the early applications for the Otto cycle engine that we know as the four stroke cycle, spark-ignited, liquid cooled engine.

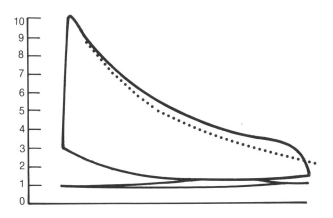

Figure 8-1. Otto Four Stroke Cycle

The engine-generator set industry rapidly expanded immediately following World War I. Many of the early engines were specially developed for the purpose. An example of such an engine is the Sunbeam, Figure 8-2. The engine had high tension ignition, ebullient cooling, automatic oil level and an electric governor. The set weighed 250 pounds and had a rating of 650 watts.

Water cooled spark-ignited engines have been the standard of the industry for many years and represent by far the highest volume internal combustion engine in the world today. The spark ignited (SI) engine has the advantage over the compression engine that, with positive ignition, it can operate at lower compression ratios, resulting in lower firing pressures. The low firing pressure substantially reduces the design weight of all components, allowing the engine to operate at higher speed for increased output. The lower firing pressure also results in both lower combustion noise and lower torsional vibration, resulting in lower installation cost. The spark-ignited engine responds well to turbocharging for increased output, taking advantage of a waist gate design for the extended speed range.

Unfortunately, the torque curve for a spark-ignited engine is relatively flat compared with a compression ignition engine and is therefore, less responsive for governing and lugging. See Figure 8-3.

Overall, the SI engine offers high output in a small lightweight package, has extremely good cold starting capability and the ability to operate under all environmental conditions from arctic to desert.

The spark-ignited water cooled gasoline engine is relatively inexpensive to manufacture, has high reliability

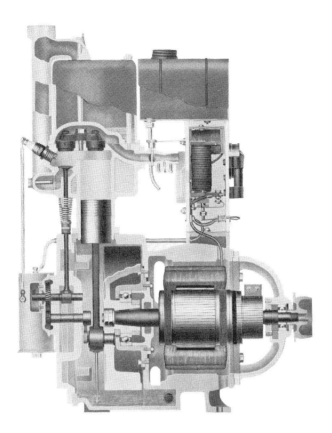 

Figure 8-2. Cross Section Sunbeam Farm-Lite Plant

and durability, and successfully accepts abusive treatment with regard to fuel, maintenance schedule, etc. Engine performance with regard to fuel economy cannot compete with a compression ignition engine. However it has been improved extensively in recent years with the use of electronically controlled ignition systems, lean burn combustion technology, and new designs for reduced friction.

The other advantage of the water cooled spark-ignited engine is that the thermal energy from the coolant system provides effective means for waste heat recovery. With recent technology the engine operates at much higher pressures and temperatures, and increases flexibility with regard to electric driven cooling fans and smaller heat exchangers. With the advent of no-lead fuel the maintenance intervals have been extended substantially with regard to lubricating oil and spark plug replacement. The disadvantage of a water cooled engine is the additional cost for the maintenance of the heat exchangers and their reliability compared with an air cooled engine.

Spark-ignited gaseous fueled engines have similar advantages to the gasoline fueled engines and are available through several thousand horsepower.

Spark-ignited gasoline fueled water cooled engines power many generator sets to approximately 100 kW.

Recent developments have applied lean burn combustion technology to spark ignited natural gas engines. The technology includes equipping the engines with turbocharging, aftercooling, electronic ignition, and electric fuel management systems. These engines provide high performance output and meet stringent emission standards without exhaust after treatment.

8.1 COOLING SYSTEM

Recent technology has satisfactorily increased the operating temperatures and pressures of the cooling system, resulting in lower cost and smaller size heat exchangers. Systems today are operating at 16 psi with top tank temperatures in the range of 250° F. With the cooling systems today, it is important to ensure required air flow, proper shrouding to prevent recirculation, elimination of excessive pressure drops, and restriction to flow in external plumbing. The water pump must not cavitate (form air bubbles in the coolant), which would cause lower heat rejection and overall performance. A plexiglass window installed at the inlet and outlet position provides a clear

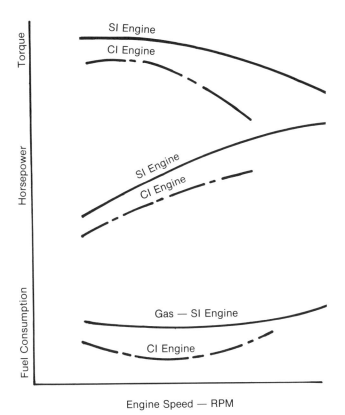

Figure 8-3. Comparison of SI Engine to CI Engine with Same cu. in. Displacement

definition of the condition of the coolant. A clear non-milky appearance is normal. All coolant systems should be properly serviced with ethylene glycol and water to prevent excessive rust buildup, and to provide increased high and low temperature operating ranges of the coolant system. Care must be taken to assure that all belt driven water pumps and fans are properly maintained and that coolant hoses are properly reinforced to avoid collapse during operation.

Heat exchangers should be designed to include the ability to digest and pass foreign debris, to avoid excessive plugging or restriction to flow. External cooling lines should be sized to properly accept the coolant flow of the engine without excessive restriction, which could cause cavitation and inadequate cooling. Most industrial engines should be equipped with high temperature shutdown devices and/or coolant level devices to properly protect the engine from catastrophic failure. See Chapter 16 for more detailed information on liquid cooling systems.

8.2 LUBRICATING OIL SYSTEM

All modern engines are built with full flow lubrication systems, in which all the oil is directed through the filter prior to entry into the critical components of the engine. All lubricating oil systems should be designed with high pressure relief valves for cold startup and be equipped with a well designed pressure regulator valve to assure adequate oil pressure under all operating conditions. External lines and filters must be sized to avoid excessive restriction to flow and be constructed with material that provides high reliability for long life requirements. Oil filters should be located so that they can be easily serviced without need for special tools or equipment and to avoid excessive oil spillage, which discourages required maintenance. Lubricating oil should be of high quality and consistent with the engine manufacturer's requirements with regard to temperature, service interval, and SAE grade and type.

Many industrial applications which operate unattended use makeup systems which automatically add oil between oil changes.

8.3 FUEL SYSTEM

The carburetor fuel system requires minimum maintenance, with the exception of a fuel filter normally installed in the fuel line prior to the carburetor. Poor performance and low HP can result if the filter is not maintained properly. Mechanical and electric fuel pumps are highly reliable items that provide a normal 3 to 5 pound fuel inlet pressure to the carburetor assembly. If on industrial applications the main fuel supply is beyond the capacity of the transfer pump, inadequate flow will result in poor performance.

Carburetor systems are being replaced with electronic fuel injection systems for improved performance and emission certifications. The fuel pump is a calibrated piece of equipment as developed by the engine manufacturer and should not be modified or substituted since this could cause serious engine damage. For more information on fuel systems see Chapters 14 and 15.

8.4 AIR INTAKE SYSTEM

All modern spark-ignited engines use a barrier dry type air cleaner which must be properly sized (CFM) for the engine, and positioned for ease of element replacement. When operating under normal conditions, the air cleaner should be positioned to avoid the intake of exhaust gas or high temperature heat exchanger air. Many industrial air cleaners are equipped with pre-cleaner to extend the life of the element, along with an air restriction gauge that indicates the element is in need of service.

8.5 IGNITION SYSTEM

Industrial gas engines are normally equipped with one of several ignition systems to include breakerless ignition, a long-life ignition system with coils at each spark plug, magneto system, or solid state distributorless system.

8.6 EXHAUST SYSTEM

The exhaust system should be sized to avoid excessive back pressure when operating under rated conditions. An exhaust system should be routed to avoid re-circulation with intake air and radiator assembly. The exhaust system should be properly supported to avoid excessive stress on the engine exhaust manifold assembly. Proper personnel protection for high temperature exposure should be provided and the exhaust system should be positioned so that rain or snow cannot enter it.

8.7 ELECTRICAL

The battery, starter and alternator system are designed by the engine manufacturer to provide reliable starting and required battery charging under normal operating duty cycles, taking accessory loads into consideration. It is important that battery cable lengths not exceed 6 ft. in order to avoid excessive voltage drop and poor starter performance. It is important that proper belt maintenance and belt tensioning be maintained to assure proper alternator performance. Battery sizes consistent with starting requirements are normally specified by the engine manufacturer.

8.8 GOVERNORS

Industrial engines are equipped with a governor, mechanical, hydraulic, electric, or velocity. Mechanical governors can be gear or belt driven. Velocity governors provide only overspeed protection and are not sensitive enough to control engine RPM for generator set applications. The other choices are based on the frequency response required for the specific application. It is important that governor linkage is designed so that any malfunction or loss of governor will return the engine to normal idle setting.

8.9 SAFETY

Many safety related items pertain to engines and engine installations including the following:

a. All fans, either mechanical or electrical, should have proper guards to avoid injury.

b. Exhaust systems should be properly shielded to avoid burns and injury to personnel operating or maintaining the equipment.

c. Linkage system should be designed to return engine to idle position when any malfunction or throttle linkage is involved.

d. Care must be taken to avoid any fuel spillage or leakage that could cause fires when exposed to high exhaust temperatures or any other source of ignition.

e. Engines should be properly installed with shutdown or warning devices that monitor low oil pressure, high water temperature, low coolant level, excessive air cleaner restriction, and low oil level.

REFERENCE

Cummins, Lyle. Internal Fire, Revised Edition, Society of Automotive Engineers.

Air-Cooled Spark-Ignited Engines

Gordon Johnson

CHAPTER 9

INTRODUCTION

One of the most useful inventions of all time is the air-cooled gasoline engine. The great majority of generator sets under 10 kW use spark-ignited air-cooled engines fueled by gasoline, natural gas, or propane. The popular applications are: portable (for construction and recreational use), marine, recreational vehicles, and standby. The factors that contribute to the overwhelming dominance of air-cooled engines in this size range are simplicity, light weight, and low cost.

In the early history of on-site power generation, a number of manufacturers designed special air-cooled engines specifically for generator set use. Today, however, most generator set engines are modifications of a standard industrial engine. The manufacturers of air-cooled engines for lawn and garden equipment have been very cooperative in making generator set versions of their engines available to generator set manufacturers. This gives the advantage of using low cost, highly developed, high production engines. The modifications most frequently furnished are constant speed governor, mounting for generator stator and rotor, and increased lubricating oil capacity.

9.1 ADVANTAGES

a. *Simplicity*. The air-cooled engine requires no radiator, water pump, or hoses. It has no coolant to leak out, freeze, or become contaminated. The cooling fan is usually integral with the engine, cooling fins are cast on the cylinder and cylinder head, and air baffles are part of the engine assembly. The ignition system is integral with the engine. The engines are usually one or two cylinders.

b. *Flexibility*. Because of its simplicity, the air-cooled generator set finds use in a wide variety of applications. A single set may be used for construction, recreational use, or standby. It can be quickly taken to fires or flood and hurricane situations. Its gasoline fuel is available everywhere.

c. *Operation*. Because of its rapid warm up and higher temperature operation, the air-cooled engine is less susceptible to sludge formation in the lubricating oil. This rapid warmup also reduces wear of cylinder bore, piston, and piston rings in short run applications.

9.2 DISADVANTAGES

a. *Noise*. In general the air-cooled engine will have a higher noise level than an equivalent water-cooled model because of its lack of a water jacket to dampen noise. The noise level, is in general, in the range of 3 to 5 dBA higher for the same speed. Because of the many sheet metal parts it is also more susceptible to rattles. Most of the portable generator sets under 10 kW (Figure 9-1) operate at 3600 rpm, which also contributes to high noise level. Other types such as those for recreational vehicles (Figure 9-2), where noise is a critical factor, are mostly 1800 rpm.

b. *Cooling air disposal*. Because of its integral cooling system it is usually more difficult to dispose of heated cooling air on air-cooled sets. (See paragraph 9.5 on cooling.)

c. *Engine heaters*. Heaters to maintain engine temperature for rapid starting in cold temperatures are much more difficult to apply to air-cooled engines. The usual choice is to place heaters in the oil sump; however, such applications are plagued with oil breakdown and coking on the heater coils, leading to heater burnout if not removed. They also do not supply sufficient heat to the fuel inlet system for good vaporization, which may result in poor starting.

Air-Cooled Spark-Ignited Engines

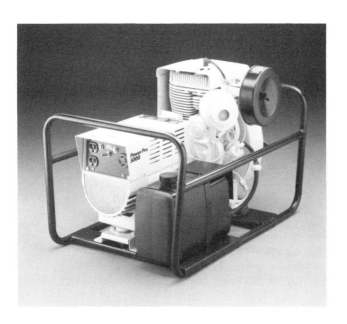

Figure 9-1. Portable Set

9.3 TWO VERSUS FOUR STROKE CYCLE

Air-cooled spark-ignited engines may be two-stroke cycle, having a firing stroke each revolution, or four-stroke cycle, having a firing stroke every two revolutions. Two-stroke engines are common in sizes below 800 watts. Two-cycle engines are the ultimate in simplicity since they do not require a valve train. They have a high power-to-weight ratio, have fewer moving parts, and are low cost. They have the disadvantages of higher fuel consumption, having to mix lubricating oil with the fuel, more frequent spark plug fouling, and high exhaust gas emissions. Four-stroke cycle engines give cleaner more efficient burning of fuel. The lubrication system is separate from the fuel; thus, the four-cycle engine does not require mixing oil with the fuel. They do require a camshaft and valve train. The trend is toward the use of four-stroke cycle engines. However, the lightweight, low cost of two-cycle engines assures that they will continue to find applications.

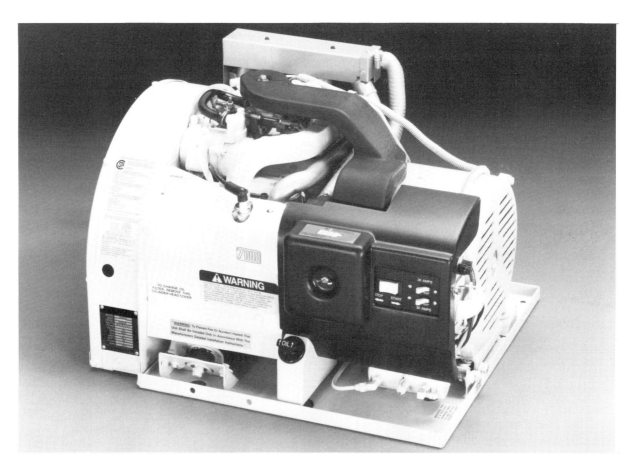

Figure 9-2. Recreational Vehicle Set

9.4 ALUMINUM VERSUS CAST IRON

The majority of air-cooled spark-ignited engines built today have die-cast aluminum alloy cylinders. Die-casting technique lends itself well to producing thin, smooth, close together cooling fins and is conducive to high volume production. Aluminum alloy conducts heat well and leads to a lightweight engine. The cylinder bore of aluminum alloy engines must be treated, usually plated or sleeved, to prevent rapid wear. Historically, aluminum alloy has been subject to distortion but new alloys have largely overcome this problem. Cast iron engines or aluminum engines with cast iron sleeves are recognized as longer life engines. The metal in the cylinder of such engines can be alloyed for minimum cylinder bore wear and is stable, giving longer valve life, better sealing and lower oil consumption. In general, the cost of cast iron engines is higher than aluminum alloy engines because aluminum alloys can be diecast. Both cast iron and aluminum engines use aluminum alloy cylinder heads. Most of the engine heat rejection is to the cylinder head. The superior conductivity of aluminum and the ability to cast more closely spaced cooling fins makes it very attractive.

9.5 COOLING

While air-cooled generator sets are designed for adequate cooling in ambient air, difficulties can be encountered when they are enclosed in a limited space. Air-cooled engines actually require much less volume of air than water-cooled engines but the temperature of the cooling air discharged from the engine may be over 200° F. The heated air may be discharged in a position where it is difficult to duct to the outside as is done with liquid cooled engines. Overheating will quickly result if heated air is recirculated through the engine. A frequent technique is to modify the cooling air flow to what is frequently called reverse or vacuum cooling. See Figure 9-3. Reverse cooling involves removing the cooling fan fins from the engine flywheel and adding a blower which pulls air across the engine and exhausts into a duct which can be directed outside. These systems are usually designed for a maximum restriction of 0.5 inches of water in the duct. The size of duct can be calculated to meet this limit and will vary with the length and number of bends. For straight out runs the size may be approximately one-quarter square foot of opening per kW rating of the generator set. The user should be aware that anything in an opening is a restriction. Louvres restrict up to 35%, screens up to 50%, and filters up to 75%.

Some cooling system designs attempt to collect the hot air coming off the standard engine cooling system. In order to be successful, these systems must allow very little leakage and the duct must have almost no restriction. Since air-cooled engines usually have heated air coming out in a number of places, the design of the duct system is likely to be difficult. In addition, the standard engine blower design does not anticipate any restriction of the outlet air. For that reason little restriction can be tolerated in the outlet duct.

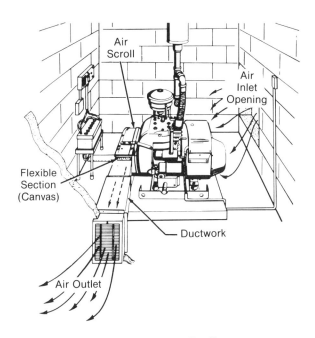

Figure 9-3. Reverse Cooling

9.6 LUBRICATING OIL

The operating temperature of the lubricating oil in an air-cooled engine is higher than for water-cooled engines, frequently as high as 275° F at full load. For satisfactory engine life this demands frequent oil changes, for some engines as often as every 25 hours. Since the oil capacity of these engines also tends to be limited and the oil consumption tends to be higher, frequent oil level replenishment is necessary. Automatic vacuum type oil level maintenance devices are inexpensive and simple to install; however, these must not be used to avoid the recommended oil change period. Low oil level shutdown devices are frequently available and are an excellent addition. They should not, however, be used to avoid frequent oil level checks.

Most single cylinder air cooled engines use splash lubrication. A simple dipper on the connecting rod splashes oil into the cylinder and over the bearings. Many two cylinder engines have pressure lubrication with an oil pump in the oil sump and oil galleries drilled in the crankcase. The industry also has a number of hybrid systems.

9.7 FUEL

One of the most frequent causes of failure of portable generator sets is stale or dirty fuel. This is particularly true of gasoline fueled engines. Since many of these sets have mounted fuel tanks, usually filled from a portable can, fuel contamination is a hazard. Good fuel filters with water separators are a necessity and since gasoline stores very poorly, fresh fuel is required. Failure to provide clean, fresh fuel will result in carburetor clogging, which will require expensive repairs. Other generator sets, such as recreational vehicle sets, draw fuel from a large frequently replenished tank. With those systems fuel problems are minimal. The generator set should have its own suction line in the tank to avoid being robbed of fuel by the main engine.

The readily available supply of gasoline everywhere usually makes it unnecessary to store large quantities for generator set use. Engineers do not recommend gasoline for automatic standby applications because of the storage problem.

Most manufacturers offer natural gas or propane conversion kits for air-cooled spark-ignited generator sets. These engines operate very well on gaseous fuel; however, they usually require derating. Gas fuel leaves few deposits; therefore, gas fueled engines usually have dramatically extended life. For more information on gaseous fuel systems see Chapter 15.

9.8 IGNITION

Air-cooled spark-ignited engines usually have self-contained magneto ignition systems which are very reliable. Both breaker and breakerless types are available. The vulnerable component is the spark plug, which is subject to fouling and failure due to its high operating temperature. Many manufacturers will recommend changing spark plugs at 100 hour intervals. Many mechanics will automatically change the spark plug of a failed or erratically operating engine.

9.9 HORIZONTAL VERSUS VERTICAL SHAFT

Most generator sets use horizontal shaft engines. However, the flexibility of the air-cooled engine never ceases. Some creative engineers have applied vertical shaft engines, very popular with lawnmowers, to generator sets. See Figure 9-4. The result has been decreased cubic volume and increased portability.

9.10 CYLINDER CONFIGURATIONS

Multi-cylinder engines have many configurations. The common two cylinder air-cooled engines found on

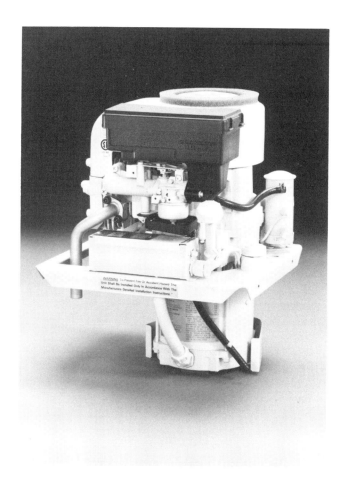

Figure 9-4. Vertical Shaft Set

generator sets use the 180° horizontal opposed configuration. See Figure 9-5. The cylinders are 180° apart with respect to the crankshaft. Most four cylinder engines use the 90° V configuration. Both configurations make the engine even firing and easy to balance. Even firing is important to generator sets to keep light flicker to a minimum. The horizontal opposed and V configurations are also reasonably simple to cool with a simple flywheel blower.

9.11 SUMMARY

Air-cooled engines are simple and remarkably reliable. Good generator set versions are available from a number of manufacturers. Air cooled engines require ample cooling air, lubricating oil maintenance, and clean fresh fuel. Emphasis on low cost and portability may necessitate compromises, using higher speed and limited life components. However, those compromises emphasize the flexibility of the air-cooled spark-ignited engine.

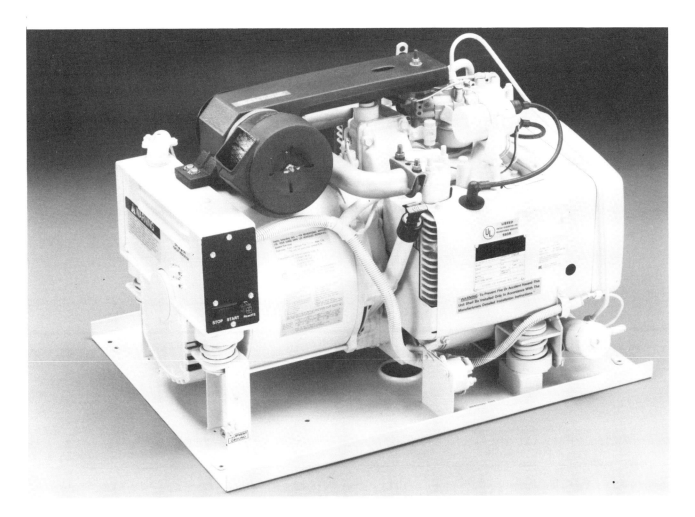

Figure 9-5. Two Cylinder Opposed Configuration

BIBLIOGRAPHY

Automotive Electric/Electronic Systems, Robert Bosch GMBH, Post Bach 50, D-7000, Stuttgart, West Germany.

Automotive Handbook, English Edition, 1976, Robert Bosch GMBH, Post Bach 50, D-7000, Stuttgart, West Germany.

Taylor, Charles Fayette. Internal Combustion Engine in Theory & Practice, Volumes 1&2, MIT Press, Cambridge, MA.

Technology Pertaining to Two-Stroke Cycle Spark-Ignited Engines, SAE PT-26, Society of Automotive Engineers, Inc., 400 Commonwealth Drive, Warrendale, PA 15096.

Liquid-Cooled Diesel Engines

Roman Gawlowski

CHAPTER **10**

INTRODUCTION

A diesel engine is a type of internal combustion engine in which the chemical energy of fuel is converted into mechanical energy. Diesel engines differ from other internal combustion engines in the method of ignition. Diesel engines use the heat of air compression to ignite the fuel. These prime movers are built in sizes ranging from a few horsepower (HP) to over 20,000 HP.

This efficient and self-contained source of power is quite versatile in its application. Diesel engines are used for electrical power in a wide variety of commercial and industrial applications. New developments have reduced the size and weight of diesel engines while increasing their power, reliability, and fuel economy. The smaller size, increased reliability, and greater efficiency of modern liquid-cooled diesel engines have dramatically increased the applications where diesel engines can be safely and profitably be used. Extensive development efforts by all engine manufacturers over many years have produced remarkable results, from Rudolph Diesel's single cylinder engine which burned coal dust, to modern multi-cylinder power plants, which burn a variety of liquid or gaseous fuels with remarkable efficiency.

The diesel, like other internal combustion engines, is basically a heat engine. Its operation consists of converting the heat energy produced by burning diesel fuel into mechanical energy, or work. The ratio of the heat equivalent of work done by an engine to the total heat supplied from the combustion of fuel is referred to as its **thermal efficiency**.

Early, naturally aspirated, engines offered a thermal efficiency of approximately 30–35%. Today's modern high-performance diesel engines, equipped with exhaust gas turbo-chargers and intercoolers, have a thermal efficiency of 40% and above. As an example, Figure 10-1 shows a 16-cylinder V configuration turbocharged and intercooled diesel engine. Diesel engines can be designed to operate on either the two-stroke or the four-stroke cycle of events in each cylinder. They are commonly referred to as two-cycle or four-cycle engines.

Most diesel engines are available in three different basic configurations:

a. Naturally aspirated

b. Turbocharged

c. Turbocharged and intercooled

A naturally aspirated engine is one that introduces the air into the cylinder at or near normal ambient pressure. A turbocharged engine uses engine exhaust to drive a turbine air compressor to increase the pressure at which air is introduced into the cylinder. While the process of turbocharging results in a substantial increase in the efficiency with which air is moved through the thermal cycle, the temperature of the air is also increased with resulting increases in combustion temperatures. The air can be cooled by using an intercooler to strip away some of the heat in the incoming air.

The intercooler, sometimes also called an aftercooler, is essentially a small radiator core that uses engine coolant to cool the air coming out of the turbocharger from temperatures in excess of 300°F down closer to the engine's coolant temperature of around 200°F. This process is also called Jacket Water Aftercooling (JWAC). Because turbocharging and then cooling the air greatly increases the efficiency of the diesel combustion process, the turbocharged and intercooled diesel engine has become commonplace. The intent of this chapter is to review the major components and characteristics of these various engine versions and to outline some installation considerations.

Liquid-Cooled Diesel Engines

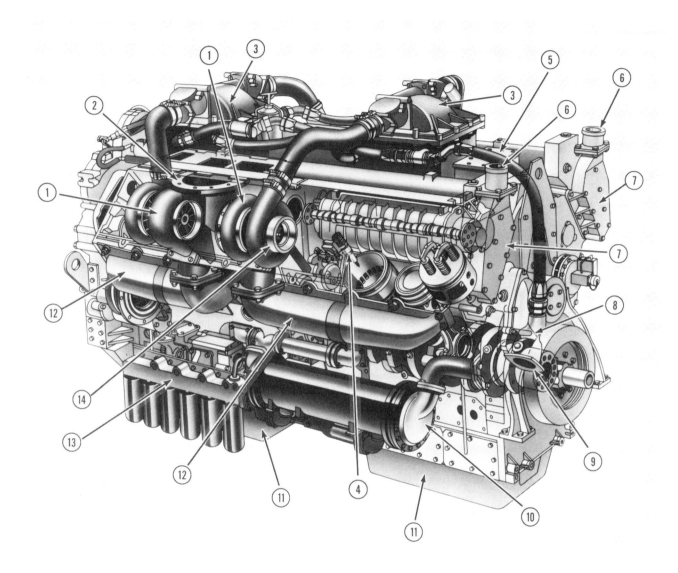

1. Turbochargers
2. Engine Exhaust Outlet
3. Intercoolers
4. Fuel Injector
5. ECM (Electronic Control Module)
6. Coolant Outlets
7. Thermostat Housings
8. Engine Coolant Pump
9. Coolant Pump Inlet
10. Lubricating Oil Cooler
11. Lubricating Oil Sump
12. Exhaust Manifolds
13. Lubricating Oil Filter Bank
14. Turbocharger Air Inlet

Figure 10-1. Modern Multi-Cylinder Turbocharged, Intercooled, Diesel Engine

10.1 DIESEL PRINCIPLE

A diesel engine is an internal combustion engine that uses the heat of compression for ignition. When air is compressed, air molecules collide more often, causing a rise in air temperature. (This phenomenon is similar to pumping a tire with a hand pump. Pushing on the pump to squeeze the air results in the barrel of the pump becoming hot to the touch.) As the piston compresses the air in the combustion chamber, the air is heated to above the combustion temperature of the fuel. The fuel is then injected into the chamber and ignites upon mixing with the heated air. In order to achieve air temperatures high enough to ignite the fuel, the compression ratio, or the ratio of air volume at the bottom of the piston stroke to the volume at the top of the piston stroke, is much higher for a diesel engine than for a spark ignited engine. In a diesel engine this ratio varies between 13:1 and 22:1, depending on engine design.

The thermal efficiency of any internal combustion engine is largely related to the compression ratio. An engine with a higher compression ratio will, on the downward stroke, expand the heated combustion gasses further, thereby converting a greater portion of the combustion heat into mechanical energy. This higher fuel efficiency translates into fuel economy 40% better than a spark ignited engine at full load, and as much as 100% better at idle.

10.2 TWO- AND FOUR-STROKE DESIGNS

There are two types of heavy duty diesel engines, two-stroke and four-stroke. Both designs have been available for many years, and both are proven and well accepted in the power generation industry.

All diesel engines have a combustion cycle of intake, compression, combustion, exhaust. The four-stroke design accomplishes the four parts of the combustion cycle with four separate strokes of the piston. The two-stroke design allows both exhaust and intake to occur at the bottom of piston travel, then completes the cycle with compression and combustion at the top of the piston travel. These differences are examined in some detail below.

10.2.1 Four-stroke cycle. The first part of the cycle is the intake of fresh air. See Figure 10-2. The intake valve opens as the piston passes the top of its travel. The piston then descends, and acting as a positive displacement pump, draws air into the cylinder through the open intake valve. This is the intake stroke. As the piston passes the bottom of its travel, the intake valve closes and the piston compresses the air as it ascends for the second or compression stroke.

At the top of the stroke, fuel is injected into the cylinder, mixes with the heated air and ignites. The expanding combustion gasses drive the piston downward. This is called the power stroke. Near the end of the power stroke, the exhaust valve opens. As the piston ascends, it once again acts as a positive displacement pump, pushing the burned gasses out of the cylinder. As the piston nears the top, the exhaust valve closes, completing the exhaust stroke. The intake valve then opens, repeating the cycle. The four-stroke power cycle is completed in two revolutions of the crankshaft.

10.2.2 Two-stroke cycle. The two-stroke cycle begins with the piston at the bottom of its travel. See Figure 10-3. At this point, the piston has uncovered the intake ports. Fresh air is blown into the cylinder under pressure from the blower. The incoming air forces any remaining exhaust gasses through the open exhaust valves. As the piston rises, it closes off the ports, beginning the compression stroke, and the exhaust valves close. The air is then compressed as the piston rises to the top of its travel. At the top of the stroke, fuel is injected into the cylinder,

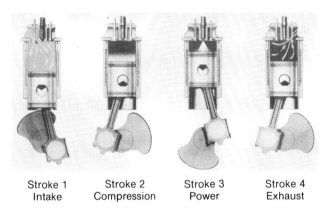

| Stroke 1 | Stroke 2 | Stroke 3 | Stroke 4 |
| Intake | Compression | Power | Exhaust |

Figure 10-2. Four-Stroke Cycle

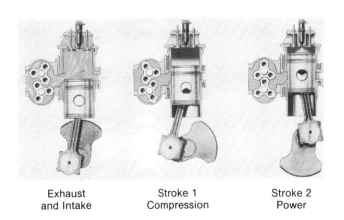

| Exhaust and Intake | Stroke 1 Compression | Stroke 2 Power |

Figure 10-3. Two-Stroke Cycle

mixes with the heated air and ignites. The expanding combustion gasses drive the piston downward, completing the power stroke. With intake and exhaust occurring at the bottom of piston travel, compression and power are required to complete the cycle. Only one revolution is required for each power cycle.

10.3 COMBUSTION CHAMBER DESIGN

The two primary types of combustion chamber systems used in modern diesel engines are the open chamber and pre-combustion chamber.

10.3.1 Direct injection. Open chamber, or direct injection systems, inject the fuel directly into the combustion chamber. See Figure 10-4. The atomized fuel is uniformly distributed and exposed to the air across the piston crown. Benefits include good air-fuel mix, smooth and complete combustion, larger valve areas, low thermal losses, and fuel economy.

Direct injection is the most common type used in heavy duty industrial diesels.

10.3.2 Pre-combustion chamber. Pre-combustion chamber systems have indirect fuel injection, where a coarse stream of fuel is forced through a single hole pintle nozzle into a pre-combustion chamber located in the cylinder head. Contacting the hot air, it partially burns, building up pressure. This pressure forces the burning mixture into the cylinder at high velocity through a multi-hole burner. See Figure 10-5.

This system is less efficient, requires timing devices, and has a high heat loss into the cooling system surrounding the pre-combustion chamber. It can have better exhaust emission characteristics and lower peak cylinder pressures. Pre-combustion chamber designs are primarily used in light duty industrial and passenger car diesel engines.

10.4 DIESEL FUEL INJECTION SYSTEM

The fuel injection system is the heart of the diesel engine. Operation of a diesel engine requires that the injection system perform the following functions:

a. Meter the quantity of fuel required by the engine and maintain this quantity constant from (1) cycle-to-cycle of operation and (2) from cylinder-to-cylinder.

b. Inject fuel at the correct point in the cycle at all engine speeds and loads.

c. Begin and end injection very quickly.

d. Inject fuel at the rate necessary to control combustion and rate of pressure rise in the combustion chamber.

e. Atomize fuel as required for efficient combustion.

f. Distribute fuel evenly throughout the combustion chamber.

Fuel under high pressure is forced into the combustion chamber through a spray nozzle; this is referred to as **mechanical injection**. The most common types utilized in present day production engines are:

a. *Multiple plunger pump system.* The multiple pump system uses individual plungers for metering and inject-

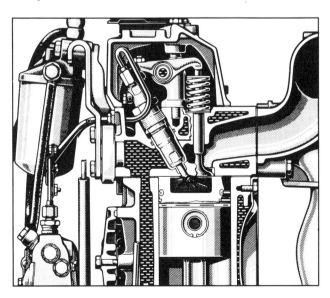

Figure 10-4. Direct Injection System

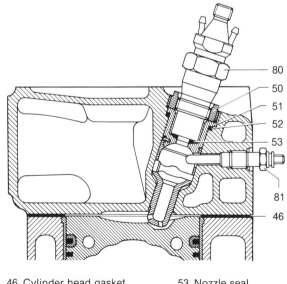

46 Cylinder head gasket
50 Threaded ring
51 Prechamber
52 Seal ring
53 Nozzle seal
80 Injection nozzle
81 Pin-type glow plug

Figure 10-5. Pre-combustion Chamber in Cylinder Head

ing fuel into each cylinder. These pumps are mounted on the side of the engine and are connected to the injectors by long, high-pressure fuel lines. Metering is controlled by a throttle acting through a fuel rack that rotates individual plunger pumps with metering helixes. Throttle and mechanical type governors move the fuel rack, changing position of the rack and metering helix on each plunger.

b. *Pressure-time (PT) system.* The PT fuel system uses injectors that meter and inject the fuel. Metering is based on a pressure-time principle. The pressure at the injector is supplied by a low-pressure fuel pump and the time for metering is determined by the interval that the metering orifice in the injector remains open. This interval is established by the engine's rotative speed, which determines the rate of motion of the camshaft controlled injector plunger. The downward movement of the injector plunger forces the metered fuel charge into the cylinder. Three components in the PT fuel pump control the pressure of the fuel at the injector: a gear pump, a mechanical governor with built-in pressure control, and a restriction type throttle.

c. *Unit injector system.* The unit injector system places both the metering and injection functions in the injector. The principle of metering with a plunger pump, helix, and fuel rack is similar to the multiple pump system, but takes place in the individual injector. Rotation of the plunger by the fuel rack changes the helix position and controls metering, while the force to apply pressure to the fuel charge comes from the engine's cam shaft. The cam acts on the plunger pump, forcing fuel past the spray tip valve and into the combustion chamber.

d. *Electronic unit injector.* The electronic unit injector is a modern development of the unit injector. As in the case of the mechanical unit injector, the cam provides the required force for injection, but rather than a complicated system of valving with a metering helix cut in the plunger, an electrically operated solenoid valve is used for metering the fuel. This solenoid valve is controlled by a microprocessor-based control called the **electronic control module** (ECM). A major advantage of electronically controlled fuel injection is the ability to control the timing as well as volume of the injected fuel charge.

e. *Distributor or rotary pump system.* The distributor or rotating type pump system uses a single plunger metering pump to meter the fuel for all cylinders. The governor and throttle control the length of pump travel and the amount of fuel delivered on each stroke. Distribution to the cylinders is controlled by a rotating disc or shaft with holes located to synchronize the metering pump delivery stroke with the various cylinders. Present distributor systems have high-pressure lines to supply fuel to the injectors. Injectors for high-pressure systems are similar to those used with multiple pump systems.

10.5 ENGINE PERFORMANCE RATINGS

Engine performance is given by the use of standard terms that define the extent of load to be applied to the engine. As with any mechanical device, the heavier the load factor, the shorter the expected service life. If an engine is to be operated as the main source of power, on a 24-hour basis, a prime power rating is used in order to extend engine life to something in the order of 10,000 to 15,000 hours. If, on the other hand the engine is to power an emergency set that will only be used when the main utility power fails, it can be presumed that an operating time of less than 250 to 300 hours per year will be accumulated. It follows then, that a set sized for prime or continuous power will give a life of 40–60 years if applied in a standby application. This then, is obviously not an efficient use of resources. For reasons of economy, it is very important to properly size the generator set to its intended application.

Typical engine ratings are defined by EGSA 101P-1988 "Engine Driven Generator Sets Performance Standard:"

Emergency standby rating is the power that the generator set drive engine will deliver continuously under normal varying load factors for the duration of a power outage.

Prime power rating is the power that the generator set drive engine will deliver when the unit is used as a utility-type power plant under normal varying load factors, operating continuously as required. This rating shall incorporate a minimum overload capability of 10%.

Industrial rating is the power that the generator set drive engine will deliver 24 hours per day when the unit is used as a utility-type power plant where there are nonvarying load factors and/or constant dedicated loads.

The key difference in the above ratings are the words "varying load factors" and "dedicated loads." In both the emergency standby and the prime power ratings, it is assumed that the loads will not be constant, but like a community utility load will peak at certain times of the day, then diminish to a lower level as demand drops. If it is intended that the generator set have a constant load, as for example the source of power for a pumping station or for a shopping center's large bank of parking lot lights, where the load will be constant while the engine is operating, the manufacturer's industrial rating should be used to assure satisfactory life from the equipment.

Figure 10-6 shows typical performance and fuel consumption curves of a turbocharged, industrial diesel engine designed to be a prime mover for an electric

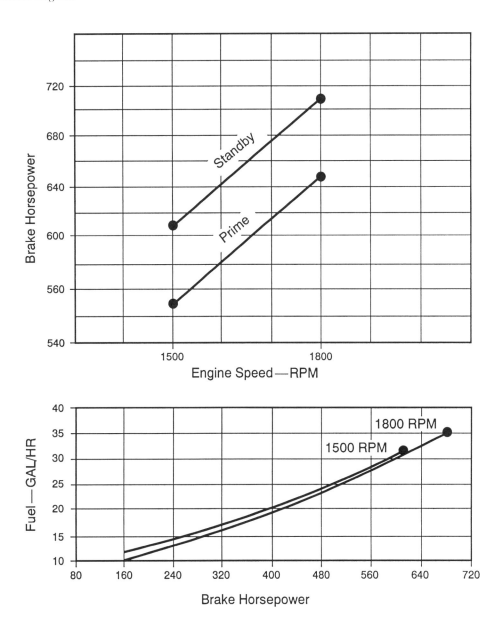

Rated power output shown represents engine performance capabilities at ambient conditions equivalent to ISO'3046, BS 5514; 100 KPA total baro press, 25°C air inlet, 30% relative humidity.

Curves also represent capabilities at the following ambient conditions: DIN 6270; 736 torr baro press, 20° air inlet, 60% relative humidity JIS D1005-1976; 760 MM. HG. baro press, 20°C air inlet, 11.4 MM, HG. vapor press.

Indicated performance is based on minimum intake and exhaust restrictions.

Fuel consumption data is based on diesel fuel no. 2 with a fuel weight of 7.11 lb./U.S. gal. (.85 kg/Litre). Fuel heating value is 18370 BTU/lb. (10210 CAL./GM).

Values are derived from currently available data and subject to change without notice.

Conversion factors: Power: kW = bhp × 0.746
Fuel: L/hr = gal/hr × 3.785

Figure 10-6. Engine Performance Curve

power generator set. Because this engine is designed to be used specifically for electric power generation, engine output and fuel consumption is given at 1500 RPM for 50 Hz, and 1800 RPM for 60 Hz requirements. As noted on this particular curve, the ratings given are in compliance with engine test specification ISO 3046.

Engines are also available for power generation for the support of military and aircraft power requirements for direct current and 400 Hz. These are usually considered special applications and are covered in special performance curves issued by the manufacturer.

In North America, engine power curves are generally drawn to comply with engine test standard SAE J1349. The SAE and ISO standards are very similar. In fact, the main difference is the power rating specified on the curve. ISO 3046 requires that a minimum power rating be shown on the curve, while SAE J1349 requires a nominal rating, which is then subject to a production tolerance of ±5%. Most manufacturers issue both SAE and ISO curves for their engines, with the standby, or maximum power rating for the ISO 3046 curve listed as 5% less than the SAE J1349 rating, in order to allow for production tolerance.

10.6 COOLING SYSTEM

A modern, efficient, diesel engine converts approximately 40% of the heat produced by fuel combustion into useful work. Another 30% of this heat is carried away with the exhaust gasses or is radiated from hot engine surfaces. This leaves 30% of combustion heat as waste heat that must be removed from the engine to prevent damage to internal engine components. This heat is carried away by circulating a liquid coolant through the passages of the engine block and cylinder head. This hot coolant emerges from the engine, is cooled, and then recirculated through the engine. The most common types of coolers are coolant-to-air (radiators) and coolant-to-water (heat exchangers).

The most common arrangement is a set-mounted radiator. In this arrangement, air is pulled from the rear of the set, passes over and through the generator, then over the engine, finally being pushed by an engine driven fan through the radiator. In some installations where it is not practical to pass large volumes of air through the engine room, the radiator may be mounted remotely and driven by an electric fan. In other cases, where there is a continuously available supply of cool water, a liquid-to-liquid heat exchanger may be used instead of the radiator. In this case, the engine coolant circulates in a closed loop consisting of the engine and the coolant side of the heat exchanger, while the "raw water" circulates in an open loop on the other side of the heat exchanger.

At this point we should define our terms. **Coolant** is a term used for a liquid medium used to carry heat away from the hot components of an engine. Coolant consists of a mixture of water and some substance intended to lower the freezing point of the water. Ethylene glycol is the "antifreeze" commonly used for this purpose. In addition, a number of chemicals are included to inhibit the accumulation of rust and scale. **Raw water** is any cool, clean liquid available to be pumped through the heat exchanger. Raw water can be sea water; water from a river, lake, or pond; or even city water from the local utility.

While the liquid cooled engine lends itself to a variety of cooling arrangements, it should be noted that the set-mounted radiator cooling system has the advantage of being self-contained. A utility water supply for cooling water is not satisfactory for an emergency generator set. The National Electrical Code, Section 700-12(b)(3), prohibits such use for emergency sets unless there is an on-site alternate supply with automatic changeover.

10.6.1 Set-mounted radiator. Set-mounted radiators are available in two types. The first type uses a fan mounted on the engine. The fan is belt driven from the crankshaft pulley. Drive belt tension is generally accomplished by adjusting the position of the fan support bracket. Because this also moves the fan itself, the fan shroud must have an elongated opening to allow for adjustment. The second type uses a fan mounted to the radiator. A radiator-mounted fan may be driven by belts from the engine crankshaft or directly driven by an electric motor. Since the radiator-mounted fan has a fixed position, the fan shroud opening can be optimized for minimum clearance. This reduced fan tip clearance results in higher efficiency and lower fan noise. However, because of the elaborate arrangement, such as adjustable idler pulleys or mountings for the electric motor, the radiator-mounted fan is usually more expensive than an engine-mounted one.

The air flow over the set and through the radiator is quite critical. The manufacturer's design requirements must be followed closely. The discharge air should flow directly outdoors through a duct that connects the radiator to an opening in an outside wall. The engine should be located as close as possible to the outside wall in order to reduce the restriction to air flow caused by the ductwork. The total restriction to air flow downstream of the radiator should not exceed 0.3 inches of water. If the application requires particularly long ducting, it may be necessary to use a remote radiator.

10.6.2 Remote radiator. A remote radiator with an electric motor driven fan can be installed in any convenient location away from the electric set. The fan may be driven by a thermostatically controlled switch and will draw electric power from the set only when needed to cool the engine.

When selecting a location for the remote radiator, follow the engine manufacturer's recommendations for the maximum height above the set as well as the maximum restriction to coolant flow. If these recommendations cannot be met, the remote radiator must be isolated from the engine cooling system by means of a heat exchanger or a hot well. With either system, the engine coolant pump circulates the coolant on the engine side, while an auxiliary pump circulates the coolant from the hot well or heat exchanger to the remote radiator.

Whether engine mounted or remote, an important advantage of a radiator cooling system (as opposed to raw water cooling) is that it is self-contained. With raw water cooling, if a storm or accident disrupts the utility power source, it could also disrupt the water supply and disable any electric set whose supply of water depended upon a utility.

10.6.3 Cooling air requirement. A radiator system mounted on the generator set is the most popular solution to the problem of engine-generator cooling and has few disadvantages. In its favor is that it is entirely self-contained. Nevertheless, it requires relatively large volumes of cooling air to be transported to and away from the unit with minimum restrictions.

For optimum fan air flow and cooling system efficiency, a shroud should be installed between the engine and the radiator. The pusher type fan location should be flush with the impeller ring and one-third of the fan width protruding into the tapered section of the shroud. The clearance between shroud and fan tip should be minimal. However, unless the shroud is mounted on the engine, the movement of the engine and fan may place a lower limit on permissible clearance. Figure 10-7 illustrates the importance of a controlled flow of air providing the most efficient cooling air movement. An average specific air volume requirement may be assumed to be:

$$V = 50 \text{ cfm per BHP} \qquad \text{(Equation 10-1)}$$

10.6.4 Cooling system design. The engine performance curve data provides us with all of the information required to select cooling components for the electric set prime mover.

Most engine manufacturers specify the air-to-boil (ATB) number. This is the main parameter in use today to characterize cooling system performance. The definition of the ATB number is:

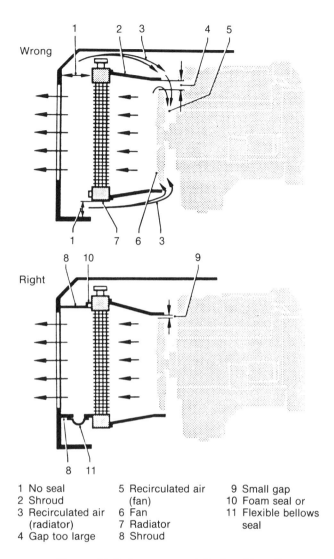

1 No seal
2 Shroud
3 Recirculated air (radiator)
4 Gap too large
5 Recirculated air (fan)
6 Fan
7 Radiator
8 Shroud
9 Small gap
10 Foam seal or
11 Flexible bellows seal

Figure 10-7. Arrangement Pusher-Type Fan and Air Flow Through Radiator

$$\text{ATB} = (212°F - \text{Radiator Tank Temperature}) + \text{Ambient Temperature} \qquad \text{(Equation 10-2)}$$

Also specified will be the quantity of heat that must be carried away by the coolant and the coolant flow capacity of the engine driven coolant pump. See Figure 10-8. Air flow for the engine driven fan will also be specified.

Given these figures, radiator cooling capacity can be determined. The capacity is generally based on a rise in coolant temperature of 8–12°F through the engine and the fact that most modern diesels operate most efficiently at engine outlet temperatures between 185–230°F.

10.6.5 Temperature control. Because these engines operate best within a rather narrow range of temperatures and the cooling system must be sized to accommodate the heat rejected to the coolant at maximum engine

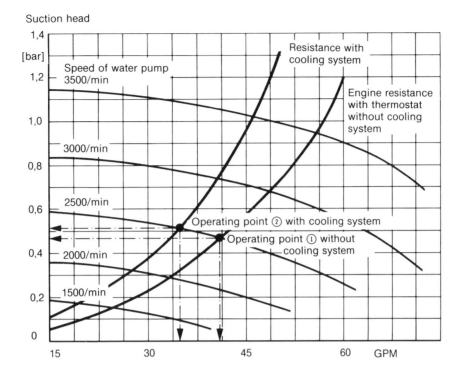

Delivery volume and suction head of the water pump

without cooling system:
The operating point is the point at which the line indicating the water pump speed and the line indicating the engine resistance intersect.
The delivery volume of the water pump can be read off vertically below the operating point ①.
The suction head can be read off the scale on the left-hand side.

with cooling system:
Add resistance of cooling system to engine resistance.
The operating point is the point of intersection of the line indicating the water pump speed and the line indicating the resistance of the cooling system added to the engine resistance.
Read off delivery volume and suction head below and on the left-hand side of the operating point ② as described above.

Figure 10-8. Water Pump Flow Rate

output, a means must be provided to maintain the coolant within this narrow operating band while the engine is starting to heat up, as well as when operating at less than maximum load. This is accomplished by engine mounted thermostats. The thermostats are used to restrict coolant flow to the radiator until the coolant within the engine reaches a preset operating temperature. At this point, the thermostats begin to open, allowing coolant to flow through the radiator, or heat exchanger, where excess engine heat is removed.

10.6.6 Quantity of coolant. The actual amount of coolant required to remove heat passing through cylinder walls depends upon horsepower capacity of the engine, design of the engine, and allowable rise in coolant temperature. As a rule, it is advisable to install sufficient coolant capacity to handle 2300–2800 BTU/HP/HR for normally aspirated engines and 1800–2300 BTU/HP/HR for turbocharged engines. The lower values apply to 4-cycle, open combustion chamber, direct injection engine types. The higher values should be used for pre-combustion chamber engines since their thermal efficiency in general is lower than that of open chamber types due to the greater heat losses from the larger pre-combustion chamber surface. Most engine builders provide values of heat rejection to jacket water, oil cooler, water cooled intercoolers, exhaust manifolds, and exhaust gas turbochargers. They should be used to determine the coolant capacity required. The quantity of coolant in gallons per minute to be circulated in a closed cooling system can be calculated using the following equation:

$$\text{GPM} = \frac{\text{BHP} \times \text{BTU/BHP/HR}}{(t_1 - t_2) \times 500} \quad \text{(Equation 10-3)}$$

Where:

$\quad$ BHP = brake horsepower
BTU/BHP/HR = specific heat rejection rate
$\quad\quad t_1$ = outlet jacket — coolant temperature, °F
$\quad\quad t_2$ = inlet jacket — coolant temperature, °F

Example: Diesel engine, 4-cycle, direct injection system, turbocharged, maximum output = 350 BHP, dry-type turbocharger and exhaust manifold, no intercooler, specific heat rejection rate = 1800 BTU/HP/HR

t_1 = 195°F; t_2 = 185°F

$$\text{Quantity of water} = \frac{350 \times 1800}{10 \times 500} = 126 \text{ GPM}$$

10.6.7 Heat rejection to coolant. The total heat rejection rate to coolant of a diesel engine must be known in order to estimate the size of the radiator or heat exchanger. For general estimates, it can be assumed that the heat rejection to the cooling water will be about 30–50 BTU/BHP/min for conventional diesel engines. Hypothetical figures for engine heat rejection are shown in Figure 10-9.

Basically, few engines use much over 40% of the fuel efficiently and many use even less. The remaining 60% appears as heat in the engine jacket water, heat in the exhaust gas, and radiated heat. Usually engine manufacturers can provide the heat balance of an engine that represents an overall picture of the heat distribution for normal applications. Figure 10-10 shows the data obtained from a typical heat rejection test in which engine load and speed have been varied. Note, that the specific heat rejection (BTU/min/BHP) does not remain constant as engine load is varied. Specific heat rejection tends to increase as the load is decreased.

Actual values can differ when additional heat must also be removed from components like water-cooled exhaust manifolds, turbochargers, and intercoolers. In certain stationary installations, such as cogeneration systems, waste heat can be recovered for heating purposes. In those special applications, close engineering cooperation between engine manufacturer and power plant builder is necessary to design an efficient and technically sound diesel generator heat recovery module.

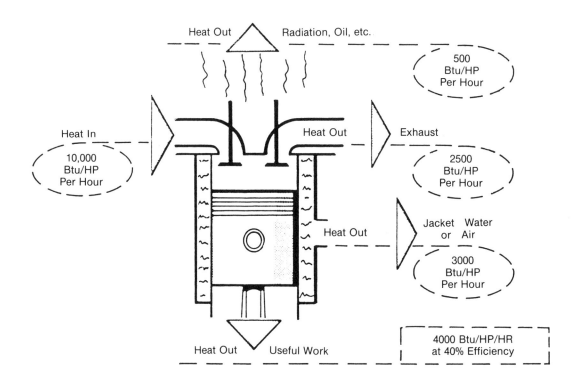

Figure 10-9. Hypothetical Figures for Engine Heat Rejection

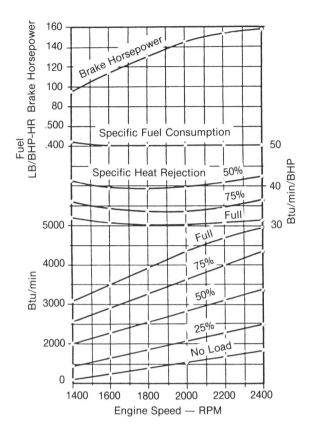

Figure 10-10. Jacket Water Heat Rejection at Varied Load and Speed

10.7 SPECIAL APPLICATION: COGENERATION

One distinct advantage of the liquid cooled engine is the fact that it can very easily be adapted to allow for the recovery of much of the heat that is normally wasted, thereby greatly improving the efficiency of the overall unit.

A typical cogeneration system uses a diesel engine to generate electricity. The waste heat produced by that process is captured and utilized for water heating, space heating, or cooling.

Cogeneration also is referred to as **total energy**. Such plants are built on-site and can operate independently of the utility power grid or in tandem so the cogenerator can buy power from the utility or sell it back. With such a system about 80–85% of the energy is recoverable. By comparison, a utility power plant is about 35% efficient. Prime cogeneration candidates include hotels, hospitals, universities, computer centers, office complexes, and food processing and printing plants. The cogeneration unit produces electricity for lighting and equipment operation, while producing thermal energy for water heating, space heating, and air conditioning from the waste heat and steam. The trend to install cogeneration power systems began during the late 1970s because of the nationwide energy crisis and by 1982 about 8% of the energy consumed by American Industry was produced through cogeneration. The figure is expected to rise to 15% by the year 2000.

Figure 10-11 outlines a cogeneration flow diagram of a typical prepackaged module. Engine builders offer engi-

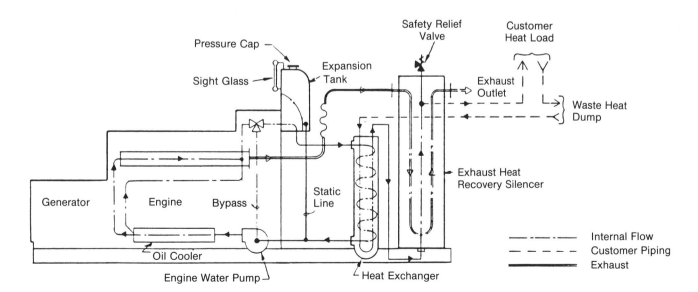

Figure 10-11. Cogeneration Flow Diagram of a Typical Prepackaged Module

neering services in the form of computerized feasibility studies to determine the proper package match with system loads.

REFERENCES

[1] EGSA 101P-1988, Engine Driven Generator Sets Performance Standard, Electrical Generating Systems Association.

[2] EGSA 101S-1988, Engine Driven Generator Sets Guideline Specifications for Emergency or Standby, Electrical Generating Systems Association.

[3] Engineering Bulletin No. 50, "Cooling System Guidelines for Radiator Cooled Applications," Detroit Diesel Corporation.

Air-Cooled Diesel Engines

Sid Bishop

CHAPTER 11

INTRODUCTION

Air-cooled diesel engines (Figure 11-1) have been used for many decades where total reliability, longevity, economy, and versatility have been required.

Air-cooled diesel power is available from as low as 2 HP single cylinder engines to over 600 HP turbocharged V12 engines, and all sizes are used in generator applications such as:

- Construction
- Rental
- Irrigation
- Agriculture
- Refrigeration
- Base power
- Standby power
- Telecommunications
- Military
- Oil field
- Recreational vehicles
- Marine

The construction of the majority of air-cooled diesel engines lends itself to simplicity in manufacture. Air-cooled diesels have higher efficiency than most other forms of reciprocating and rotary engines, will endure environmental conditions from arctic cold to desert heat, and use a cooling medium that is always available.

11.1 ADVANTAGES

a. *Simplistic in design.* An air-cooled diesel requires no radiator, water pump, or hoses. There is no coolant to leak out, freeze, become contaminated, or cause erosion within the cooling system. Therefore, it requires less maintenance, is less prone to catastrophic failure, and reduces downtime and costly repairs to a minimum.

b. *Diesel fuel.* Reduces the amount of fuel that has to be handled. The fuel is safe to handle and usually, where other major equipment is used, is the main fuel used on-site.

c. *Flexibility.* Because of its simplicity and reliability, the air-cooled diesel generator set can be used for powering diverse applications. Navigational light houses, a critical long run application, frequently use air-cooled diesels. Disaster crews working in flood, fire, or earth-

Figure 11-1. Two-Cylinder Air-cooled Diesel

Air-Cooled Diesel Engines

quake areas use lightweight portable air-cooled diesel generators. Air-cooled diesel power is less prone to failure where catastrophies such as fire, earthquake, or flood occur, because of its stand-alone capability. Similarly, the air-cooled diesel generator set (Figure 11-2) has proven to be more reliable under battle conditions.

d. *Operations.* Air-cooled diesel engines have rapid warm up and higher operating temperatures, which reduce cylinder bore, and piston and piston ring wear. The high temperature reduces sludging of the lubricating oil, especially in short run applications. Further, because of the faster warm up period, the air-cooled engine will accept its full load capability faster than liquid-cooled engines.

e. *Installation.* Air-cooled diesel power plants require very little different treatment than their liquid-cooled counterparts, but are less prone to vibration failures. Both require clean, cool air for combustion and cooling, good ventilation, proper exhaust piping, adequate mounting, and guards applied over hot and rotating parts. The air-cooled diesel requires less than half the cooling air of its water-cooled equivalent.

11.2 DISADVANTAGES

a. *Noise.* The noise level may be somewhat higher with air-cooled diesel engines, due to the elimination of the secondary portion of the cooling system — the water jacket used by liquid-cooled engines. Treatment of the sheet metal work by some air-cooled engine manufacturers helps to offset this disadvantage.

b. *Heat recovery.* Due to the difficulties in capturing the waste heat from air-cooled engines, they are less likely to be considered for a heat recovery system.

c. *Cooling.* The enclosures, the inlet cooling air, and the hot air outlet ducting often have less flexibility in the cooling system for reversing the flow of air. Liquid-cooled engines usually have the availability of push or pull radiator fans to suit the cooling arrangements (see 11.4).

Figure 11-2. 20 kW Generator Set

11.3 ALUMINUM VERSUS CAST IRON

Generally engines built with cast iron crankcases and cylinders are considered to be of the heavy duty type, whereas engines built from aluminum are for light duty applications. However, some engines have a satisfactory compromise in the use of both these metals, and can be considered of the heavy duty type. This latter type unit usually has an aluminum crankcase, with cast iron cylinder barrels and sleeved bearings to give the same integrity as the engine having a cast iron crankcase. With modern aluminum alloys and casting techniques, there is little distortion of the alloy crankcase.

Although aluminum crankcase engines have an obvious weight advantage, the modern iron foundry can now produce what is known as *thin wall* castings, which maintain the structural integrity of the crankcase but considerably lower the weight. Therefore, aluminum does not have the weight advantage it once enjoyed. See Figure 11-3.

11.4 COOLING

A high percentage of liquid-cooled engine failures occur in the cooling system. It is obvious there is considerably less expense in maintaining, repairing, and servicing the air-cooled engine, especially where the cooling fan is integral with the flywheel of the engine. The air-cooled system is designed for each model engine to serve in all climates without change to the system, whereas radiators on liquid cooled engines are often sized for the site conditions.

There are two basic arrangements for air-cooled engines. The first is by a fan attached to or integral with the flywheel. The fan fits into a volute, which produces air flow into a chamber at one side of the cylinders. The pressurized air then passes over the cylinder barrel and cylinder head fins, cooling them. The second is by a belt driven axial flow fan, which pushes air into a chamber at one side of the cylinders. The pressurized air then passes over the cylinder and cylinder head fins, cooling them. The flywheel fan cooling system is almost 100% reliable because there is no drive to fail. This type of cooling can be used for generator sets up to 25 kW. Air-cooled diesel engines are less prone to plugging of the cooling system by dust or debris than their counterpart, the liquid-cooled engine, due to the wider air passages. See Figure 11-4.

The heated cooling air discharged from the engine is approximately 100°F (56°C) over ambient with the engine running at full load. It is, therefore, necessary to ensure that the heated discharged air is not allowed to recirculate through the engine or into the combustion system or overheating may occur. This is especially important to consider when installing engines in confined spaces such as housings. It should be remembered that the engine will discharge at least as much heat in kW as the generator output it is driving and therefore, good ventilation is required.

Engine manufacturers have very defined recommendations regarding air flow for cooling, which should be followed strictly to ensure satisfactory conditions. See Figure 11-5 for an example of cooling duct recommendations.

It is essential that the intake for the air cleaner is positioned to receive the coolest air possible. Otherwise the engine can de-rate itself, accelerate heating, and possibly cause early failures.

A common mistake made with housings are the restrictions caused by louvers, screens, or filters. In all cases, the number of open square inches must be considered. With louvers, the air usually has to turn through 90°, which reduces air flow, so a greater amount of free air space should be provided. Screens or mesh should act as safety guards, and filters must be sized correctly and be easily replaceable. In preference to filters in extremely dusty conditions, barriers should be positioned to force the air to change direction and velocity, causing heavy particles of dust to drop into a collection point. Where generator sets are installed in housings, an auxiliary fan can be

Figure 11-3. Engine with Thin Wall Iron Crankcase

Figure 11-4. Axial Belt Driven Fan and Flywheel Fan Cooled Engines

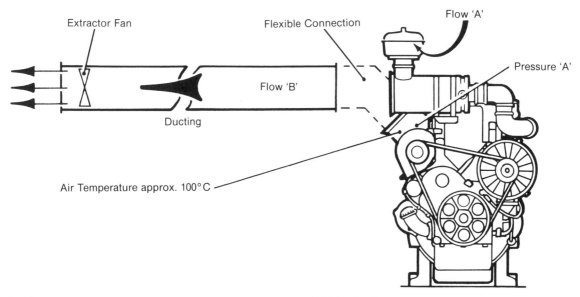

a Ducting Areas

The recommended minimum cross-sectional areas of air outlet ducting on air cooled engines are given in the table. These areas apply only to ducting lengths of up to 1.5m (5 ft.), and should a greater length be required the figure given must be multiplied by the following factors:

Ducting length 1.5 to 3.0m (5-10 ft.) — Multiply by 1.4

Ducting length 3.0 to 7.6m (10-25 ft.) — Multiply by 2.25

Ducting length 7.6 to 15.2m (15-50 ft.) — Multiply by 3.5

b Extractor Fan

If ducting lengths above 1.5m (5 ft.) are required, but it is not possible to install ducting of the correct cross-sectional area, an extractor fan having a capacity equal to, or slightly greater than flow 'B' must be fitted in the duct. The installation must be so arranged that the extractor fan is started at the same time as the engine, and continues to run until the engine is brought to rest. If the fan motor is driven from the engine, or the engine driven alternator in the case of generating sets, the power absorbed by the motor must be deducted from the rated output of the engine or the alternator.

If the ducting is restricted in any way, i.e., area restriction, bends, louvres, etc., these points must be taken into consideration when evaluating fan requirements.

Figure 11-5. Cooling Data

employed to assist ventilation and prevent radiated heat build up.

As a safeguard, engine driven generator sets should be equipped with a high temperature warning, and shutdown and belt breakage devices (where belt driven fans are used) to avoid catastrophic failure.

11.5 LUBRICATION

11.5.1 Lubricating system. Modern high speed air-cooled diesel engines are designed and built with a full flow oil filter system to prevent unwanted particles entering the critical components of the engine.

In most designs, the lubrication system (Figure 11-6) pressurizes the oil to the major bearings in the engine, using splash or mist to lubricate the piston, barrel, cams, and gears. To ensure that correct pressure is maintained, pressure relief valves are included in the system. The more durable engines have 250 hour oil change periods. Some air-cooled engines have self regulating pumps, which eliminates the pressure relief valve and reduces the power required to drive the pump, making the engine more efficient. See Figure 11-7.

Generally the manufacturer sizes oil filters for satisfactory operation and life. If the application requires additional filtration, the manufacturer should be consulted. Improper filter selection could void the warranty, or early failures could occur due to excessive pressure drop or similar problems.

11.5.2 Extended run lubricating oil system. The four most predominant extended run lubricating oil systems are:

a. *Oil make up*, as the engine uses it, from a reservoir controlled by a float device. The oil change period is usually restricted to that of the manufacturer's standard oil change period, because the contaminants remain in the crankcase and are recirculated.

b. *Oversize oil pans and oil reservoirs*. Again, this is limited due to contaminants being recirculated.

c. *Dry sump system*. The oil is carried in a reservoir dependent on length of run (sometimes up to six months, 24-hour/day operation). Oil is drawn from the reservoir by the engine's lube oil pump and returned via a drain line. The system is arranged to pick up oil at one end of the reservoir and return it at the opposite end. This causes the contaminants to fall out, thus preventing them from being recycled.

d. *Wet sump system*. Similar to the dry sump system, but with a small auxiliary pump that pulls the oil from the tank and passes it into the engine. The return line (pipe) to the tank is connected just below the normal high level mark so that oil is maintained at the near normal level mark in the oil pan. This system is used where the oil pump capacity creates a difficulty in returning the oil sufficiently fast to prevent over-fill of the engine.

Both systems c and d (Figure 11-8) are customized to suit the length of run between oil changes.

11.5.3 Lubricating oil. Lubricating oils must be of the correct SAE Grade and API (American Petroleum Institute) rating, as recommended by the engine manufacturer for temperature and service intervals. Ignoring the manufacturers' recommendations can cause:

a. Voiding the warranty

b. Early failure due to various damaging conditions

The lubricating oil selected may need to be matched to the site conditions, especially where heavy loads with high temperatures or light loads with low temperatures are encountered. (Consult the engine manufacturer's representative.)

11.6 FUEL SYSTEMS

The most popular fuel system (Figure 11-9) employs a fuel lift pump, a filter, a fuel injection pump, high pressure fuel lines, and high pressure injectors into the cylinders. These components are highly reliable under normal working conditions and will withstand a reasonable amount of abuse with regard to performance conditions or working load. Due to the very fine limits of manufacture, they cannot tolerate foreign matter in the form of fine abrasives or water. Fine abrasives cause rapid wear, and water will cause rust and possibly erode injector nozzles or blow the tips off, causing poor performance and expensive repair. It is recommended that the manufacturer's filtration system (which should have some form of water separation) be used to reduce the risk of water entering the fuel injection system.

Most fuel systems have a "leak off" line from the fuel injectors (sometimes fuel pumps); it is important that this line is of adequate size, kept clear, and preferably returned to the fuel tank on generator sets. If the return line is connected into the filter or main fuel line, it is possible for gases to enter the fuel system, causing erratic running or intermittent poor governing.

Experience has shown that fuel filters/separators should be on the pressure side of the fuel transfer pump to prevent air from entering the fuel system, which can cause erratic running, poor governing, or even total shutdown.

Diesel fuel can deteriorate when stored over an extended period of time without use or replenishment, and should

Air-Cooled Diesel Engines

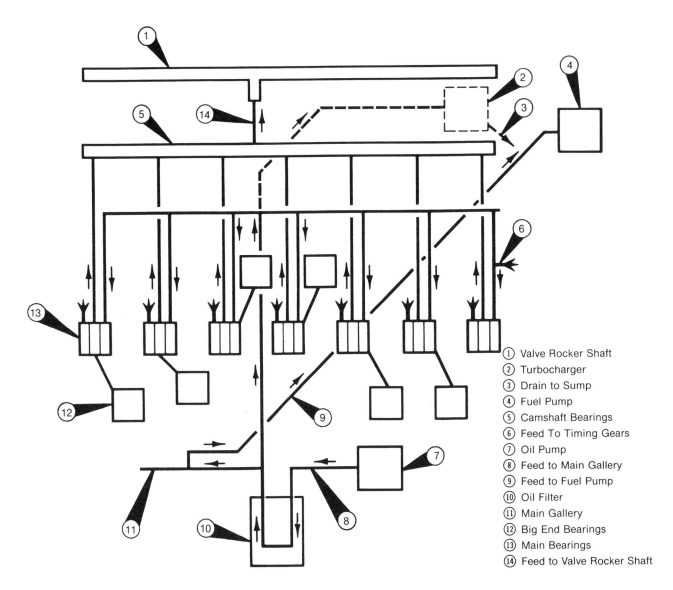

Figure 11-6. Lubricating Oil System

occasionally be checked for water and fungi. Standby systems are the most likely applications where the fuel will be exposed to these conditions.

The fuel transfer pump, when used, must maintain adequate pressure and flow, to satisfy the total performance of the engine.

11.7 AIR INTAKE SYSTEM

If the manufacturer's standard product is not used, the system must be sized per the manufacturer's instructions.

The air cleaner should be suitable for the climate conditions under which it is expected to work. Extremely dusty conditions may require a two- or three-stage air cleaner with safety element, whereas marine applications may only require a simple single-stage cleaner. Generally, paper element cleaners are part of modern air-cooled diesel generator sets, but remote situations may require an oil bath cleaner because replacement elements are not easily available. The cleaner should always be positioned for easy maintenance and to induce the coolest air possible, even if it means ducting the air to the cleaner.

11.8 EXHAUST SYSTEM

To ensure the best performance from the engine generator set, the engine manufacturer's recommendations should be followed.

To prevent heat recirculation, exhaust runs within a housing should be kept as short as possible and lagged with an insulating material. Materials containing asbestos should be avoided.

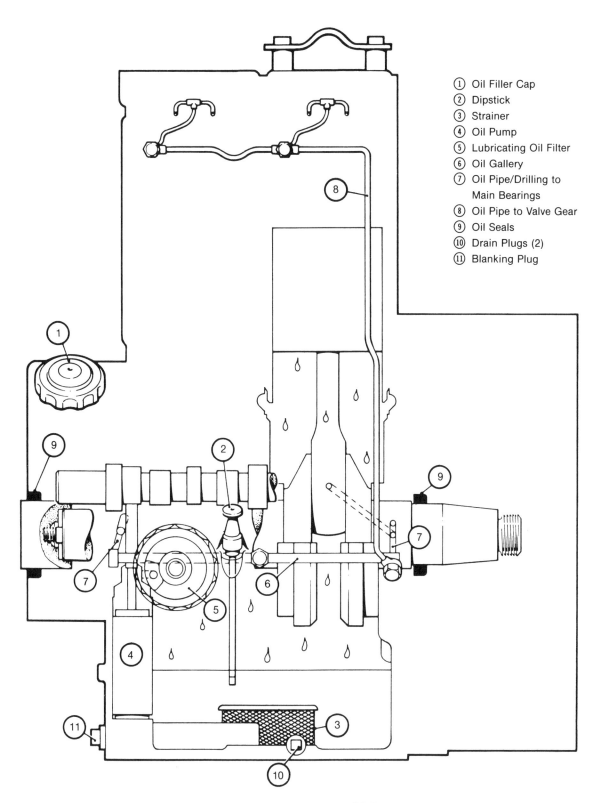

Figure 11-7. Self Regulating Oil System

Air-Cooled Diesel Engines

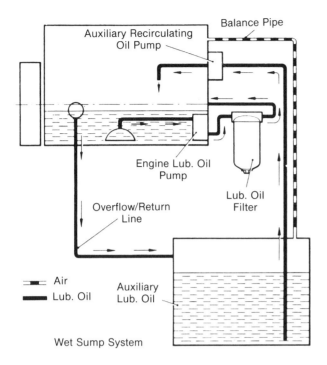

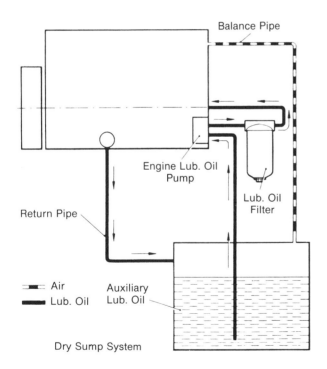

Figure 11-8. Long Run Oil Systems

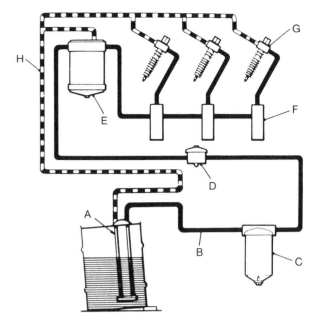

A Fuel System Using Fuel Drums.:

A - Fuel Probe
B - Low Pressure Fuel Supply
C - Remote Mounted Filter
D - Lift Pump
E - Engine Mounted Filter
F - Fuel Injector Pumps
G - Injectors
H - Injector Leak Off

Figure 11-9. Fuel System

Ensure that the exhaust system is well supported and that there is a flexible connection at the exhaust manifold, to avoid manifold breakage and transmitting unnecessary vibration to other equipment. Where the exhaust passes through the wall of a housing, a sleeve should be employed to prevent heat transfer and damage due to expansion and contraction.

A condensate trap should be used to prevent moisture from getting back into the engine and to ensure that the exhaust outlet is terminated so that rain cannot enter it.

11.9 STARTING SYSTEMS

Starting systems must be designed to give adequate cranking to the engine under all expected temperature conditions. Most starting systems are electric but in certain hazardous conditions either air or hydraulic starting may be employed.

Electric systems use some form of charger to keep the battery charged. The charger must be suitable for the application. For example, standby requires a long term charging system designed to prevent overcharging but maintain the battery for sudden use. In the case of a trailer mounted generator, an engine mounted battery charger is normally employed for fast recharging. See Chapter 24 for more information on batteries and battery chargers.

Air systems have a reservoir tank, often recharged by a belt driven compressor, from the engine or from a main compressor on site. Hydraulic systems have reservoirs recharged by a belt driven pump from the engine or by a hand pump.

11.10 GOVERNORS

All air-cooled diesel engines have governors of the flyweight type, either built into the engine in various forms, or into the fuel injection pump. The degree of governing is usually within 5% droop for fixed speed generator set engines, which adequately covers the majority of applications. When more accurate or isochronous governing is required, the modern engine manufacturer employs an electronic governor and uses the standard mechanical governor as a safety or overspeed device.

11.11 RATINGS

The rating defines the output of an engine or generator set under specific conditions. The generator can only put out what the engine is capable of. Therefore, it is important that the correct manufacturer's rating for the application be used.

11.12 SAFETY

Diesel generator sets can be dangerous. The installer should take precautions to avoid personal injury. All rotating and high temperature components should be guarded. All electrical connections should be adequately insulated or covered. All floors should be kept clean of oil or grease (use oil pans) and other fluids. Ensure that wiring cannot be chafed or damaged by exposure to heat from the engine. Make sure that control circuits, especially interlocking devices function correctly. Ensure that components are sized correctly for the application; do not skimp. Avoid fuel spillage or leakage that could ignite. Post adequate signs warning of danger.

11.13 SUMMARY

Air-cooled diesel engines are usually simple to maintain, are extremely reliable, and are economical power sources. They are available from a number of manufacturers in a wide variety of sizes and models to suit most generator set applications up to 200 kW. Smaller units are portable and rotate at speeds suitable for two- or four-pole generators.

Gas Turbines

Sam Laux

CHAPTER 12

INTRODUCTION

While the diesel engine has dominated the industrial engine market for generator set applications up to 1000 kW, the gas turbine engine has achieved significant market acceptance, particularly in the range above 3 MW, and has made inroads into generator set markets as low as 150 kW.

As a prime mover, the gas turbine represents the latest and most advanced engine technology. Continuous combustion, vibration free, non-reciprocating motion and very low weight to horsepower ratio are advantages not enjoyed by any other direct-fueled engine.

12.1 PRINCIPLE OF OPERATION

The simple cycle gas turbine operates on a continuous cycle, which normally consists of the following events. See Figure 12-1.

a. Compression of air from atmosphere (Sta. 2 to Sta. 4)

b. Increase of the air temperature by constant pressure combustion of fuel (Sta. 4 to Sta. 5)

c. Expansion of hot gasses through a turbine (Sta. 5 to Sta. 6)

d. Discharge of gasses to atmosphere (Sta. 7)

12.1.1 Compressor. Compressors are configured as multi-stage axial, as suggested in Figure 12-1, or centrifugal (either single or multi-stage), as suggested in Figure 12-2. Generally, small gas turbines utilize centrifugal compressors, while larger engines are configured with multi-stage axial compressors of higher compression ratio.

12.1.2 Combustion. Gas turbines may utilize single, double, or multiple chambers to handle the combustion process. The choice is primarily dictated by the space requirements of the engine design. Larger engines normally utilize small multiple combustors, arranged around the engine periphery, while small engines generally use single combustion cans, whose general arrangement commonly requires change of air flow direction during the cycle.

12.1.3 Power turbine. The third basic element in the gas turbine is the expansion turbine, or power producing element, of the engine. The pressurized hot gases from the combustion chamber provide the energy for the turbine. The temperature of the gases entering the turbine range from approximately 1300°F–2000°F under continuous full load. The first part of each turbine stage is a nozzle assembly. The nozzles restrict, accelerate (an expansion process), and direct the flow into the turbine wheel. After passing through the nozzle and entering the rotor, the hot gas continues its expansion process through the turbine wheel blading, and imparts rotative force to the rotor shaft.

Approximately two-thirds of the total power developed by the turbine is used to drive the compressor and engine accessories. The remaining shaft horsepower is the useful output of the engine, available for driving the generator.

12.2 ADVANTAGES

There are several inherent advantages of the gas turbine engine as a prime mover for generator sets.

12.2.1 Performance. The very high rotational speed of the gas turbine contributes a rotational inertia to the engine-generator system, which provides superior generator

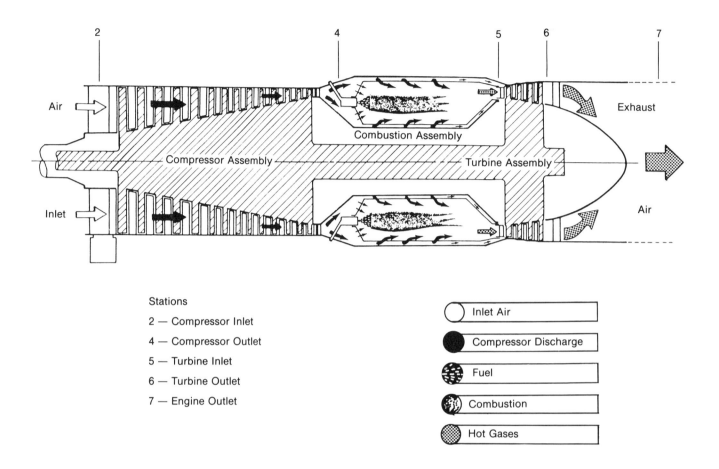

Figure 12-1. Power Section Air Flow Schematic

set response to load transients, and excellent steady-state frequency regulation.

12.2.2 Reliability. The simplicity of the pure rotary motion of the gas turbine and minimal contamination to the engine lube are primary reasons for the demonstrated superior reliability of this power plant. Inherent MTBR (mean time between removals) of 30,000 hours have been recorded over several million operating hours.

12.2.3 Size and weight. Gas turbine engines, particularly the aero-derivative variety, enjoy horsepower to weight ratios of four horsepower per pound and higher. This startling weight advantage, combined with the minimal space required for the engine package allow smaller, lighter generator set packages than any other prime mover.

12.2.4 Multi-fuel and dual fuel capability. The primary fuel for dual fuel operation is usually natural gas, waste gas, or process gas, depending on cost and availability. Use of a second fuel, which can be stored on the premises, is most often dictated by requirement for emergency operation, or by the economical advantage of an interruptible service gas supply contract. The system for providing fuel transfer under load is generally more complex, but many such systems, using dual fuel nozzles, are in operation.

12.2.5 Recoverable heat. Most of the thermal energy of the fuel which is not converted to shaft horsepower is available in the exhaust stream, at relatively high temperatures. This feature has led to the recent rapid market growth of cogeneration systems, in which this exhaust heat energy is utilized for heating, cooling, or generation of process steam.

12.2.6 Low emissions and noise. Since the combustion process takes place in an excess of air in the gas turbine cycle, the products of combustion are inherently cleaner. In recent years, new technology involving the

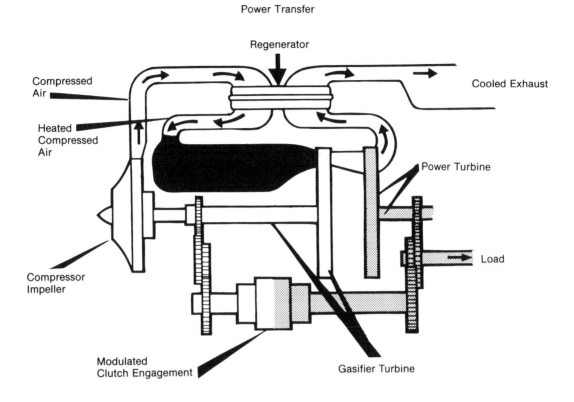

Figure 12-2. Double Shaft Turbine

injection of water or steam into the combustor has provided reduction of emission levels sufficient to meet nearly all existing local codes.

In general, the noise produced by the turbine is of a relatively high frequency, and is therefore easily attenuated in an enclosure. In addition, gas turbines of the regenerative type enjoy the inherent attenuation of the regenerator system, and have been used in both commercial and military generator applications without further silencing.

12.2.7 Low maintenance. The inherent freedom of the gas turbine engine from those maintenance activities associated with water or glycol cooling systems, and with contamination of the engine lube as experienced in reciprocating engines, are reasons for the demonstrated low maintenance requirements of the gas turbine, as compared with reciprocating diesel or gas engines.

12.3 DISADVANTAGES. There are disadvantages associated with the application of the gas turbine engine when compared with other prime movers for generator sets.

12.3.1 Cost. Generally, the gas turbine engine is more costly than its diesel counterpart. This is true because the efficiency of the turbine is directly related to the maximum temperature that can be tolerated; thus, expensive, high temperature alloys are commonly used. Cost is also affected by production volume, which is generally lower for turbines, making them more expensive.

12.3.2 Specific fuel consumption (SFC). Modern gas turbine engines have demonstrated SFC performance in the range of 0.4 lb/HP. While this fuel consumption is acknowledged to be good and represents considerable improvement over earlier performance, reciprocating diesel engines have demonstrated superior SFC and are constantly improving.

12.3.3 Sensitivity to ambient conditions. Gas turbine output horsepower is more sensitive to changing inlet air temperature and pressure than either two- or four-cycle diesel engines. For example, a change from 85° F and 500 ft to 90° F and 3000 ft will cause a power loss of 13% in a turbine engine as compared to 9% in a four-stroke diesel, 2% in a two-stroke diesel, and no loss at all in a turbocharged two-stroke diesel engine.

12.3.4 Training of service personnel. Effective, in-depth training for gas turbine engine service personnel has long been in place in the aircraft industry. Exclusive industrial engine training is now operational, but it will be a considerable time before the local service station attendant can be called to service a gas turbine powered generator set. A service agreement with a qualified contractor should be arranged before purchase.

12.4 SINGLE SHAFT AND SPLIT SHAFT

In a modern turboshaft engine the turbine, or power producing section, serves two purposes: (1) it drives the compressor, and (2) it provides shaft power to drive the generator or other driven machine. In a single shaft engine, these functions are not separated; the power producing turbine section must rotate at the same speed as the turbine that drives the compressor. Generally, a single turbine serves both functions. See Figure 12-1. In a split shaft engine, separate turbines are utilized for the two functions, allowing them to rotate at different speeds. See Figure 12-2. This is an advantage when starting the engine because the gasifier turbine is free to accelerate the compressor rapidly to a maximum power producing speed, unhindered by the inertia of the power turbine and the driven machinery, which is then free to accelerate at its own rate. Single shaft engines are most commonly used for generator set applications, however, because this configuration mechanically connects the generator to the high inertia of the gasifier, minimizing the effect of load transients on generator speed.

12.5 AERO-DERIVATIVE VERSUS INDUSTRIAL

Two basic derivations of industrial gas turbine engines are used as prime movers for generator sets: (1) aero-derivative engines, which are developed through modification of engines designed and produced to power aircraft, and (2) industrial engines, which are designed and produced for industrial application. In general, aero-derivative engines enjoy a much more extensive development process in order to meet the exacting requirements of the aircraft industry. Industrial turbines are usually heavier, since weight is less important to engines that do not fly. It is sometimes claimed that this heavier design is also more rugged; however, the static structure, which is the prime recipient of the additional weight and ruggedness, does not contain those components that are generally life limited for either the industrial or the aero-derivative design.

12.6 COOLING

Engine jacket cooling systems employing water or glycol are not required for the gas turbine engine, which helps to reduce installation and required maintenance costs. Normally, turbine engine cooling is accomplished in two ways: (1) dissipation of engine heat through lubricating oil to atmosphere by means of an oil/air heat exchanger, and (2) dissipation of heat from the engine surface by means of direct transfer to the engine compartment cooling air flow. This simplicity of engine cooling contributes to the relative ease and low cost of installing today's gas turbine engine in a generator set application.

12.7 ENGINE LUBRICATION

Generally, lubrication of the gas turbine engine is a simple process because lube oil in these engines does not become contaminated by the products of combustion to the extent that lubricating contamination occurs on the cylinder walls of reciprocating engines. Most gas turbines can operate with either synthetic or mineral lubricant. It is accepted practice to have a common lubrication system serving the engine, gearbox, and generator with a common sump, reservoir, and lubricating oil cooler. Specific lubrication recommendations by the engine manufacturer should be followed meticulously to ensure realization of the durability designed and tested into the particular turbine engine. As mentioned previously, a significant quantity of engine waste heat is rejected to the engine lubricant, requiring an effective means of rejecting this heat to ambient. External lubricating oil coolers can be sized to meet the requirements of the engine only, or include the requirements of the reduction gearbox and generator. Specific recommendations by the engine manufacturer should be followed.

12.8 FUELS AND COMBUSTION

Since the gas turbine engine combustion process is open and continuous, and occurs in an excess of air (oxygen), the presence of exact stoichiometric ratios is not a combustion requirement, and almost any fuel can be burned without major engine modification. Among fuels currently being used successfully in gas turbine engines are: natural gas, propane, diesel fuel, lo/Btu gas, JP-5, and kerosene.

12.9 AIR INLET AND EXHAUST SYSTEMS

All gas turbines today have higher airflow and greater sensitivity to inlet pressure loss than their diesel equivalents. It is, therefore, necessary to size the air cleaner and ducting to provide the lowest possible pressure drop so as not to penalize engine performance. In order to minimize loss of performance due to accumulation of dirt or erosion in the compressor components, air filtration or cleaning is required. The application, as well as environmental and maintenance requirements will dictate the type of air cleaner or filter that should be employed. In relatively clean installations, such as generator sets, the barrier type cleaner with a pleated paper filter element can be used effectively. The element is designed for an initial inlet system pressure drop of approximately 6 in. of water and is replaced with a clean one when the total pressure drop exceeds 10 in. For dirtier environments, the inertial separator type of air cleaner is desirable in order to reduce maintenance.

Exhaust system design must depend on the specific requirements of the application. Exhaust pressure loss should be limited, and must be considered, along with its attendant power loss, when determining system capacity, particularly when heat recovery steam generators or exhaust silencers are to be used.

12.10 GENERATOR SET RELATED IMPROVEMENTS

In recent years, two improvements to the gas turbine engine cycle have been introduced that enhance generator set operation.

a. The regenerative cycle (Figure 12-2) in which a porous, rotating disk absorbs heat from the turbine exhaust and returns this energy to the cycle by heating compressor discharge air prior to combustion.

b. Power transfer, in which a clutch can engage, providing fixed-shaft operation for generator set performance, or disengage, allowing the starting advantages of the split-shaft engine.

12.11 SUMMARY

Already a dominant force in larger generator set installations and in the burgeoning cogeneration industry, the gas turbine engine is now gaining acceptance in smaller generator applications, such as the 150 kW military generator set employed with the Patriot Missile System by the United States Army. It is predictable that the gas turbine engine will continue to increase in acceptance as a prime mover for generator sets, particularly when turbines begin to be mass produced for autos and trucks, greatly reducing their cost.

Governor Fundamentals

David Fredlake

CHAPTER 13

INTRODUCTION

The use of the word *precision* as it is used in connection with speed governors implies that there are governors that are not precise, or that there are varying degrees of precision in governor control. This is, of course, true. It is also true that for many purposes, such as limiting controls in the automotive field, relatively crude and insensitive governors are perfectly adequate for the job required of them. An outstanding example of this type of governor is the one used to control the throttle on small engines such as those used to power lawn mowers. In this case, a flat vane is placed in the blast from the cooling fan and so pivoted and connected to the throttle valve that an increase in the air blast, resulting from increased speed, moves the vane against a light spring in the direction to close the throttle (Figure 13-1).

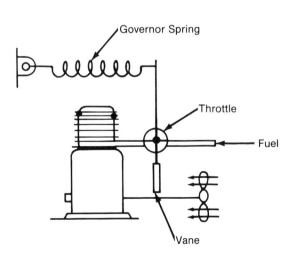

Figure 13-1. Air Vane Governor

This represents the ultimate in inexpensive governing, yet it is completely adequate for the application. Our interest, however, lies with the job requiring high accuracy of control and the availability of considerable force to operate the throttle. Such a problem does not yield to such a simple and inexpensive solution.

13.1 SPEED SENSING TECHNIQUES

It is apparent that a speed governor must include at least two components: a speed sensing element and a device to operate the throttle. In the simplest governors, these may be the same. Where considerable force is required, however, an additional device called a **servomotor** is required. This servomotor, capable of exerting the required force, is controlled by a speed sensing element which, in itself, may have very little energy available at its output. We will discuss a number of means used for sensing speed changes and actuating the gate or throttle.

13.1.1 Centrifugal pump. A speed sensor of long standing consists of a centrifugal pump that takes the dead-end or zero flow pressure as an indication of speed (Figure 13-2).

13.1.2 Constant displacement pump. Another uses a constant displacement pump discharging through a fixed orifice; the pressure drop across the orifice is indicative of speed (Figure 13-3).

13.1.3 DC generator. A number of electrical schemes have been used, including a small dc generator with a permanent magnetic field for which the output voltage is proportional to speed (Figure 13-4).

13.1.4 Permanent magnet alternator, rectified. A variation of this scheme that eliminates commutator and brushes uses a small permanent magnet alternator, the output of which is rectified, using the dc voltage as before (Figure 13-5).

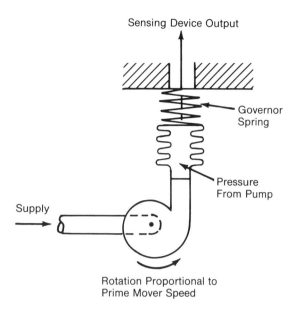

Figure 13-2. Centrifugal Pump Governor

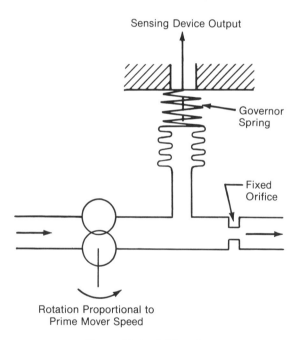

Figure 13-3. Orifice Governor

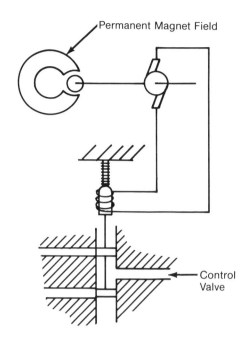

Figure 13-4. Permanent Magnet DC Generator

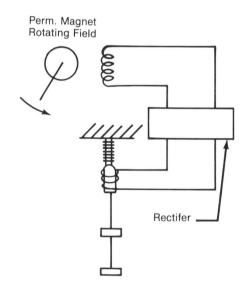

Figure 13-5. Permanent Magnet AC Generator

13.1.5 Permanent magnet alternator, frequency sensitive. A third electrical scheme uses the permanent magnet alternator, but feeds its output into a frequency sensitive network that provides a signal approximately proportional to the deviation in speed from an equilibrium value (Figure 13-6).

13.1.6 Centrifugal ballhead. All of the above have their advantages and disadvantages that determine their fields of application, but the speed sensing device preferred for its sensitivity, ruggedness, and the usefulness of its output is the centrifugal ballhead. This mechanism is probably the oldest, and certainly one of the simplest, speed sensing devices. It depends for its operation upon the fact that a force is required to compel a mass to follow a circular path, as mentioned earlier. This force is proportional to the square of the speed of rotation, and to the first power of the distance of the mass from the axis of rotation. In its best known form the ballhead consists of a pair of weights, usually spherical, at the ends of two arms pivoted near the axis of rotation in such a way that the flyweights can move radially in a plane through the axis.

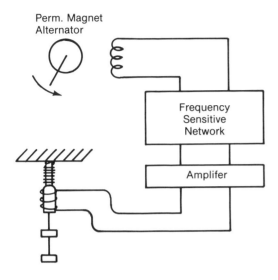

Figure 13-6. Frequency Sensing Governor

Additional links are attached to the arms and a collar about the axis to form a parallelogram configuration. Thus, when the weights move outward, the collar moves up (Figure 13-7). Since the centrifugal force always acts at right angles to the axis of rotation, it exerts a torque about its pivot equal to the product of the force times the vertical distance of the ball below the pivot. This torque is opposed and, if no other forces are present, must be balanced at equilibrium by the torque of gravity, which is

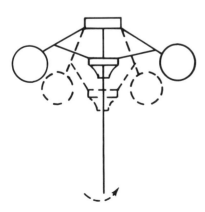

Figure 13-7. Centrifugal Ballhead

equal to the weight of the ball times the horizontal distance to the pivot (Figure 13-8). Thus as the speed increases, the centrifugal force increases and the ball moves outward, decreasing the centrifugal force torque arm and increasing the gravity torque arm until equilibrium is reached. This results in a unique equilibrium position of flyball and collar for each speed of rotation.

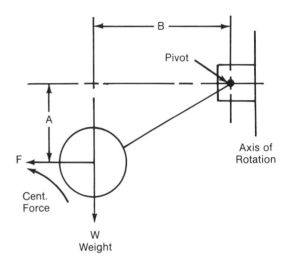

Cent. Force Torque = FA
Gravity Torque = WB
At Balance FA = WB

Figure 13-8. Torque Balance

In the direct mechanical governor this collar is connected to the throttle so as to close it as the flyweights move outward. Early steam engine governors were of this type. It should be noted that the unique relationship between speed and position of ballhead and collar no longer exists if friction is added to the system. This is for the reason that as speed increases from an equilibrium condition, the centrifugal force torque must reach a value equal to the gravity torque plus that due to friction before movement results. Similarly on decreasing speed, the centrifugal torque must go down to a value equal to the gravity torque minus that due to friction. The result is a so-called **dead band** or region in which the speed may wander without producing a corrective motion of the throttle. Efforts to minimize this dead band resulted in the huge cast iron flyweights common to early mechanical governors.

From the above, it is apparent that for greatest sensitivity the force available from the ballhead upon a small speed change should be large relative to the force required for control. There are two approaches to this problem: the friction of the entire system and the force required for control may be minimized, or the force available from the ballhead upon a small speed change may be increased. The second is accomplished when the size of the ballhead is increased or when a small ballhead is run at high speed. The first implies careful construction to reduce friction, and removal of the throttle load to a servomotor controllable by a ballhead output of low power level. Over the years the centrifugal ballhead has progressed through a great number of design configurations to reach its current form (Figure 13-9).

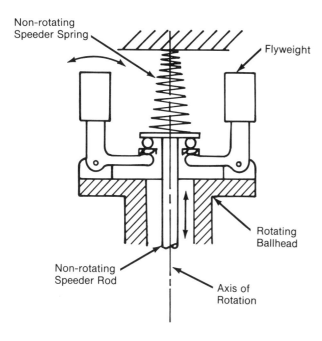

Figure 13-9. Modern Ballhead

13.2 MECHANICAL GOVERNOR BASICS

As used for internal combustion engines, the mechanical governor usually consists of two weights with their centers of mass approximately the same distance from the axis of rotation as the pivots about which they swing. So-called **toes** are arranged substantially at right angles to the body of the flyweight in such a way that as the weight moves toward and away from the axis of rotation, the toes convert this motion to an axial movement of a pilot valve or speeder rod through a suitable thrust bearing. The centrifugal force is opposed and balanced by the force exerted by a compressed speeder spring instead of gravity. In some designs the toes are on the center line, and since they move in an arc, some sliding with resultant friction takes place. In other designs the toes are offset and contact the thrust bearing on a line at right angles to their plane of movement, so that the arcuate movement is converted into a slight rotation of the thrust bearing with a minimum amount of sliding friction (Figure 13-10).

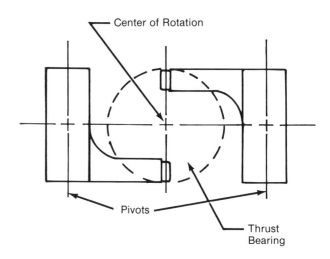

Figure 13-10. Thrust Point

Friction is further reduced in the more sensitive designs by the use of antifriction bearings and cross spring pivots. These precautions are taken to reduce the dead band that results from friction.

It should be noted that complete elimination of friction from the ballhead produces an undesirable result. This is because the flyweight-speeder spring combination, forming a pendulum system, tends to oscillate after a disturbance, giving a false indication of speed deviation. It is desirable to have sufficient damping in the system so that it will return to equilibrium with no more than one very small overswing. It appears this might best be done by utilizing a dashpot to provide viscous damping for a frictionless ballhead, since this would give the required damping without producing dead band. In smaller governors, lubrication is inherent in the design, and this construction is unnecessary, since the amount of friction required for satisfactory damping is so small that the best we have been able to accomplish in the reduction of friction in the pivoted ballhead still leaves enough for adequate damping and is, at the same time, sufficiently low to provide an acceptably small dead band.

As stated above, the force necessary to restrain the flyweight, in this case by loading of the flyweight toe, increases with the square of the speed and directly with the radial distance of the flyweight center of gravity from the axis of rotation. Thus, at a given speed, the force will increase as the flyweight is permitted to move out. The flyweight scale is defined as the rate at which this force increases with movement, referred to the flyweight toe. Thus, if at some speed a movement of 0.010 inch at the toe produces a force change of 0.2 lb., the flyweight is said to have a scale of 20 lb/in., which for the usual ballhead with two flyweights is 40 lb/in. ballhead scale.

For a given ballhead this is different at each speed, varying proportionally with the square of the speed.

It should be noted that as the flyweights move out, the speeder spring is compressed. Since the speeder spring also has a scale, this results in an increase in the force opposing flyweight movement. If this increase is less than that of the ballhead for the same movement, instability results. In other words, if a condition of equilibrium exists (which merely requires ballhead and spring forces to be equal), and a slight displacement of such a combination takes place, the new forces of ballhead and spring are not equal and the net force is in the direction to continue the motion. Such a ballhead and speeder spring combination will, upon varying speed over the necessary range, snap quickly from one extreme position to the other. This characteristic may be designed into an overspeed trip, but it is useless in a regulating governor.

If, on the other hand, the scale of the speeder spring is appreciably greater than that of the ballhead, a stable system results. At each speed within the range of movement of the flyweights, the governor finds a unique equilibrium position. This is the design used for regulating governor ballheads. By properly choosing the ratio of speeder spring and ballhead scales, a wide range of sensitivities or movement of pilot valve for a given speed change can be had. The best choice, however, depends upon the nature of the system that the governor is to control.

13.3 HYDROMECHANICAL GOVERNOR BASICS

Thus far, no mention has been made of an additional force factor that must be taken into account in ballhead design, namely, the hydraulic reaction force exerted on the piston type pilot valve due to flow of oil through the ports. This force, for the usual range of port openings, is in the direction to re-center the valve and varies approximately linearly with the valve opening. Thus, this force also has a scale which, in the case of the directly connected valve, is additive to the spring scale. If not taken into account, this may produce a ballhead of much lower sensitivity than anticipated. The magnitude of this force is a function of the flow through the valve and the pressure drop across it. It is the principal limiting factor in the determination of the size of valve that can be satisfactorily actuated by a ballhead of given size.

This implies that control of large oil flows, as required in water wheel governors, could not be handled directly by a ballhead of reasonable size. This problem is solved in cabinet actuators and water wheel governors by the use of an additional valve. This valve, capable of controlling large oil flow, is arranged so that it takes a position proportional to the speeder rod position, but is actuated by a hydraulic force many times greater than that available from the ballhead (Figure 13-11). The dashpot illustrated will be discussed later.

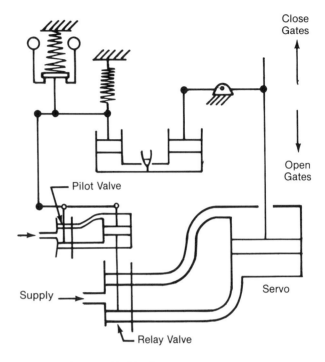

Figure 13-11. Control by Pilot Valve

As mentioned, the ballhead force and scale vary with the square of the speed. This suggests that to design a speeder spring for a wide range operation may be something of a problem, since it is desirable to have a nearly linear relation between speed and speed setting and since for proper operation, the ratio between spring scale and ballhead scale should not vary too much. The problem has been satisfactorily solved for speed ranges of 6 to 1 and, in special cases, even higher, by the trumpet shaped or conical spring. This spring has a non-linear load deflection curve (Figure 13-12), since it is wound so that at light loads all turns are active, resulting in a low spring scale when the ballhead scale is low, while at high loads the larger turns close out, providing a stiffer or higher scale spring for the higher speeds.

Despite the obvious changes that have taken place during the evolution of the ballhead from its crude beginnings, the qualities that recommend it remain substantially the same: it is rugged and simple mechanically, has a useful output at a reasonably high power level, is relatively insensitive to temperature changes, and is simply adjusted for speed setting. For these and other reasons, this type of speed sensing device is widely used in governors in conjunction with a servomotor to provide a high power level output.

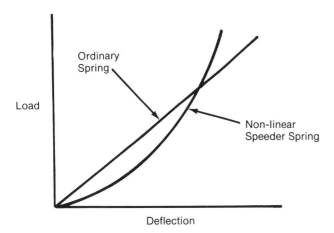

Figure 13-12. Spring Comparison

13.4 SERVOMOTORS

Various types of servomotors are worthy of consideration. In the case of hydraulic turbine gates, it is obvious that the energy required for their operation makes direct control by the speed sensing element impractical. In the early water wheel governors, the power required was taken mechanically from the combined water wheel through clutches selectively operated by the ballhead. As control power requirements became greater and higher speeds of gate movement were made necessary by the demand for better regulation, the practical limit of the simple clutch system was reached. Even for power requirements well within its range, this mechanism has certain disadvantages as a control servo, such as the difficulty of securing and maintaining a small dead band, and, most important, the fact that its speed is proportional to prime mover speed regardless of magnitude of speed error. The latter characteristic makes stable operation difficult because of the tendency to overcorrect for small errors. It is desirable to have the speed of servo movement proportional to speed error at least for small errors. An approximation of this can be secured by intermittent operation of the clutch, decreasing the percent of time engaged as the unit speed approaches the proper value. However, this means added complication and more clutch wear.

With electric motors so common, their use as servo units is an obvious possibility. As in the case of the clutch controlled servo, the power required is a vital factor in determining the suitability of electric motors as servos. For water wheel gates that involve large forces and rapid movements (that is, considerable power), the size and cost of the electric servo becomes objectionably large. Furthermore, control of such a motor, continually started, stopped, and reversed, is a serious problem. A further objection is that the speed of response is slower than desired due to relatively high inertia.

For relatively low power requirements, or where high speed of response is not essential, the electric servomotor finds entirely satisfactory application, but for our purposes it does not compare favorably with the hydraulic servo.

The hydraulic servo has a great many desirable characteristics to recommend it and comparatively few objectionable features. Without storage accumulators, the combination of hydraulic servo and pump driven by the controlled prime mover offers a relatively inexpensive, lightweight, compact means of providing power for control. With accumulators, power can be supplied for a short time (long enough to make one or two full servo strokes) at a very high level without the necessity for increasing pump size (Figure 13-13). This is important where pumps may be electric motor driven or where a limited volume of oil makes heating a problem. Control of a hydraulic servo is readily accomplished with little expenditure of energy by means of a piston type valve. If relative rotation between plunger and sleeve of such a valve is used, frictional opposition to motion is substantially eliminated, so that speed sensitive elements of low energy output may actuate directly a valve controlling a fairly large and fast servo. Furthermore, it is inherent in such a control that the speed of servo movement decreases as the valve approaches its equilibrium position. The mass of the moving parts is so low compared with the force available that the response of the hydraulic is extremely rapid. If oil is used as the hydraulic fluid,

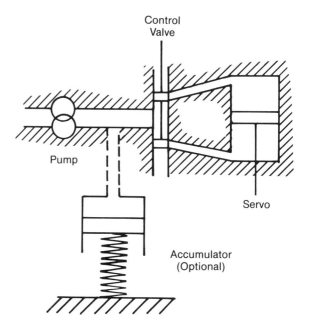

Figure 13-13. Double Acting Piston

lubrication is inherent. The possibility of oil leaks, difficulties with flow at extremely low temperatures, and the necessity for keeping the oil clean may be considered disadvantages, but they are at most no greater than those inherent in other systems, and are for our purpose far outweighed by the advantages.

Having given consideration to the relative merits of the various governor components discussed above, most governors are built around the hydraulic servo of the reciprocating piston type and the centrifugal ballhead actuating a piston type pilot valve. Other types of hydraulic servo such as the vane type unit capable of movement of less than one revolution, and the hydraulic motor capable of continuous rotation, have found special application, but the simplicity and dependability of the reciprocating piston have thus far not been surpassed for general use. Three types of reciprocating piston servos are used:

a. *The double acting piston*, requiring a pilot valve with two control lands simultaneously regulating flow of oil to and from opposite sides of the piston. It is connected to its load by a relatively small diameter piston rod projecting through a seal at one end (Figure 13-13).

b. *The single acting spring loaded piston*, in which hydraulic force under control of a single land pilot valve overcomes spring and external load in one direction. The spring discharges the oil from the cylinder and overcomes external load in the other direction (Figure 13-14).

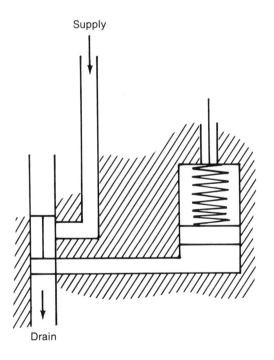

Figure 13-14. Single Acting Piston

c. *The differential piston*, in which the piston rod becomes large enough to reduce the effective area on its side of the piston to one-half that of the other side. In this case, oil at supply pressure is maintained in the small area end of the cylinder and oil controlled by a single land valve flows in and out of the other side (Figure 13-15).

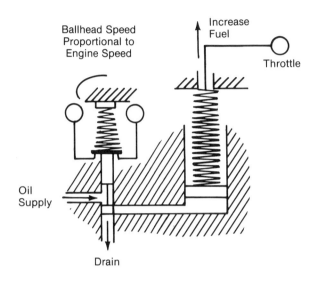

Figure 13-15. Differential Piston

In all cases, the design is such that substantially identical forces are available in both directions of motion.

The basic principle of operation of the centrifugal ballhead and the general form in which it is used in most governors was described above (Figure 13-9). However, ballheads appear in a number of variations; the design is dictated in detail by the specific requirement. The same basic principles apply to all. The servomotor is so simple in principle that it requires no further description than has already been given.

13.5 HYDRAULIC GOVERNOR

Let us now consider the simplest form of hydraulic governor, in which the ballhead and pilot valve described control a simple reciprocating piston servo. Assume an adequate oil supply is available. For the sake of discussion, assume that the governor is to control a diesel engine. The ballhead is driven at a speed proportional to that of the engine and the servo is connected to operate the fuel racks (Figure 13-16). It should be noted that the simple combination of ballhead and directly connected pilot valve has only one equilibrium position: the position in which the valve is closed, neither admitting oil to nor discharging oil from the servo cylinder.

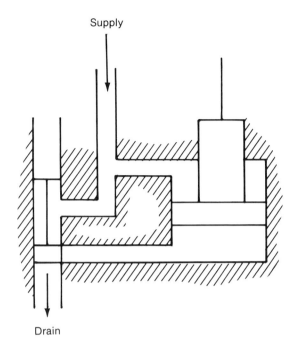

Figure 13-16. Servo System

For a given setting of the top end of the speeder spring, the ballhead and valve will take this position at only one speed; in other words, such a speed sensitive device is inherently **isochronous**. Unfortunately, such a system is also inherently unstable. This is because the engine speed does not instantly assume a value proportional to the rack position due to the inertia of the rotating mass. Therefore, if the engine is below the governor speed setting, the pilot valve is positioned to move the servo to increase the fuel. By the time the speed has increased so the valve is centered and the servo stopped, the fuel has already been increased too much and the engine continues to speed up. This opens the pilot valve the other way and fuel begins to decrease. As before, when the speed gets to the right value, the fuel control has traveled too far, the engine underspeeds, and the whole cycle repeats. Some means for stabilizing such a system must obviously be added to the two components to secure a satisfactory governor.

The simplest method of securing stability in the system is to add means that will provide speed droop in the governor, which in turn results in regulation in the governed system. The distinction between these terms will become clear as we proceed.

13.6 SPEED DROOP GOVERNOR

For some reason, understanding speed droop seems to be difficult to grasp. It is merely the governor characteristic that requires a decrease in speed to produce an increase in throttle or gate. Since an increase in throttle or gate is required if the prime mover is to carry more load, it follows that increased load means decreased speed. The reason speed droop should stabilize an otherwise unstable system is seen by a comparison of the differential equations of the two systems. However, since this explanation is not universally satisfactory, consider the following analogy.

Assume that the machine to be controlled is an ordinary automobile with foot throttle. The driver of the car will function as the servo in operating the throttle. The speedometer, slightly modified, will be the speed indicator. The speedometer needle will be replaced with a solid disc, half red and half green, with the dividing line where the needle pointer usually is. The face of the speedometer will have an opaque disc with a very narrow slot near the edge to view the indicating disc. This slot may be positioned at will as an adjustable speed setting, but for this example it will be at the location corresponding to 50 mph. The driver will attempt to maintain this speed by reacting to the speed indication as follows: whenever the slot appears red (indicating an overspeed), the throttle will be completely closed; if green, completely opened.

Suppose the car is moving at a steady 50 mph as it approaches a slight upgrade. The speed will drop, the slot will show green, and the driver will tramp the throttle to the floor. The car will speed up, the indication will turn quickly from green to red as it goes past 50 mph, and the driver will completely close the throttle. The speed will drop quickly below 50 mph and the cycle will be repeated. This is a system without speed droop.

Now assume that the slot is increased in width so that its low edge is at 45 mph and its high edge at 55 mph. When the slot shows all red, which will occur at 55 mph, the driver will close the throttle completely; when all green (at 45 mph), the throttle will be fully open. For intermediate speeds or ratios of green to total slot width, the throttle must take a proportional position. With such an arrangement, the slight decrease in speed caused by the upgrade does not change the indication to all green, but only increases the amount of green. This results in a small increase in fuel, not a complete throttle opening. The car will take the grade at a slightly reduced but stable speed, and the throttle will not be completely opened unless the grade is sufficiently severe to require full power at 45 mph. Such a system has speed droop.

Note that the simple mechanical flyball governor, in which the throttle is operated directly by the ballhead, has inherent speed droop, since the only way the throttle opening can be increased is by the flyweights moving inward, which requires a decrease in speed.

Speed droop in a simple governor can be provided by a mechanical interconnection between servo (throttle) movement and governor speed setting such that, as fuel is increased, the speed setting is decreased. Such a device may consist simply of a lever of suitable ratio between servo and speeder spring (Figure 13-17). The equilibrium relationship between speed setting and servo position for such a system may be represented by a line sloping or "drooping" downward to indicate decreased speed setting with movement of the servo in the increase fuel direction.

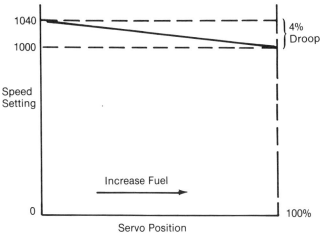

Figure 13-18. Droop Curve

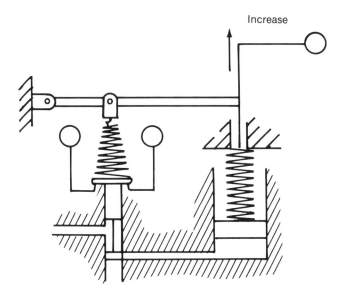

Figure 13-17. Simple Droop Governor

For each position of the manual speed adjustment, there will be a sloping line as shown in Figure 13-18. In this example, the speed adjustment is assumed to be such that if the servo is at its extreme position in the "increase fuel" direction, the governor speed setting is 1000 RPM. As the servo is moved to the opposite limit of its motion, the speed setting is increased to 1040 RPM. The speed droop is usually expressed as the percent change in speed setting for full servo stroke in reference to the speed setting at the maximum fuel position. In this case the governor is adjusted for 4% speed droop. This characteristic is a function of governor design and adjustment only.

Regulation, however, is dependent not only upon the speed droop setting, but also the percentage of governor servo stroke required to move the throttle between no load and rated load positions. The steady speed rise resulting from decrease in load from rated value to zero is expressed as a percentage of rated speed. If the governor discussed above was connected to the throttle of its engine so that only 50% of the servo travel was required to move the throttle between no load and rated load, the regulation would be 2% although the speed droop is 4%.

Speed droop in a hydraulic governor is not always attained by operation on the speeder spring. It can also be secured by changing the position of the flyweights (and therefore their speed for a given speeder spring setting) to center the control valve at its equilibrium position. This might be done by using a floating lever connection between speeder rod, servo, and pilot valve (Figure 13-19), or by having the servo move the bushing (Figure 13-20).

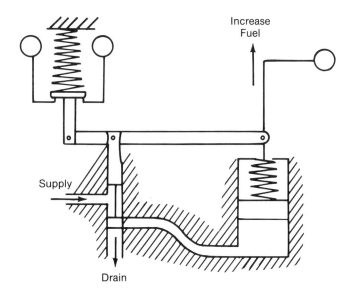

Figure 13-19. Floating Lever

Governor Fundamentals

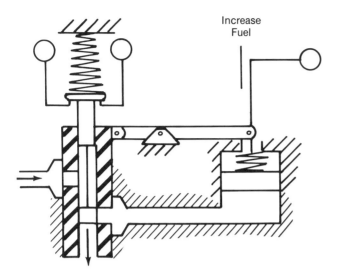

Figure 13-20. Moveable Bushing

13.7 ISOCHRONOUS GOVERNOR

It is sometimes desirable to have an isolated prime mover run isochronously (speed constant regardless of load within the capacity of the prime mover), or perhaps the allowable regulation is not sufficient to provide stability. In such cases, **transient speed droop**, or **compensation**, is used. This calls for the introduction of a temporary readjustment of speed setting with servo movement to produce the stabilizing speed droop characteristic, followed by a relatively slow return of speed setting to its original value.

This can be accomplished in a number of ways. Present day hydraulic governors use a variety of schemes. Perhaps the most common (Figure 13-21) involves a floating lever connecting speeder rod, pilot valve, and receiving piston that are urged by spring force toward an equilibrium position. As long as the receiving piston is in this equilibrium position, centering the valve requires that the flyweights are always in the same position. For a fixed speeder spring setting, this means that the ballhead must always, for final equilibrium, be running at the same speed. The receiving piston is displaced from its equilibrium position by a flow of oil initiated by movement of a transmitting piston, which moves with the throttle- or gate-actuating servo. Thus, if there was no leakage from this compensating hydraulic system, the receiving piston would move as though rigidly connected to the servo, and permanent speed droop as described above would result since the centering spring is ineffective. However, if an adjustable leak in the form of a needle valve is provided between the compensating hydraulic system and an oil sump, the centering spring slowly returns the receiving piston to its initial position after a disturbance by forcing

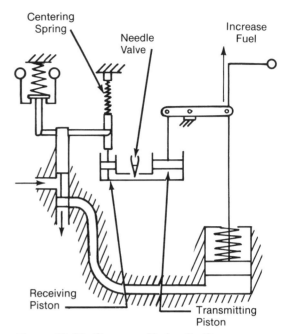

Figure 13-21. Common Hydraulic Governor

oil out or drawing oil in through the needle valve as required. As this occurs, the speed setting of the governor slowly returns to its original value, although the servo and throttle will remain at the new load position. Such a governor is isochronous, although it is provided with the necessary transient droop for stability. This scheme is used in current water wheel applications.

Another method of producing the same result involves the direct (Figure 13-22) application of pressure to the pilot valve plunger, adding to or subtracting from the speeder spring force to effect a change in speed setting.

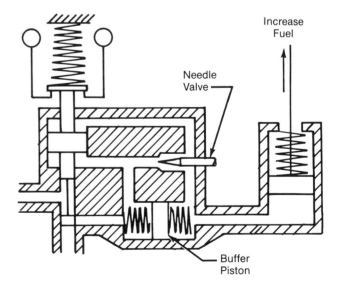

Figure 13-22. Use of Buffer Piston

In the adaptation of this method, the oil actuating the servo is required to deflect a buffer piston against a centering spring load. This produces a pressure differential across a receiving piston rigidly attached to the pilot valve plunger. A needle valve permits equalization of pressure across the pilot valve receiving piston to restore the initial speed setting. In operation, as oil flows to the servo, the buffer piston is moved against the force of its centering spring. This results in a higher pressure on the lower side of the receiving piston, which produces an upward force on the pilot valve. In effect, this decreases the force that the flyweights must balance, resulting in centering of the pilot valve at a lower speed, thus providing speed droop. As the displaced oil is permitted to leak through the needle valve, the buffer piston returns to its equilibrium position, the differential pressure disappears, and the speed setting reverts to its original value. This method, although requiring a more expensive pilot valve construction, has several operating advantages. Not the least of these is that exact return of the buffer piston to center is not essential to exact return of the system to normal speed. Only equalization of pressure is required.

It is obviously possible to combine temporary and permanent droop in a single governor (Figure 13-23). This is done in many standard units that anticipate parallel operation requiring prescribed division of load.

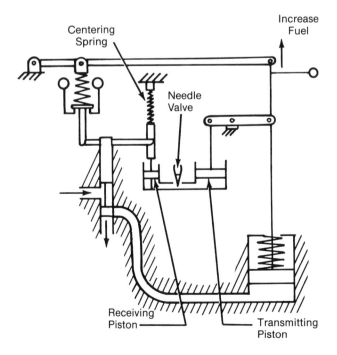

Figure 13-23. Droop System

Other methods of securing stability have been used. Among these is the derivative control. This is a device sensitive to the first (or higher) time derivative or rate of change of the controlled variable. In the case of a speed control, a first derivative system is a device sensitive not only to speed, but also the rate of speed change (first derivative), which is acceleration. Acceleration sensitive governors were used long before they were called derivative controls, and an extensive mathematical analysis of them in various applications has been made. Acceleration sensitivity alone is unsatisfactory, since it will not accurately maintain speed. The minimum requirement in an isochronous control so stabilized is, therefore, a speed sensitive element and an acceleration sensitive element. Properly designed, it is possible to secure stability in such a mechanical device, but it has not been looked upon favorably.

If wander over a narrow band is to be avoided, the stabilizing effect of the acceleration sensitive element must be present for very small accelerations or decelerations. This is very difficult to accomplish mechanically without the use of objectionably large mass or very expensive and delicate construction. Furthermore, mathematical analysis, confirmed by tests, reveals that with rare exceptions, use of the derivative control can accomplish nothing to regulate an on-speed condition that cannot be secured as well, and at much lower cost by proper design of the compensating systems described above. If the disturbance consists of a large change in speed setting rather than a change in load, the acceleration sensitive element offers some advantage, but it is doubtful if the improvement obtained justifies the complication and cost involved.

13.8 PROPORTIONAL TYPE ELECTRIC GOVERNOR SYSTEMS — BASIC ELECTRIC SPEED GOVERNOR

A basic proportional electronic speed governor is composed of three parts (see Figure 13-24):

a. A speed sensing device

b. A speed control that generates a speed reference, compares the speed reference to the actual speed, and outputs a signal proportional to fuel

c. An actuator to position the fuel metering mechanism of the prime mover. (Proportional actuators can be either electrohydraulic or all electric.)

The important thing to note about these three basic parts is that they compose a closed loop system. Figure 13-25 illustrates the closed loop concept.

Governor Fundamentals

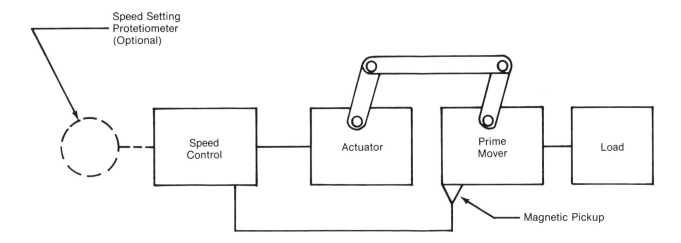

Figure 13-24. Single Unit Speed Control

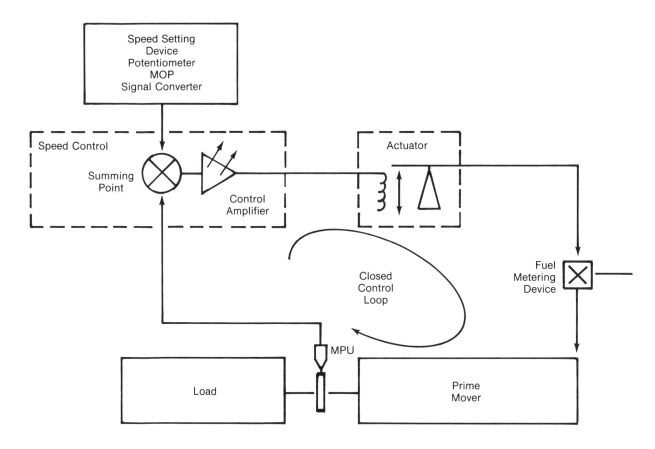

Figure 13-25. Closed Control Loop

Prime mover speed is normally sensed by one of two methods. The most common is a magnetic pickup mounted in close proximity to a ferrous metal gear, either on, or being driven directly by, the prime mover. The other method is to measure the frequency of a generator being driven by the prime mover. This generator may be the primary generator, an auxiliary PMA (permanent magnet alternator), or PMG (permanent magnet generator).

The speed signal is converted to an analog dc level where the voltage is proportional to the actual speed of the prime mover. This dc voltage is compared to the speed reference voltage. If a difference or error exists, the output from the amplifier causes the actuator to move to minimize this error. This movement modulates the fuel to either slow the engine or speed it up to bring the actual speed into agreement with the speed reference. The summing point is shown in Figure 13-26.

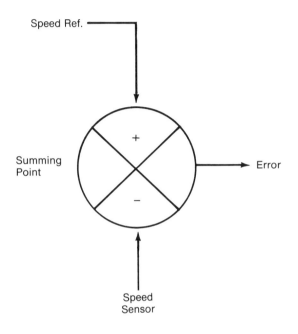

Figure 13-26. Summing Point

A magnetic pickup is a permanent magnet, mono-pole device. It transmits a voltage signal each time the flux path across the pole is disturbed by a passing gear tooth. The shape and spacing of the gear teeth determine the wave form of the output voltage. The number of teeth on the gear determines the number of pulses per revolution of the gear. Thus an 80 tooth gear turning at 3600 RPM would produce a frequency of 4800 Hz, as calculated by the following equation:

$$\text{Frequency (Hz)} = \frac{\text{Number of Teeth} \times \text{Speed (RPM)}}{60}$$

(Equation 13-1)

The perferred wave form is sinusoidal, with one zero crossing per cycle. Most controls require a minimum of 1.5 Vrms from the magnetic pickup to the control. This voltage can be measured at the control, with the leads from the magnetic pickup connected to the control input. The output of the magnetic pickup should be greater than 5 Vrms at idle speed.

The speed sensor is a frequency-to-voltage converter. This circuit converts the generated frequency to a dc voltage proportional to the speed of the prime mover. In some controls a transformer is required for electrical isolation.

The set speed reference is normally obtained by use of a potentiometer generating a reference voltage in the control. This speed setting potentiometer can be either an integral part of the electric control or external, in the form of a hand or motor operated potentiometer.

13.8.1 Loop dynamics. To obtain a proper transient response from the control, several other circuits are incorporated into the amplifier section of the speed control. They are the proportional, integral, and differential circuits. The name **PID control** is sometimes used to describe these features. They are used to tune the control to allow for delays in carburetion, plenum chambers, ignition, momentum, etc. The differential circuit determines how quickly the amplifier recognizes that speed is drifting away from the set speed. The integral circuit will integrate (or ramp) to actual speed. The proportional circuit will determine the magnitude of the change in the output signal of the amplifier to the actuator. Because of system and component tolerances, the electric control should be tuned on the actual system for optimum performance. To make the tuning easier, two adjustments (called **gain** and **stability**) are provided on most controls. See Figure 13.27. On some controls the stability adjustment may be labeled *reset*. For most installations, setting the gain as high as possible, while keeping the system stable with the stability adjustment, results in satisfactory control. These adjustments should be made and then checked by momentarily upsetting the system. This can be accomplished by physically moving the actuator or disturbing the electric control, and then adjusting for a minimum overshoot and minimum stabilizing time.

13.9 ELECTRICAL VERSUS MECHANICAL

So far, we have discussed a basic speed control. You might question the use of an electric control when a mechanical control could essentially do the same job. This is true for a basic speed governor. However, it is much easier to tune an electric governor for transient response and to make changes to preselected speeds. The big advantage of the electric governor over the mechanical governor lies in the ability to easily modify the speed reference by various means, in order to accomplish such things as controlled acceleration and deceleration, load control, synchronizing phase and speed during paralleling, load sharing after paralleling, etc. The applications are endless.

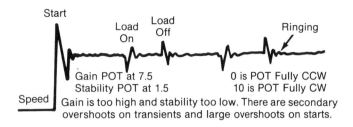

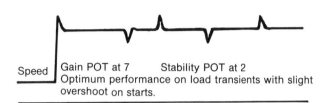

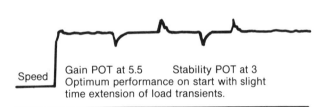

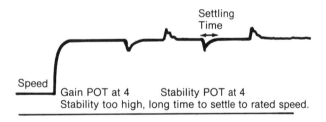

Figure 13-27. Engine Performance Curve

13.10 LOAD SHARING

Two basic methods are used in paralleling several generators. The first is droop, where speed decreases with load. The second is isochronous, where speed remains constant. Droop control is sometimes referred to as **base loading**.

13.11 ISOCHRONOUS SPEED CONTROL

An isochronous speed control maintains the prime mover at a constant speed, no matter how much load is on the machine. The prime mover will run at the same speed at no load and full load. The only way speed can be changed is by adjusting the desired speed or speed setting.

Figure 13-28 illustrates isochronous operation by plotting load versus speed.

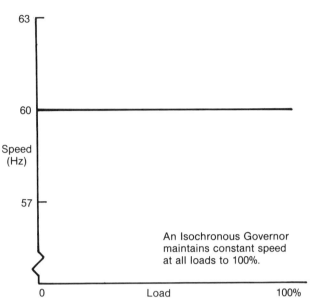

Figure 13-28. Isochronous Speed Control

13.12 DROOP SPEED CONTROL

In speed droop, speed is reduced as the load on the machine increases. Droop is usually represented as a percent of rated speed. If at no load the prime mover is running at rated speed and at full load the machine's speed is 5% below rated, the system is said to have 5% speed droop. Figure 13-29 illustrates droop operation by plotting load versus speed. Droop can be used for many different reasons. Droop is used in mechanical governors to allow load sharing. It is also used for added stability in single engine applications where constant speed is not critical. Droop may be derived as a function of governor terminal shaft position, since shaft position is approximately proportional to real load. The droop signal in an electric governor is derived from actuator command voltage (or kW load) on the generator. The more accurate method is to measure the real kW load and develop a speed bias signal as a direct function of load. Electric load sharing governors include droop as an optional operating mode.

Droop paralleling is used on governors where interconnection of all controls is not possible, such as when paralleling to a large electric grid network, referred to as an **infinite bus**.

Consider the operation of a 60 Hz generator set paralleled with an infinite bus. Assume a 5% droop (3 Hz) for the example.

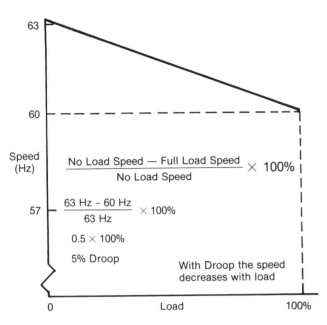

Figure 13-29. Droop Speed Control

Notice in Figure 13-30 that as the speed setting is increased, the speed droop line intersects the bus frequency at higher load levels. Thus the infinite bus maintains frequency and the governor speed setting establishes the generator's load level. If the actual local load is more than the generator's load setting, the infinite bus will provide the balance. If the local load is less than the generator's load setting, the excess power goes into the infinite bus system.

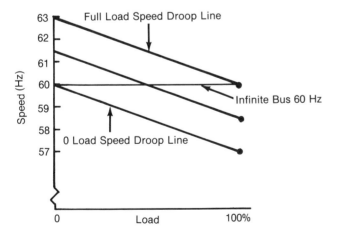

Figure 13-30. Droop Parallel Operation with an Infinite Bus

13.13 ISOCHRONOUS-DROOP PARALLEL OPERATION

The infinite bus can be replaced with another generator set that has a governor operating in the isochronous mode. However, be sure that the isochronous generator set can accept the expected load swings. For example, two generator sets are each rated for 100 kW full load. One governor is run in isochronous mode. The other governor is set for droop, paralleled, and its speed setting adjusted for 100 kW load generation. The total system load in this case must be more than 100 kW and less than 200 kW for satisfactory operation. If the total load on the system is less than 100 kW, the speed setting of the droop unit would have to be lowered to prevent "motoring" the isochronous unit. Obviously, constant operator attention may be necessary. Figure 13-31 illustrates the operation of these two generators.

13.14 DROOP-DROOP PARALLEL OPERATION

If all generator units in a system have governors in droop operation, the generators can be set to share load equally, but the frequency of the system will vary with load. To achieve load sharing, the speed settings must be matched and the droop must be set the same on all units. See Figure 13-32.

The sequence of setting the speed for this operation should be considered. Assume one set is loaded to near 100%, and the speed setting has been raised to give 60 Hz operation. The sequence for adding the second set of equal rating is illustrated in Figure 13-33.

a. Adjust the speed of the oncoming unit to 60 Hz to allow the breaker to close. After the breaker is closed, the oncoming unit remains at near zero load.

b. Increase speed setting of oncoming unit to assume half the existing load.

c. Readjust both speed settings to achieve desired loaded frequency.

Note that the parallel operation of generator sets in droop is straightforward but there are two major characteristics that may be undesirable.

a. Bus frequency changes with load changes.

b. Each time a set is added to the bus, the speed settings must be reset to establish a new load balance and reestablish system frequency.

The droop governor system could be automated through governor speed setting motors, but the hardware would be cumbersome and complex. A more desirable solution is to use electric load sharing governors.

13.15 ISOCHRONOUS LOAD SHARING

Load sharing speed control was developed to alleviate the need for readjusting speed on a drooped system as loads changed. The only way two simple speed controls

Governor Fundamentals

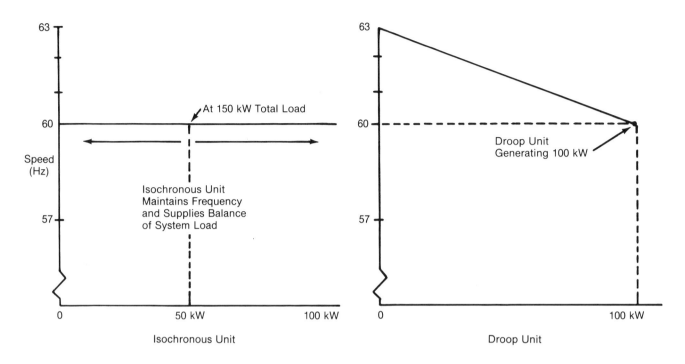

Figure 13-31. Paralleled Generator Sets (one in Isochronous and one in Droop)

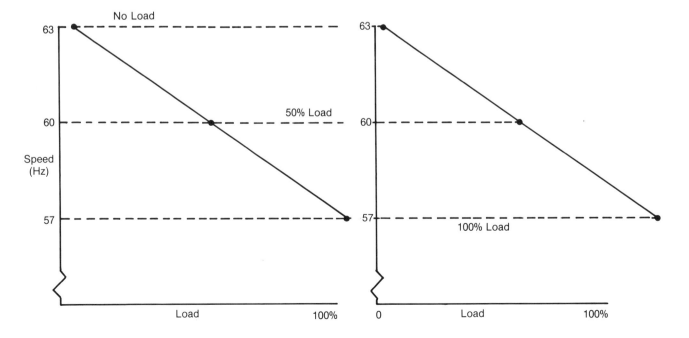

Figure 13-32. Paralleling Operation

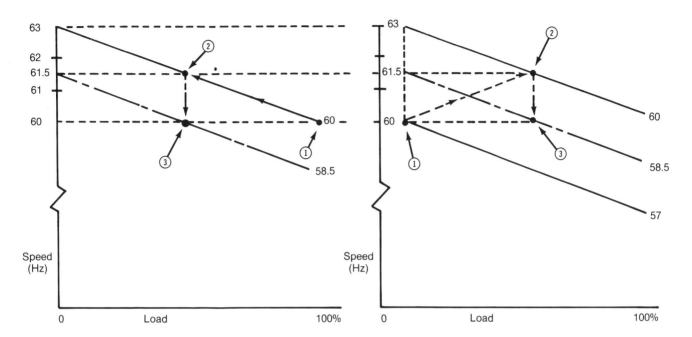

Figure 13-33. Paralleling Sequence

can share loads while their generators are tied together is by operating in droop. As discussed earlier, the load and the amount of system droop determine the machine's speed. To remain at synchronous speed, the operator must manually readjust the speed reference (or load reference) as load varies. A means of balancing or equalizing load between machines without manually readjusting the speed reference is needed. This would allow the machines to be set up and operated at synchronous speed. Two machines, paralleled and operating isochronously, will not share load equally. It is impossible to set both machines at exactly the same speed before paralleling. After they are tied together, the machine that is running just slightly faster will pick up all the load. The machine running just slightly slower will shed all its load.

The operating mode that is needed is isochronous load sharing. If several prime movers are driving generators that are to share a common load, one control needs to know what the others are doing. The parallel (or load sharing) lines are the communication link between two or more load sharing speed controls. Figure 13-34 illustrates a typical load sharing system.

Isochronous paralleling governors are normally used with isolated systems where compatible controls exist, i.e., in systems where load sensing devices in the controls may share load information through paralleling lines.

The elements of the load sharing governor include a basic electric speed governor and a load sharing section. The speed governor system consists of the magnetic pickup, the electronic speed control, and the actuator to position the fuel metering mechanism. The load sharing section measures kW load and provides a correcting signal to the speed governor. If the speed governors are set close to the desired bus frequency, the correcting signals required for good load sharing will be very small.

The basic operation of the load sharing system in Figure 13-35 is described below. The two generator sets are synchronized and closed onto a common bus. Their speeds are locked at synchronous. Each gen-set is equipped with a load sharing speed control. Their load sensors are looking at their respective generators. The load sensors are adjusted so that at full load (5A on CTs and 115 Vac on PTs) the load gain voltage is set at 6 volts. The load gain voltage is fed into the top of the load bridge. The load bridge is constructed so that voltages across each half of the bridge are equal. The midpoint of one side of the bridge is connected to one input of a differential amplifier. At this point the voltage should be one-half the load gain voltage, or 3 volts. The midpoint of the other side of the bridge is connected to the other input of the same differential amplifier. It is also connected to the load sharing lines. Again, when the system is balanced, the voltage at this point should be one-half the load gain voltage. The output from the differential amplifier is fed to the summing point of the control amplifier. When the inputs of the differential amplifier are not equal, an additional bias voltage is placed on the summing point of the control amplifier. This bias influences the actuator's position. When the load sharing line voltage is not equal to

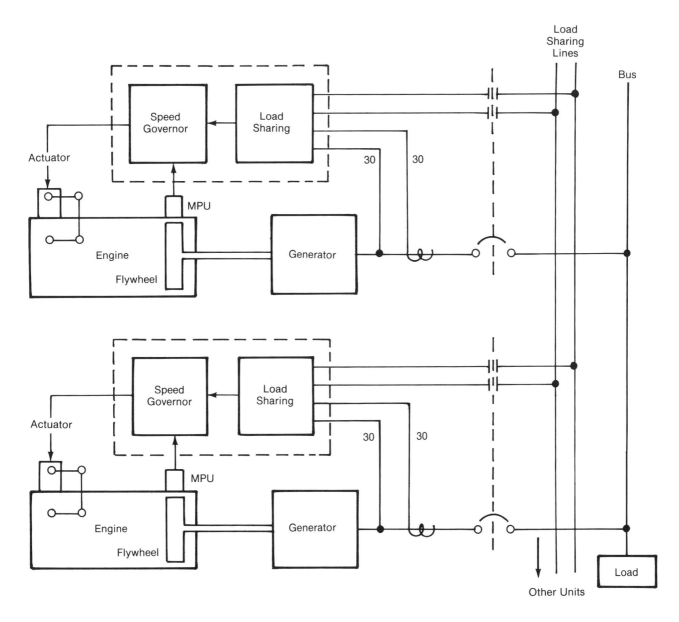

Figure 13-34. Isochronous Load Sharing Governor System

one-half the load gain voltage, the differential amplifier influences the control amplifier, causing the actuator to move to equalize the inputs of the differential amplifier. The load sharing line voltage will always try to equal one-half the load gain voltage.

Returning to the generator sets, if one control senses an increase of load on its generator, its load gain voltage increases. The voltage it sends out on the load sharing lines increases. The other control senses this increased voltage on the load sharing lines and moves through its differential and control amplifiers to open up its actuator to carry some of that load. The first unit senses that the other unit is picking up some of the additional load. Its load gain voltage will decrease. This action continues until the closed loop system returns to a balanced state.

The key to the system is the voltage on the load sharing lines. The load signal voltage on the load sharing lines is an average voltage of all controls connected to the lines. By comparing the load sharing lines' voltages to its own load signal voltage, each control's load sharing section calculates an output. The output is used by the control amplifier to raise or lower the generator output to make the load signal voltage equal to that of the load sharing lines. This maintains the desired bus load. When the load signal voltage of all load sharing controls are equal, the load is balanced proportionally among the generators.

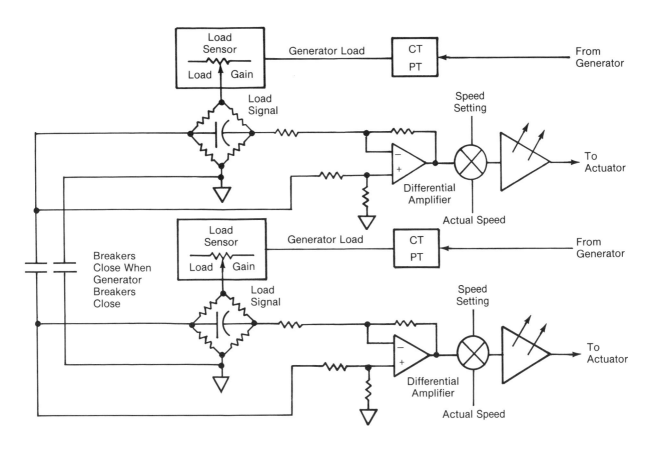

Figure 13-35. Load Sharing System Schematic

If the load sharing lines are left unconnected, the differential amplifier always sees a balanced bridge condition and will not influence the control amplifier. The control then becomes a simple speed control and will maintain speed wherever it is set.

If the load sharing lines are connected to a similar governor and the load gain voltages are equal, i.e., the loads are balanced, no current flows in the load sharing lines. Again the load signal and bridges remain balanced. The differential amplifier will not influence the control amplifier.

If the load gain voltage on different units is different, current flows in the load sharing lines. The bridges become unbalanced and a correcting signal is sent to the respective control amplifier. The magnitude of the load sharing line voltage is proportional to load unbalance, and is in a direction to rebalance the load. The operation of the system is stated in terms of adding a generator set to the bus. Assume one unit is carrying the load at 60 Hz. The oncoming unit is matched in frequency. (The frequency will be preset to 60 Hz and the breaker is closed when the units are in phase.) When the breaker is closed, auxiliary breaker contacts connect the load sharing lines.

After breaker closing, the bridges are unbalanced and correcting signals (increasing on the oncoming, decreasing on the existing) are sent to the respective control amplifier sections. As the loads are balanced, the correcting signal of the differential amplifier decreases to near zero. The load is maintained in balance with very small continuing correction signals. This operation is common to all electric load sharing governors.

The load sharing line voltage does not need to come from another load sharing speed control. Any source may be used to inject a voltage into the load sharing lines. Since the load sharing lines' inputs only see a differential input voltage, it is possible to deceive the control into picking up or shedding load as desired. This feature can be exploited for a variety of system applications.

13.16 DIGITAL GOVERNING

Digital governors have been applied to large gas turbines and steam turbines since the early 1980s. The increased complexity, the need for automation, and the need for improved performance made digital governing a practical solution for most turbine applications.

Governor Fundamentals

By the year 1990, almost all new diesel and natural gas engines, regardless of their application, were using some form of digital control. The need for this was determined by air quality standards. Secondary factors contributing to the use of digital controls on diesel and natural gas engines include automation and fuel economy.

Digital governors for diesel engines are available today in two different forms:

a. Electronic fuel injection.

b. An electronic module that replaces control (discussed previously in Section 13.8).

Electronic fuel injection refers to a complete system of controlling the fuel where the digital speed control regulates the duration of time fuel is injected into each cylinder. An electronic fuel injected system consists primarily of the digital control box, with one electronic fuel injector per cylinder. An electronic module gathers data from engine speed, crank shaft position, turbo boost pressure, fuel pressure, air temperature, etc. and calculates the exact amount of fuel required for each cylinder; it then sends a pulse with modulated signal to each electronic fuel injector to meter the required amount of fuel.

The other form of digital control consists of a digital speed control module working with a conventional electric actuator. This system was first applied on low to medium speed diesel and gas engines and generator set marine and gas compression applications. Typical features include automatic and manual engine-mapped variable dynamics, enhanced steady-state dynamics, self diagnostics, precise start fuel limiting to prevent overfueling during start-up, load limiting, and a 4–20 mA remote speed setting signal.

Both of the digital controls described above use a hand-held programming device that has several advantages over the conventional potentiometers used in analog controls. The programmer allows for more accurate calibration. Once calibrated, the programming device can be removed so no settings can be changed by unauthorized personnel. The programmer can also be used as a service tool to diagnose problems more quickly and accurately.

Ultimately, digital controls can have a widespread use in all engine related markets. The digital controls make it possible to establish completely automated systems, tying a variety of engines located miles apart to a master control station.

Fuel Systems, Diesel

Marjorie Bernahl

CHAPTER 14

INTRODUCTION

The recommended diesel fuel system has three major components: the main supply tank, the transfer pump, and the transfer (day) tank. See Figure 14-1. Manufacturers frequently supply the transfer pump as a part of the transfer tank assembly. The transfer tank is sometimes omitted, but few installations can operate satisfactorily without it. The three-component system provides operating fuel adjacent to the engine. With fuel at this location, the engine fuel transfer pump can operate within its capability and reliably supply the fuel injection pump.

The National Electrical Code (NEC) Section 700-12(b)(2) requires that, for an emergency system, not less than two hours fuel supply be on-site. Many engineers size the transfer tank for the required two hours operation. The main supply tank should contain enough fuel for an extreme emergency, but it should not be so large that fuel aging occurs. Twenty-four hours supply is a frequent choice for emergency applications.

14.1 TRANSFER TANK

14.1.1 Sizing. Two hours operation at full load is a good guide for sizing the transfer tank. Standard transfer tanks range from 10 to 600 gallons. The engineer selecting the tank should be familiar with state and local codes, which may limit the size.

14.1.2 Connections. For proper installation, a transfer tank should contain six fittings (see Figure 14-1) for connection to:

a. *Supply line from main tank*. Line that transfers the fuel from the main tank to the transfer tank. Refer to section 14.2 Pumping system, for correct application and sizing.

b. *Supply line to generator set*. Line transfers the day tank's fuel to generator set. Refer to engine generator specifications for correct application and sizing.

c. *Return line*. Line that allows unused fuel to return to day tank through top or bottom fitting. The return line can also be plumbed to the main tank so that the transfer tank fuel will not rise in temperature. This rise in temperature could cause a decrease in the efficiency of the engine-generator operation.

d. *Drain*. Plugged fitting at bottom of tank to allow drainage of tank. If a rupture basin is used, a means should be provided for draining the transfer tank through the rupture basin. Strict attention should be paid to local codes.

e. *Vent*. Connection for a vent so transfer tank has atmospheric pressure and to allow fumes to vent outside of building. Again, special attention should be paid to all codes. NFPA 30 requires 1¼" size. UL tank vents vary according to capacity.

f. *Overflow*. This line allows the tank to be fail-safe if the transfer tank is above the main tank. In the event of an overflow, the fuel will transfer to the main tank by gravity.

14.1.3 Construction. A number of construction qualities are important for long life in a transfer tank. The recommended thickness of the tank itself is 14 gauge for 25 gallons and less, 12 gauge for 50–500 gallons. Also, an epoxy coating on the tank's interior is needed to prevent oxidation. For extra protection from leakage, a rupture basin or double wall can be used to encase the transfer tank.

NFPA 37-1990, Stationary Combustion Engines and Gas Turbines, requires a wall, curb, or dike (double basin

Fuel Systems, Diesel

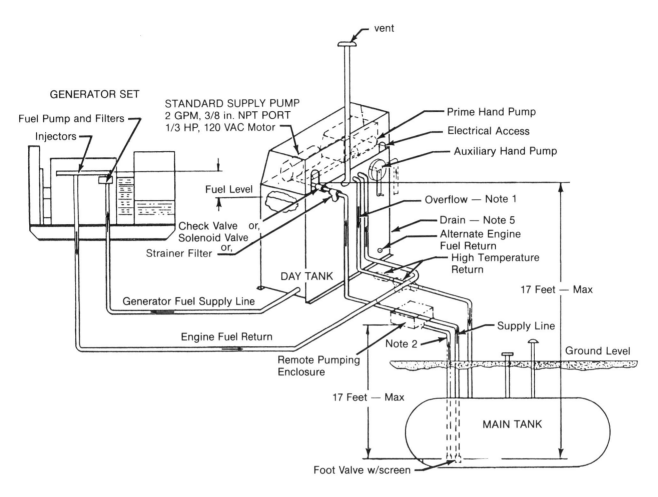

NOTES:
1. Overflow port must be above main tank to prevent spillage in case of an overflow condition (see option pipe stand adaptor).
2. If a duplex system is required, it is recommended that each pump be supplied by a separate line, or an accumulator be used at the day tank. (Consult factory.)
3. For additional safety features, see day tank system below main tank (Figure 14-2).
4. Black iron pipe, schedule 40, should be used for all fuel piping.
5. Water and sediment should be drained from tank on a periodic basis (petcock or manual ball valve).
6. All standard fittings 1 in. NPT, except vent 1-1/4 in.) and drain (3/8 in.).

Figure 14-1. Day Tank System — Above Main Tank

or double wall); rupture basin leak detector; and high level alarm with automatic pump shut off. State and local codes should be referenced as to UL listing on electrical and mechanical requirements.

14.2 PUMPING SYSTEM

Two types of pumping system are possible. The transfer tank can be placed either above or below the main tank.

Designing pumping systems can be a very complex problem. The following is a general guide to help the designer in the proper installation of his day tank pump. In critical or borderline applications, an experienced hydraulic engineer should always be consulted.

14.2.1 Lifting system. With the transfer tank above the main tank, a pump is used to transfer the fuel from the main tank to the transfer tank. (See Figure 14-1.) The engine will then pump the fuel out of the transfer tank when needed. The pump should be sized to keep up with the demand of the generator set. The most common pump size is 2 gallons per minute (GPM). Transfer pumps can go as high as 23 GPM. Another type of pumping arrangement is the duplex system, which uses two pumps. There are two advantages to this arrangement. First, it can be used as a redundant back-up system in the event the first pump fails. Second, if the fuel demand exceeds the capacity of the first pump, the second pump will deliver the additional fuel as needed. The duplex system is also more energy efficient than a single large pump.

14.2.2 Pump lift. Another factor to consider is the vertical lift of the pump. A pump will lift fuel by displacing air from suction to discharge line. This creates low pressure in the suction line allowing the higher atmospheric pressure (14.7 psi at sea level) to lift liquid into this vacuum. If a perfect vacuum could be created and maintained, fuel theoretically could be lifted to 34 feet. Since a perfect vacuum cannot be created, the lift a pump can actually achieve is approximately 50% of theoretical lift or 17 feet (7.4 psi).

To determine the total available lift, the following factors need to be considered:

a. The vertical distance the pump needs to lift fuel is the main factor in lifting capabilities. This measurement should be taken from the bottom of the main tank to the pump's inlet port.

b. The total length and size of piping is crucial due to the internal frictional losses. This will reduce lift and must be considered (see Table 14-1). All calculations are based on 60°F temperature. Frictional resistance will increase as temperature decreases.

c. Fittings in the line will disrupt flow and create friction. These fittings include elbows, tees and unions (see Table 14-2). Valves also need to be checked for possible pressure drops.

d. Elevation above sea level is important, since the atmospheric pressure acting against the pump's vacuum is reduced, thereby reducing lift (see Table 14-3).

Table 14-1. Frictional Head Loss (in Feet) for 100 Feet of Standard Weight Pipe at 60°F at Sea Level — Diesel Fuel

GPM	Pipe Size						
	3/8	1/2	3/4	1	1-1/4	1-1/2	2
2	15.2	5.5	1.1	.5	.2		
4	55.5	20.3	5.1	1.4	.5	.2	
7		61.0	15.3	4.6	1.2	.5	
10			26.3	8.5	2.5	.9	.2
19				28.5	7.5	3.5	1.2

Table 14-2. Frictional Loss in Pipe Fittings in Terms of Equivalent Feet of Straight Pipe

Pipe Size (in.)	Ball Valve	45° Elbow	Std. Elbow	Std. Tee	Check Valve	Angle Valve	Globe Valve	Diaphram Valve
3/8	.28	.70	1.4	2.6	3.6	8.6	16.5	
1/2	.35	.78	1.7	3.3	4.3	9.3	18.6	40
3/4	.44	.97	2.1	4.2	5.3	11.5	23.1	
1	.56	1.23	2.6	5.3	6.8	14.7	29.4	
1-1/4	.74	1.6	3.5	7.0	8.9	19.3	38.6	
1-1/2	.86	1.9	4.1	8.1	10.4	22.6	45.2	
2	1.10	2.4	5.2	10.4	13.4	29.0	58.0	

Table 14-3. Lifting Capacities at Various Elevations

Elevation (feet)	Atmospheric Pressure (psi)	Available Lift (feet)
Sea Level	14.7	17
1000	14.2	16
2000	13.6	15.5
3000	13.1	15
4000	12.6	14.5
5000	12.1	14
6000	11.7	13.5

Example 1

Given:
- Vertical distance — 12 feet
- Total length of pipe — 100 feet
- Pipe size — 1″ in diameter
- Pump size — 2 GPM
- Fittings in line — 3 elbows, no valves
- Elevation above sea level — 3,000 feet

Solution: Referring to Table 14-2, an elbow equals 2.6 feet of pipe (2.6 × 3 elbows = 7.8 feet). The corrected length of pipe is now 107.8 feet. Referring to Table 14-1, 100 feet of 1″ pipe at 2 GPM has a head loss of .5 feet. By dividing 107.8 by 100 and multiplying by the .5, the actual head loss is .54 feet. Therefore, the total lift needed for this system is the vertical distance plus .54 feet, or 12.54 feet.

Since the pump is safely capable of lifting 15 feet at a 3,000 foot elevation (see Table 14-3), the previous example will perform satisfactorily. However, if a 3/8″ diameter pipe had been used, the head loss would have been 16.48 feet. Adding the vertical distance to this figure equals 28.48 feet. The pump would not be able to lift the fuel.

If the plumbing system cannot be built under a 17 foot lift limitation (at sea level), a remote pumping station must be used. This is placed between the main tank and the transfer tank. The proper placement is determined by

Fuel Systems, Diesel

the pump lift calculations and the following pump head calculations.

14.2.3 Pump head. The pump's head is the theoretical vertical distance a pump will push fuel. Day tank standard pumps have 231 feet of head (100 psi) at nominal flow rate. Because of electrical convenience, the pump is normally located with the transfer tank. But when pump lift demands are exceeded, a remote pumping station is required. This allows us to utilize the head (pushing) capabilities of the pump. See Table 14-4 for typical discharge pressure capabilities.

To determine the total available head, three factors need to be considered:

a. The vertical distance the pump needs to push the fuel is the main factor in head capabilities. This measurement should be taken from the output port on the pump to the transfer tank's uppermost piping connection.

b. Length and size of pipe need to be considered in the same manner as the lift calculations.

c. Fittings also are calculated in the same manner.

The following example is based on 60°F temperature. Elevation does not need to be considered in head calculations.

Table 14-4. Pump Discharge Pressure (psi)

Motor H.P.	Nominal Pump Size (GPM) at 1725 RPM					
	2	4	7	10	19	23
1/3	100	60	2			
1/2		100	20	2		
3/4			40	20		
1			100	40	20	2
1-1/2				80	40	40
2				125	60	60
3				150	100	125

NOTE: Pump discharge volumes (GPM) can decrease by as much as 25% when higher pressures are required. Please consult factory for borderline consumption rates.

Example 2 (see Figure 14-2)

Given:	Vertical distance	150 feet
	Total length of pipe	175 feet
	Pipe size	3/4" diameter
	Fittings	1 check valve, 1 solenoid valve
	Pump	7 GPM

Solution: Referring to Table 14-2, a 3/4" elbow equals 2.1 feet of pipe (2.1 × 2 = 4.2). The check valve equals 5.3 feet of pipe. Also, the solenoid valve has a 3 psi drop, (consult manufacturer) or 6.93 feet (3 × 2.31). The total adjusted length of pipe is: 175 + 4.2 + 5.31 + 6.93 = 191.4 feet. Referring to Table 14-1, 191 feet of 3/4" pipe at 7 GPM interpolates to 29.2 feet of head loss (1.91 × 15.3). Therefore, total equivalent height is (150 + 29.2) 179.2 feet. (Note: The resulting pressure at transfer tank is (231 feet − 179.2 feet) divided by 2.31 = 22 psi). Since the pump will push fuel to a height of 231 feet, this system will work.

14.2.4 Pump prime. Maintaining the prime on a pump is of critical importance. Fuel must be maintained with no air pockets in the suction side pipe. Foot valves at the main tank or check valves at the transfer tank can be used to prevent fuel flowing back to the main tank and losing prime.

Pump cavitation (air in the line) can occur gradually and will eventually ruin a pump. Therefore, a visible discharge hose from pump to day tank is an important feature. It allows a visual inspection to be sure of the proper transfer of fuel without loss of prime. Also, the pump motor should be thermally protected to safeguard against burn-out and possible fire.

Vertical piping loops or traps should be avoided when designing a pumping system. Air pockets can become trapped in the high point of the vertical loop, resulting in pump cavitation.

A hand pump is recommended for initial priming to avoid undue wear on the fuel pump. If the fuel pump must be used for initial priming, do not run for more than 60 seconds. Fuel should be flowing within that time.

A fuel strainer on the inlet side of the pump is also recommended. Foreign particles entering the pump chamber will diminish it's life expectancy. The strainer should be checked periodically to avoid particle build-up which would limit pumping capabilities.

14.2.5 Pumping Notes. Proper engineering practices should always be used when calculating pump head and especially pump lift. By following these guidelines, costly repair due to improper installation can be avoided.

NOTES:

a. 1 psi = 2.31 feet of head is the conversion for water. As a general rule, this is a safe conversion for #2 diesel fuel.

b. For more precise calculations refer to the equations and conversions listed below:

$$\text{Head in feet} = \frac{\text{PSI} \times 2.31}{\text{Specific gravity of fuel}} \quad \text{(Equation 14-1)}$$

$$\text{PSI} = \frac{\text{Head} \times \text{specific gravity of fuel}}{2.31} \quad \text{(Equation 14-2)}$$

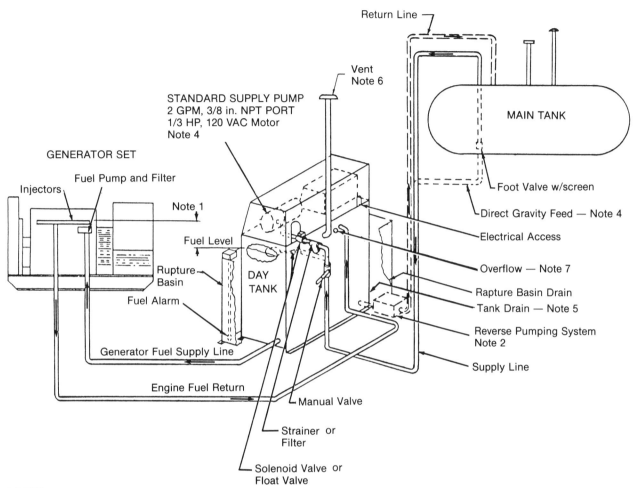

NOTES:
1. Black iron pipe, schedule 40, should be used for all fuel piping.
2. Reverse pumping system transfers fuel back to main tank when a high level condition exists.
3. All standard fittings 1 in. NPT, except vent (1-1/2 in.) and drain (5/8 in.).
4. Many state and local codes require main tank fittings to be top mounted with a pumping system. Solenoid valve still required to prevent syphoning effect.
5. Water and sediment should be drained from tank on a periodic basis (petcock or manual ball valve). Tank is planned to allow draining through rupture basin.
6. Day tanks are not intended to be pressurized vessels. Do not use vent line as a stand pipe.
7. Due to the inherent danger of a gravity feed system, we strongly recommend the safety features shown on this page.

Figure 14-2. Day Tank System — Below Main Tank

Fuel Systems, Diesel

Specific Gravity of #2 diesel fuel — .88 at 60°F.

Weight of #2 diesel fuel — 7.3 #/gal

c. All calculations are based on a 60°F temperature. Allowances must be made for extreme temperature variances.

14.2.6 Gravity system. With the main tank above the transfer tank, the pump is not always needed. A gravity feed may be sufficient (see Figure 14-2). This is a potentially dangerous situation, since a 10,000 gallon tank can be above a 50 gallon transfer tank. Therefore, we recommend avoiding this situation. When this situation is impossible to avoid, the following safety items are recommended.

a. High level alarm (refer to Section 14.3)

b. Critical high level alarm (refer to Section 14.3)

c. Rupture basin and fuel in rupture basin alarm. The rupture basin surrounds the transfer tank and allows time for corrective action before flooding occurs.

d. Solenoid valve or float valve to stop flow of fuel to transfer tank.

e. Filter or strainer to maintain clean fuel. This will ensure proper seating of valves.

f. Manual shut-off valve to stop flow of fuel in the event of an overflow.

g. Reverse pumping system, which will pump the overflow fuel back to the main tank.

14.3 ELECTRICAL CONTROL

14.3.1 Single system. The transfer tank system is controlled by electrical switches. This is a vital system component. The main function is to maintain the fuel in the tank by monitoring the level and controlling the pump.

A level indicator is necessary for visual inspection of the fuel level in the tank. Also recommended are warning lights for critical tank levels. Along with warning lights, it is important to have a means for remote annunciation (i.e., relays) so the transfer tank can be wired as an integral part of the generator system. This can be accomplished by wiring it to such accessories as flashing lights, sirens, a computer, or micro-processor.

The levels are:

a. *High level.* Signals overfilling or syphoning by the main tank.

b. *Low level.* Signals that a pump(s) has malfunctioned, is unable to keep up with demand, or low level in the storage tank. (This level should be set to provide ample time to react before a shutdown occurs. A stand-by hand pump should be ready in this case.)

c. *Low level shutdown.* Signals insufficient fuel to operate generator and should be wired to shut down the engine-generator.

d. *Low fuel in main tank.* Allows a means to monitor the main tank at the transfer tank.

e. *Fuel in rupture basin.* Signals an overflow or leak into an optional rupture basin.

14.3.2 Duplex system. The duplex system controls two pumps. The usual set-up is to have the lead pump turn on at the 88% level, with the secondary pump turning on at the 75% level. The two pumps should be alternated as the lead pump, on each cycle. This allows for even wear on pumps.

If a completely redundant system is needed, a secondary critical high level float switch can be added. This back-up switch will signal there is a high level condition and stop the pump in the event the main controller fails.

14.3.3 Testing. A major maintenance consideration is to be able to test the motors and pumps and also test all the warning signals for proper operation. This is of extreme importance because the transfer tank does not function all the time, but when it does, it must work. Therefore, these tests should be part of a periodic maintenance schedule.

Fuel Systems, Gaseous

J. A. Lang

CHAPTER 15

Gas engines can be designed to operate on various gaseous fuels, including natural gas, HD-5 propane, digester (sewage) gas, and landfill gas.

Each engine installation must be designed with a particular fuel in mind, because the octane rating and Btu content of different fuels can vary greatly (see Table 15-1). A gas analysis should be submitted to the engine manufacturer if a non-commercial grade fuel is to be used.

Table 15-1. Fuel Specifications

Type of Fuel	Octane Rating	Btu/Cu Ft LHV
Natural Gas	115–120	900
HD-5 Propane	95	2400
Digester Gas	115–120	600
Landfill Gas	115–120	400–450

15.1 NATURAL GAS ENGINES

Natural gas engines are often used for prime power applications. Generally the gas is supplied by utility owned lines that run to the installation site. The major components in the natural gas fuel system are the pressure regulators, piping, and the carburetor (see Figure 15-1).

15.1.1 Pressure Regulators.
Pressure regulators (see Figure 15-2) are designed to control the pressure of the gas as it enters the engine. Through an arrangement of a diaphragm and springs, the pressure of the natural gas coming to the engine is lowered and controlled. This provides a steady supply of gas to the carburetor.

The fuel system uses a high pressure line regulator, mounted in the main fuel line, and a low pressure engine mounted regulator. The line, or high pressure, regulator lowers the gas pressure in the lines that feed the engine mounted gas regulators to a lower predetermined value, depending on the application. The engine regulator sets the gas pressure to the carburetor. From the engine mounted regulator the gas flows into the carburetor, where air is mixed with the gas. The mixture then flows into the engine to be burned. Gas pressure to the engine regulator is typically 5–10 psi (0.34–0.68 bar) for naturally aspirated engines, 22–25 psi (1.5–1.7 bar) for turbocharged engines, and 35–50 psi (2.4–3.5 bar) for lean burn configured engines. Low gas pressure will starve the engine of fuel and reduce engine output. Excessively high pressures could damage the regulator, allowing excessive fuel to flood the cylinders. This could lead to detonation and serious engine damage.

If at all possible, avoid feeding any gas operated equipment from the supply line between the line regulator and the engine regulator. The supply pressure to the engine could be disrupted. If there is no way to avoid such an installation, add a second line regulator close to the engine and increase the pressure from the first line regulator by 10 psi (0.68 bar) to compensate for the pressure loss.

A second line regulator may also be necessary if the initial line regulator is located far from the engine.

On turbocharged engines, a balance line must run between the pressurized air at the carburetor air horn and the atmospheric vent in the regulator. By using this line, the gas pressure will increase as the air pressure increases, maintaining gas over air pressure differential at idle. Use a 7/16 in. (11 mm) ID line for the balance line.

a. *Pressure drop between regulators.* The maximum pressure drop across a line regulator is generally 50–75 psi (3.5–5.2 bar). Consult the regulator manufacturer for specific information.

15.1.2 Carburetors.
The carburetor on all gaseous fueled engines is designed to mix gas and air into a

Fuel Systems, Gaseous

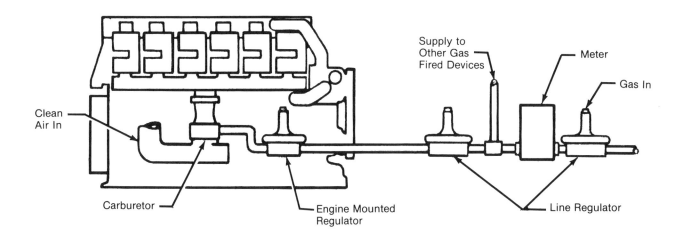

Figure 15-1. Gas Fuel System Installation

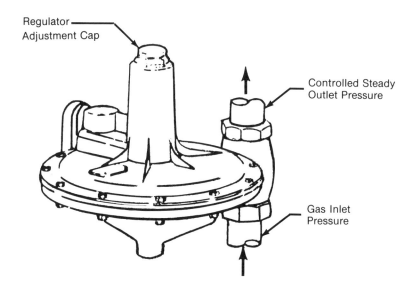

Figure 15-2. Pressure Regulator

combustible mixture. Through a series of orifices, springs, and diaphragms, gaseous fuel and air are mixed to provide the proper fuel-to-air ratio for efficient engine operation. Carburetors are used to provide a constant fuel-to-air ratio under varying and constant loads.

The gas-to-air mixture must be established at start-up. Consult the engine manufacturer for instructions on gas carburetor settings.

15.1.3 Piping. Piping to the engine site is generally supplied by the utility. Installation piping will have to be fitted from the meter or main feed line to the engine. This piping should be absolutely clean and scale free. If possible, blow out the lines with clean, dry compressed air before mounting the regulator. Piping should be black iron or steel to avoid reacting with the sulfur in the fuel. **Never use galvanized metal or zinc alloy piping.**

The pipe size used should be the same size as the engine fuel inlet. All threaded pipes should be sealed to prevent leaks. A shut-off valve should be installed along the pipe span just before it enters a building.

WARNING

All gas installations in closed areas or buildings should have a positive shut-off valve to prevent gas leakage when the engine is shut down. Consult all applicable local, state, and federal building codes for each installation.

Natural gas piping should never run near furnaces, heating pipes, electric wiring, or exhaust manifolds. The high temperatures could cause an explosion. Natural gas piping should always be insulated for added protection.

Always incorporate a flexible pipe connector in the piping system. Locate this flex connector as close to the engine regulator as possible.

15.1.4 Fuel treatment. Fuel filters and/or scrubbers are often recommended to remove dirt, rust, scale, water, and chemical contaminants from the fuel. If debris remains in the fuel, the regulator orifice and gas jets in the carburetor could clog up and reduce engine performance. In areas where the gas has a large sulfur content, specially treated scrubbers and cleaners should be used. Sulfur in the fuel would combine with the water formed in the combustion process to make sulfuric acid. The acid causes rapid corrosion of engine components.

15.1.5 Gas mixing. Some gas distribution systems add a propane-air mixture to the natural gas to compensate for low pressure conditions. This is called **gas mixing**.

Gas mixing lowers the octane level of the fuel. To determine the octane drop, take the percentage of propane added to the system and multiply it by the difference between the two octane levels. For example, if the normal natural gas supply is established at 115 octane and 25% of 95 octane HD-5 propane is added, the resulting octane number is decreased by 25% of the difference between 115 and 95, or 5. The resulting octane level would be 110. If the propane level is increased to 50%, the octane level would drop to 105. Depending on engine load, gas mixing can lead to damaging detonation. If detonation is experienced, the load must be reduced or the spark retarded, or both, when the engine is adjusted for natural gas. In areas where gas mixing is common, a second set of marks on the magneto will help the operator retard the spark quickly. The addition of propane-air mixture is proportioned so carburetor readjustment is not necessary.

15.2 HD-5 PROPANE FUEL SYSTEMS

Propane is a high Btu, low octane fuel that is generally sold and stored as a liquid. Propane is converted to a gaseous state at the engine, in a separate piece of equipment called a vaporizer. The liquified gas runs through the vaporizer and is warmed and converted to vapor. Propane fueled engines are most often used for non-prime power installation.

CAUTION

The other LPG fuel, butane, has too low an octane rating (approx. 80) to be a satisfactory fuel in modern engines.

The gas-to-air mixture must be established at start-up. Consult the engine manufacturer for instructions on gas carburetor settings.

This type of fuel system is very similar to the natural gas system; however, several additional components are needed.

15.2.1 Filter. A small liquid stage filter should be used to protect the vaporizer, carburetor, and engine against fuel tank and line scale.

15.2.2 LPG vaporizer. This fuel is liquified by compressing it while it is in the gaseous state. The high pressures caused by compressing the gas convert it into a liquid. The liquid fuel is then stored. Before the fuel can be used by the engine, it has to be reconverted to a gas in the fuel vaporizer. Heated air or water runs through the vaporizer and warms the fuel to a gaseous state. From there, the gas flows to the carburetor, where it is mixed with air. A vaporizer must be used on most LPG applications. Only very small engines can run off the gas vapor that forms in the fuel tank. See Appendix Section A.9 for propane vaporization data. Many local codes require an outside location for the fuel vaporizer. Contact your local distributor for the type of vaporizer that will be best for your installation.

Exterior propane fueled engine installations can usually be equipped with a combination vaporizer and pressure regulator (converter) mounted on the engine. These engine mounted converters use heated engine coolant to vaporize the liquid propane.

15.2.3 Fuel tank. An approved LPG fuel tank must be used to store the fuel. Shut-off valves and pressure gauges are usually incorporated in the tank. Consult the tank manufacturer for more specific information.

Due to the volatile nature of the fuel, the tank should never be located inside any structure. (Most local codes require that the tank be installed outside a building.) Always plan the installation to keep the fuel tank away from open flames, sparks, or electrical connections.

Since propane gas is heavier than air, the tank enclosure should not contain any piping trenches or floor drains that would permit escaping gas to get under the building or into city storm drains or sewer systems.

15.2.4 Piping. As with piping for natural gas, never run an LPG line near heat sources, exhaust manifolds, or electric wires. This particular fuel will vaporize at −44° F (−42°C). Any excess heat or loss of pressure could lead to rapid fuel vaporization within the system. Try to keep the pipe spans as short as possible. Piping can be insulated for added protection.

A balance line may be used between the carburetor air horn and the atmospheric vent in the regulator. The balance line should be 7/16 in. (11 mm) ID.

15.2.5 Carburetors. As with the natural gas engines, a propane carburetor is nothing more than a gas and air mixer. The springs, orifices, and diaphragms will determine the gas-to-air ratios. Carburetor adjustment is outlined in the engine operation and service manuals.

15.3 DIGESTER (SEWAGE) GAS FUEL SYSTEMS

The natural by-products of modern sewage plants can often be used as fuel for gas powered engines. Digester gas is economical and its use leads to a certain degree of independence from local utilities.

A sewage disposal plant will produce from 0.8 to 1.0 cubic foot of gas each day for every person served by the system. For large cities, this amount becomes significant. Great fuel savings can be realized by supplementing the natural gas supply with sewage gas.

Sewage gas installations require considerable experience and specialized detail with respect to controls and fuel handling. In some instances, the gas may require special filtering or pre-treatment. In other cases, efforts to use exhaust heat exchangers to reclaim waste heat for processing may be handicapped by corrosion due to the chemical composition of the gas.

Most of these problems can be solved if the prospective user works closely with the design engineer in planning the installation.

Cooling Systems, Liquid

Lyle Christiansen

CHAPTER 16

INTRODUCTION

Liquid cooled engines are cooled by transferring the waste heat to a coolant pumped through the engine. The coolant must be cooled outside the engine for reuse or discarded after passing through the engine.

Liquid cooled engines in most cases need a cooling system added to the basic engine. The only exceptions are those engines that flow coolant through the engine and discard it. Examples of the latter are direct sea water cooling and direct city water cooling. Two general types of heat exchanger systems are used: water and air. Both types remove the waste heat from the engine coolant so that it can be reused. The coolant used in the engine is treated water in most applications. Water used as a coolant should be soft or have as few scale forming minerals as possible. Also the water should be treated with a corrosion inhibitor and freeze protection if needed. (See Figure 16-1 for freezing point table.) The engine and heat exchanger manufacturers should be contacted for specific coolant requirements.

16.1 WATER COOLED SYSTEMS

These systems utilize a heat exchanger to transfer the waste heat to the cooling water. (See Figure 16-2 for a typical heat exchanger.) Water cooled systems offer quiet operation, high pressure capability, small package size, and no need for cooler ventilation. The system does require a source of water and a location to dump the water. Freeze protection can be a problem.

In most cases, the engine coolant flows through the shell side of the heat exchanger. This is because of the lower heat transfer rate of the usually inhibited and cleaner fluid. On small heat exchangers, the bundle or core is fixed and cannot be removed to clean the shell side.

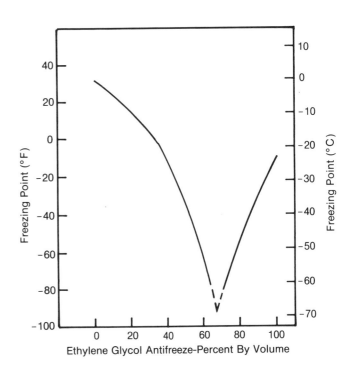

Figure 16-1. Freezing Points of Aqueous Ethylene Glycol Antifreeze Solutions

Large 8 to 10 inch diameter units may have removable bundles for cleaning. The cooling water or raw water flows through the tube side of the exchanger. This allows cleaning of the unit if the raw water causes fouling.

Figure 16-3 shows a typical water cooled heat exchanger installation. For applications using only one cooling water pass through the cooler, counter flow plumbing is desirable (raw water inlet on coolant outlet end of cooler).

Cooling Systems, Liquid

Figure 16-2. Heat Exchanger

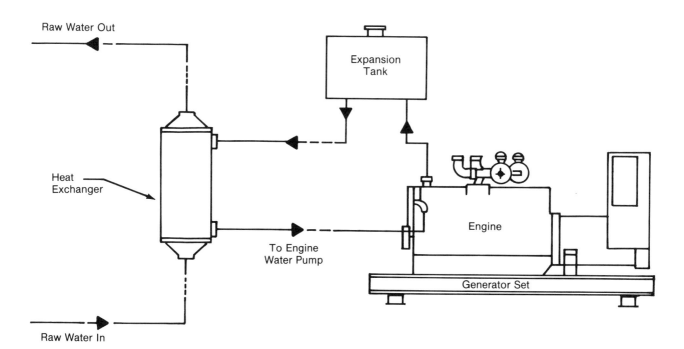

Figure 16-3. Heat Exchanger Installation

To properly size a heat exchanger, the manufacturer will need to know:

a. Coolant flow and type*

b. Maximum coolant temperature

c. Desired heat rejection

d. Raw water flow and temperature

16.2 FAN COOLED SYSTEMS

These systems use a heat exchanger, normally called a radiator, to transfer the waste coolant heat to the ambient air. A radiator system does not require a source of water, nor a drain system, and freeze protection is easily provided. The system does require large air flows, is generally noisier, requires considerable space, and a source of power for the fan. A typical radiator has top and bottom tanks, core, support structure, and in the case of remote radiators, an air moving device. The tanks direct the engine coolant through the core, provide for expansion, and allow for deaeration in some systems. The core typically consists of tubes and fins bonded together to form a heat exchanger to transfer the waste heat from the engine coolant to the air. The support structure holds the components in a durable assembly and the air moving device moves air over the core to pick up the waste heat.

16.2.1 Construction. Most radiator variations are found in the core construction. Tubes can be either round or oval and are typically made of either copper or brass. Fins are either plate type or serpentine and are typically made from steel, aluminum, copper, or brass. The tubes are inserted into the plate fin or assembled in alternating layers for the serpentine fin to form a core. The tubes are either mechanically or metallurgically bonded to the fins for strength and to increase heat transfer. The tubes are either soldered, welded, or roller expanded to headers that are then joined to the tanks. Some radiators feature tubes that are sealed in the headers individually with rubber grommets for ease of replacement in the field. On some radiators, the headers are round pipes that also act as the inlet and outlet tanks. For most radiators, the headers are joined to the tanks by soldering (soldered construction) or a bolted joint (bolted construction).

Most industrial engine mounted radiators have a water flow from the top to the bottom, or downflow on vertical core radiators. Less used but available are radiators with tanks on the sides and side to side flow or crossflow. The former design is most common in car and truck applications where a low profile is needed. The large number of variations in radiator design covers the need for low cost, performance, durability, ease of repair, and thermal durability. Based on the application and design requirements, a manufacturer should be able to combine the radiator variations into an acceptable package.

16.2.2 Remote radiators. These radiators are usually located some distance from the engine and have an integral means of moving air through the core. This is usually a fan driven by an electric or hydraulic motor. Typical remote radiators have either a vertical or horizontal air flow direction. Site conditions and prevailing winds determine which is best for the job. Figure 16-4 shows a typical remote radiator.

Figure 16-4. Remote Radiator

The proper sizing of a remote radiator requires:

a. Desired heat rejection

b. Coolant flow and type

c. Maximum allowable coolant temperature

d. Ambient air temperature and elevation at site

---

*Coolant type refers to whether the coolant is pure water or 50% water and 50% glycol. The addition of glycol reduces the heat transfer rate of the heat exchanger.

e. Any external factors that might influence the air flow, such as ducting, placing of other coolers in the airstream, louvers, or other obstructions that might restrict air flow or change the ambient air temperature to the radiator

This information and the vendor's performance chart will allow selection of a remote radiator.

16.2.3 Engine mounted. These radiators are mounted in front of the engine and use a fan driven by the engine to force air through the core. (Figure 16-5 shows a typical engine mounted radiator.) The fan is usually supplied with the engine. If the radiator is also supplied with the engine, the rated ambient capability of the system should be obtained from the engine supplier. A radiator purchased separately must have the ambient capability calculated. The following information is needed in addition to that listed for a remote radiator:

a. Fan speed

b. Fan performance at rated speed

c. Air direction (blower or intake)

Figure 16-5. Engine-mounted Radiator

d. Amount of radiated heat from engine or other equipment in close proximity to the radiator, such as compressors, generators, or other sources of radiant heat that will affect the temperature of the air at the fan.

The amount of radiant heat from the various sources is used to calculate the air temperature rise for air drawn over these sources on blower fan applications. For these blower fan applications, ambient air temperature and air temperature to the radiator are not the same. Most engine and radiator suppliers provide the system ambient capability from the above data. Some provide worksheets to allow the end user to do the calculations.

16.3 RADIATOR SYSTEM CONSIDERATIONS

Engine manufacturers or job specifications may have specific requirements that the cooling system must meet. Some common requirements are:

a. Specific fill times

b. Expansion volume in the top tank

c. Air handling

d. System pressure drop

e. Draw-down

f. Designs to reduce pump cavitation

g. Filling ability

h. Built-in sight glass

i. Low liquid level sensing

These specific requirements are met through a combination of radiator and system design changes in top tank design, addition of tank baffles, core changes, or surge tanks. Horizontal core radiators will, in most cases, require an additional surge or expansion tank to meet the above needs. Adding vent lines, as well as the location and size of vent lines and hose sizes to and from the radiator, will also influence some of the above requirements.

Many of the above requirements are considered under the general requirement for system deaeration. Deaeration is the removal of gas trapped in the system at initial filling or due to system leaks such as head gasket leaks. Each engine manufacturer has some requirement regarding deaeration. The specific needs should be presented to the radiator supplier at the time of radiator selection. (A review of deaeration is included in the Appendix.)

Most systems include a pressure cap to keep the system pressurized during operation. This pressure keeps the coolant from boiling and also aids in preventing pump cavitation. From a radiator life standpoint, the pressure should be as low as possible. The coolant boiling point is usually the limit that controls cap pressure. (Figure 16-6 shows the boiling point of coolant at various pressures.)

A maximum coolant temperature is specified for each engine system. The higher the temperature, the smaller the radiator needed. The engine manufacturer should be

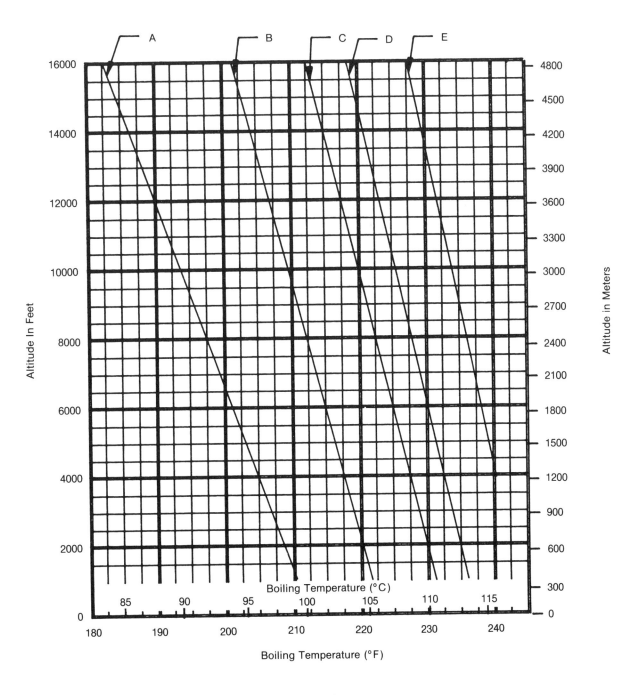

A — No Pressure Cap
B — 4# Pressure Cap (27.6 KPA)
C — 7# Pressure Cap (48.3 KPA)
D — 9# Pressure Cap (62.1 KPA)
E — 12# Pressure Cap (82.8 KPA)

Figure 16-6. Water Boiling Temperature At Altitudes Above Sea Level

Cooling Systems, Liquid

contacted for the maximum approved coolant temperature. This temperature is usually well above the thermostat opening temperature.

Many engine manufacturers require a system cooling test before any approval is given. When running the test, the thermostat should be blocked open to ensure that the radiator is controlling the temperature. Do not remove the thermostat in place of blocking it open.

16.4 ADDITIONAL SYSTEMS

Some applications require that the radiator be located considerably above the engine. If this height exceeds the allowable height specified by the engine manufacturer, the use of one of the following two systems is advisable.

A combination of heat exchanger and a radiator can be used to solve the problem. The heat exchanger is located near the engine and the coolant is run through one side of it. The radiator is located as desired and is plumbed to the other side of the heat exchanger. Figure 16-7 shows a typical system. Because of the addition of the heat exchanger, the radiator will have to be larger than one that is used directly to cool the engine. An auxiliary pump will be required in the radiator circuit to provide the desired flow to the radiator. The heat exchanger is able to handle the radiator induced head because it has a higher pressure rating than most engines. This is a more costly system than the single cooler system but does allow the use of a radiator well above the engine.

The other system used to solve the elevation problem is the hot well/cold well system. (Figure 16-8 shows a typical hot well/cold well system.) The heat exchanger is replaced with a large partitioned tank. Tank size must be large enough to contain all the water in the system unless check valves are used to keep the fluid in the radiator. This system requires more space because of the size of the well but is typically lower in cost depending on fabricating costs.

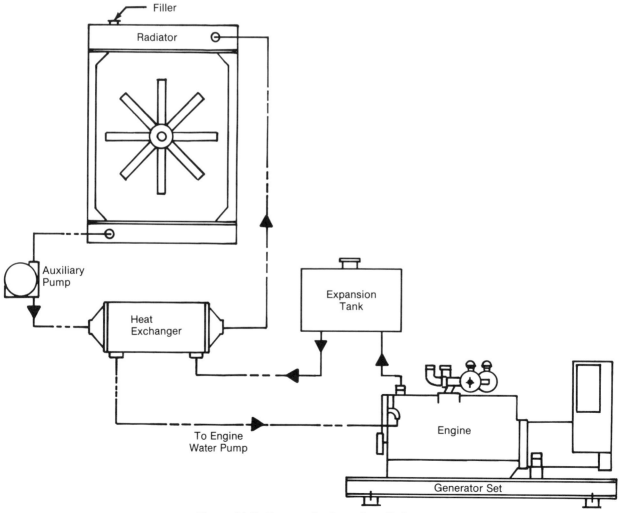

Figure 16-7. Remote Radiator Installation

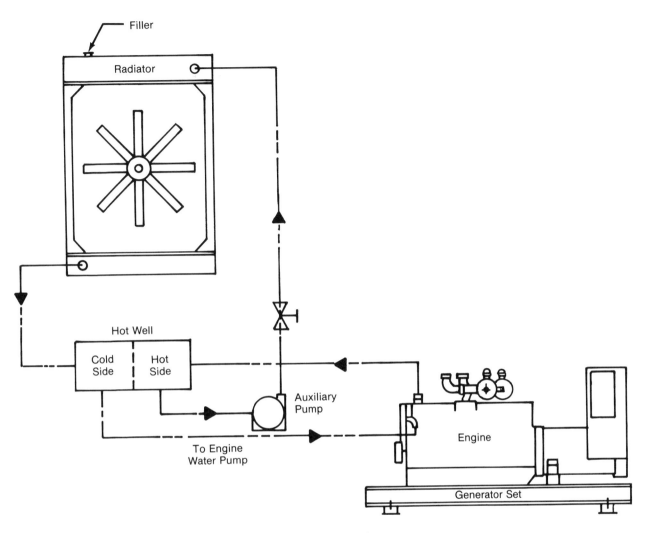

Figure 16-8. Hot Well System

Sizing of either system requires the same information as that needed for a remote radiator. Without the additional heat exchanger, the radiator size should not increase with the hot well/cold well system.

16.5 EXHAUST GAS HEAT EXCHANGERS

When engine waste heat is recovered for use in other systems, the heat recovered from the exhaust must also be considered in radiator sizing. Large quantities of heat can be recovered from the exhaust for use in heating and other processes. The proper sizing of an exhaust gas exchanger requires the following information:

a. Exhaust gas flow

b. Exhaust gas temperature

c. Water flow to heat exchanger

d. Water temperature

e. Desired gas temperature out of the heat exchanger

16.6 FUEL COOLER

Another type of cooler used on some applications is a fuel cooler. This cools the unused fuel that is returned from the engine. The need for this type of cooler is covered in job specifications and by engine manufacturers. To properly size a heat exchanger, the supplier will need to know:

a. Desired heat rejection

b. Ambient temperature and elevation at site

c. Fluid flow and type of fluid to be cooled

d. Maximum fluid temperature allowed

e. Desired type of cooler

16.7 DEAERATION

Deaeration is the removal of entrained gas from the coolant in an engine cooling system. This gas can be entrained in the system by any of the following:

a. During initial fill, air can be trapped at high points of the system, such as the heater. Also, because of the geometry of the engine or its vehicle position, air can be trapped in the engine water passages during initial fill.

b. Air can be pumped into the system through head gasket leakage of the water cooled air compressor found on many vehicles.

c. Combustion gases can enter the system through head gasket leakage or leakage around the fuel injectors or precombustion chambers.

d. High water velocities and rapid changes of water in the top tank can entrain air in the system.

e. Air can be entrained in the system through leakage at water pump seals. This is not serious; in today's systems, most radiators employ a pressure cap to keep the total system at a positive pressure.

f. Vapor (steam) can be created in the system at the water pump inlet. If the coolant is allowed to flash into steam at the pump inlet because of the pressure and temperature conditions, the pump will cavitate. This will reduce flow in the total engine and allow boiling on the water side of the engine.

16.7.1 Why deaeration?
In a system that has no deaeration provisions or improper provisions, the possible results are cylinder head cracks, water side corrosion, and erosion. The cylinder head cracks are a result of thermal stress in the heads. On the combustion side of the head, the gas temperatures can exceed 1500°F. On the water side, excessive boiling could allow the water side passages to go dry; thermal stresses will then be such that the heads will crack. This excessive boiling is accelerated with aerated water and will certainly happen if the flow through the head ceases or is reduced substantially.

Gas bubble formation on cylinder liners will cause local hot spots. Collapse of the bubble allows coolant to contact the hot metal and local erosion takes place. If this is severe, failure of the engine parts can occur.

Entrained gas in the system will also accelerate water side corrosion. The corrosion products that form on the water side reduce the water side heat transfer, resulting in high metal temperatures and ultimately cracked heads.

16.7.2 How to deaerate.
There are a number of means of providing a deaeration system, but they all basically function by bypassing approximately 5% of the total coolant flow into a calm area where the entrained gas is allowed to separate from the coolant. The deaerated coolant is then routed back into the system. Figure 16-9 shows a means of gas separation in the top tank of the radiator itself and Figure 16-10 shows a remote mounted surge tank system that accomplishes the same end.

As shown in Figures 16-9 and 16-10, the coolant flows from the engine to the radiator via line A. In the system shown in Figure 16-9, the coolant is discharged directly into the core side of the baffle. In the system shown in Figure 16-10, the coolant is discharged directly into the top tank. Here the flow is separated, with approximately 95% of the flow going to the core and 5% going through line C. The coolant in lines A and C is considered as aerated coolant. As the coolant reaches the calm area of the top tank, or surge tank, the gas settles out of the mixture and comes to the air-water interface. The deaerated water returns to the system through line D, which is connected directly to the inlet side of the water pump. Connecting line D directly to the pump inlet ensures that the pump will always operate with a positive head on it. The positive pressure at the pump inlet reduces boiling at the pump inlet or impeller, which could result in reduced flow in the system.

There are other means of deaeration that are not as sophisticated as the systems shown in Figures 16-9 and 16-10. One such system employs a partial baffle in the top tank and is quite often used on sheet metal radiators. In this system, the coolant is discharged directly under the baffle in the top tank and is allowed to spill out into the top tank at the baffle ends. This system minimizes turbulence in the top tank and therefore, eliminates reaeration in the top tank. In this type of system, the deaerated water is returned to the system via the radiator tubes.

16.7.3 How much deaeration?
Before we discuss how much deaeration is required, we should introduce two terms that have not been used in the discussion thus far. These terms are **draw-down** and **air handling**. Draw-down is the amount of water that can be removed from the system without seriously affecting the system performance. Air handling is defined as the amount of air that can be injected into the system at a constant rate without seriously affecting system performance. With these terms defined, we can now examine how much deaeration is required in the system.

The amount of deaeration required is specified by the engine manufacturer. This varies from manufacturer to manufacturer, and with any manufacturer, it may vary with the type of engine. With requirements on air handling, draw-down, time to deaerate, and size restrictions, a supplier can make design proposals to meet the required deaeration. In most cases, an actual test is used to verify that the system works.

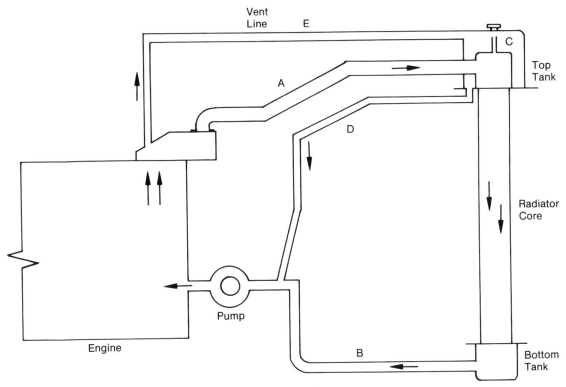

Figure 16-9. Diagram of Deaeration System

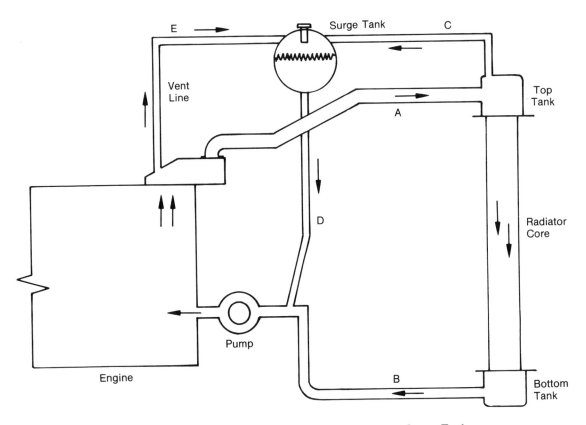

Figure 16-10. Diagram of Deaeration System with Surge Tank

16.8 SUMMARY

The preceding discussion of why, how and how much deaeration outlines the general basic principles of deaeration. Although the various manufacturers have different parameters for acceptability, they recommend deaeration for increased engine life.

Engine Exhaust Systems

Don Panetta

CHAPTER 17

INTRODUCTION

It is essential to the performance of a generator set that the exhaust system be properly designed. The most important factor is that the installed system not exceed the engine manufacturer's maximum exhaust back pressure limit. Pressure drop includes losses due to piping, silencer, and termination. Excessive back pressure could be from one or more of the following:

- Exhaust pipe diameter too small, or pipe run too long
- Too small an exhaust silencer or incorrect silencer design
- Too many sharp bends
- Excessively rough pipe or restrictions in pipe cross section

17.1 EXHAUST PIPING

Exhaust piping should conform to NFPA 37, Stationary Engines and Gas Turbines (Ref. 1), and any applicable local codes. In general, exhaust temperatures will be approximately 1000°F (538°C) measured at the engine exhaust outlet, except for infrequent brief periods. Standards for low heat appliances apply; for example, NFPA 211, Chimneys, Fireplaces, Vents and Solid Fuel Burning Appliances (Ref. 2). If these temperatures are exceeded, special consideration must be given to the higher temperatures. Exhaust pipes passing directly through combustible roofs, walls, or partitions, shall be guarded with metal ventilated thimbles to protect the combustible material from excessive temperatures.

Some of the different methods of calculating back pressure for piping are presented on the following pages.

17.1.1 Empirical formula. Calculate backpressure by:

$$P(psi) = \frac{L \times S \times Q^2}{5,184 \times D^5} \quad \text{(Equation 17-1)}$$

$$P(kPa) = \frac{L \times S \times Q^2 \times 10,000}{0.0027787 \times D^5} \quad \text{(Equation 17-2)}$$

Where

- P = Backpressure (psi) (kPa)
- psi = 0.0361 × inches water column
- kPa = 0.0098 × mm water column
- L = Length of pipe (feet) (meters)
- Q = Exhaust gas flow (cfm) (m^3/min)
- D = Inside diameter of pipe (inches) (mm)
- S = Specific weight of gas (lb-ft^3) (kg/m^3)

$$S(lb\text{-}ft^3) = \frac{39.6}{\text{Exhaust Temperature} + 460°F}$$

$$S(kg/m^3) = \frac{352.05}{\text{Exhaust Temperature} + 273.16°C}$$

To obtain equivalent length of straight pipe for each long radius 90° bend:

$$L = 33 \times \frac{D}{X} \quad \text{Standard Elbow (Radius = Diameter)} \quad \text{(Equation 17-3)}$$

$$L = 20 \times \frac{D}{X} \quad \text{Long Elbow (Radius Greater Than > 1.5 Diameter)} \quad \text{(Equation 17-4)}$$

$$L = 15 \times \frac{D}{X} \quad 45° \text{ Elbow} \quad \text{(Equation 17-5)}$$

$$L = 3l \quad \text{Flexible Connections} \quad \text{(Equation 17-6)}$$
$$l = \text{Connector Length}$$

Where X = 12 in or 100 mm

The radius of 90° bends with radii 1½ times the pipe diameter help to lower resistance.

Exhaust silencer back pressure must be added to pressure calculated for system piping to estimate total back pressure on the engine (Ref. 3).

17.1.2 Nomograph. The nomograph, Figure 17-1, may be used to calculate pipe back pressure.

Exhaust silencer back pressure must be added to pressure calculated for system piping to estimate total back pressure on the engine (Ref. 4).

17.2 DRY EXHAUST SYSTEMS

When designing an exhaust system with schedule 40 pipe, remember that the ID of this pipe can sometimes be smaller than the nominal pipe size indicates, especially on some of the larger diameters, e.g., 16 in. normal has approximately 15 in. ID. The weight of schedule 40 must be compensated for when designing supports for this type system. Schedule 40 pipe is very heavy (10 in. = 40.5 lbs/ft) and could have an inner surface that is rough and scaly. An engine will produce approximately one gallon of water for every gallon of fuel. Standby engines are run about an hour a week for check out. This amount of water could, over a time period, create a scaling problem in the pipe and restrict the engine exhaust flow, increasing the back pressure. This could reduce performance and even damage the engine. A smooth inner liner of stainless steel eliminates the possibility of rust and reduces the flow resistance of the exhaust gases. A double wall construction or insulated system reduces the possibility of condensation within the exhaust system.

A section of flexible exhaust line should be installed within two feet of the engine exhaust outlet. The flexible section limits the stress on the engine exhaust manifold or turbocharger that results from engine motion on its vibration mounts and temperature induced pipe expansions. The flexible section should be at least 12 in. (30.5 cm) long. When threaded flexible exhaust connectors are used, a 6–8 in. (15.3–20.4 cm) length of pipe should separate them from the exhaust manifold. This will make it easier to remove the flexible section, if necessary, without putting excessive strain on the engine manifold.

In any piping system where temperatures are a factor, thermal expansion must be evaluated to avoid excessive load on supporting structures. Steel exhaust pipe expands 0.0076 in. per foot of pipe for each 100° F rise of exhaust temperature (1.13 mm per meter for 100°C). Long pipe runs are sectioned with bellows joints. Each section is fixed at one end and allowed to expand at the other. Supports are located to allow expansion away from the

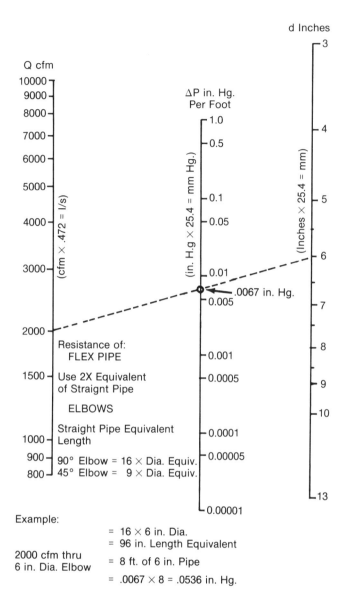

Figure 17-1. Nomograph for Exhaust Back Pressure

engine, avoid strains or distortion to connected equipment, and allow equipment removal without additional support.

A tee type condensation trap with drain plug or cock should be installed to prevent condensed moisture in the engine exhaust system from draining back into the engine when it is shut down. (See Figure 17-2, Dry Exhaust System.)

17.3 PRESSURE TIGHT PIPE

Another means of running exhaust piping is with a UL listed low-heat appliance factory-built pressure-tight pipe

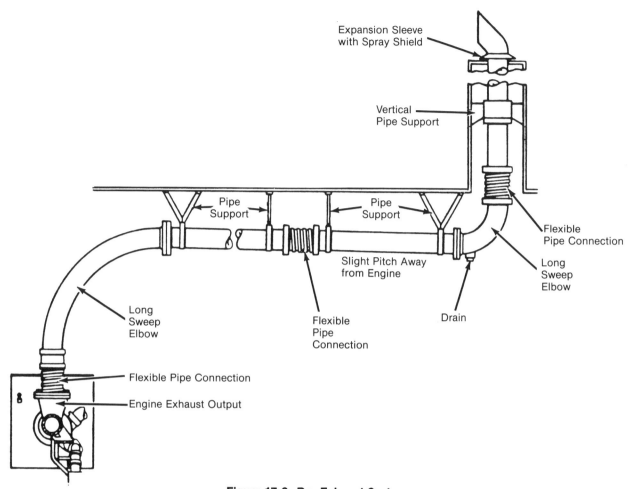

Figure 17-2. Dry Exhaust System

that is in accordance with NFPA 37 and NFPA 211 (Figure 17-3). This is a light-weight double wall pipe having a weight factor of 0.9 × diameter per foot. This double wall pipe has a very smooth inner liner constructed of 0.035 inch number 304 stainless steel. The pipe is available with a one inch air space or can have a factory applied fiber insulation of 1", 2", or 4" and an outer jacket of 0.025 inch aluminized steel.

Where heat or personal contact is of concern, the fiber insulated product can reduce outer jacket temperatures by as much as 84% of the exhaust gas temperatures (Table 17-1). These parts are assembled by means of an overlapping vee band and a pre-mixed high temperature sealant. The product is designed to hold up to 60 in. of water (4.4 in. Hg) positive pressure. With all parts factory-built, the flow resistance factors remain consistent and the piping system can be designed more accurately with regard to the maximum allowable back pressure. These factory-built systems are complete with regard to parts such as supports, bellows to allow for expansion of the piping system, elbows, tees, and ventilated roof thimbles. The manufacturer will size and design the exhaust system for desired back pressure and skin temperatures and will issue a ten year warranty on the product.

17.4 GENERAL

With today's concern for energy and where the exhaust gases could be used in a heat recovery type application, a relief valve should be designed into the exhaust system (Figure 17.4) not only to protect the exhaust piping but all items within the system, silencer, heat recovery equipment, etc.

When designing any type exhaust system, take care to locate the exhaust outlets so that no one will come in contact with hot surfaces. Exhaust gases are poisonous, and should be directed away from any occupied area. Be certain that exhaust gases cannot be drawn into any air intake vents, windows, doors, or other openings or enclosed spaces where gases could accumulate.

Although economically tempting, a common exhaust

Engine Exhaust Systems

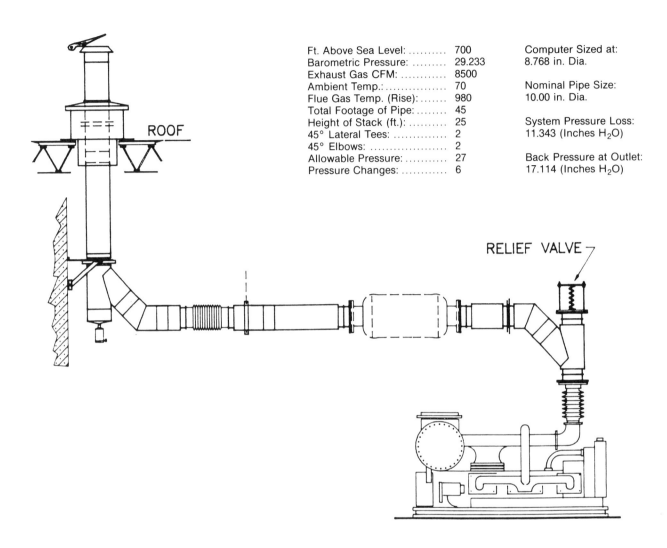

Figure 17-3. Factory-Built Pressure-Tight Exhaust System

Table 17-1. Insulation Thickness Required vs. Flue Gas Temperature,
Outer Pipe Surface Temperature and Pipe Size at 70°F Ambient Temperature

	120°F Surface Temperature				140°F Surface Temperature				160°F Surface Temperature			
	IPS Diameters				IPS Diameters				IPS Diameters			
ACTUAL FLUE GAS TEMPERATURES - °F	6"	12"	24"	36"	6"	12"	24"	36"	6"	12"	24"	36"
1000	—	—	—	—	4	4	—	—	4	4	4	—
900	—	—	—	—	4	4	—	—	4	4	4	4
800	4	4	—	—	4	4	4	—	4	4	4	4
700	4	4	—	—	4	4	4	4	2	2	2	4
600	4	4	4	—	2	2	4	4	2	2	2	2
500	2	4	4	4	1	2	2	2	1	1	1	2
400	1	2	2	2	1	1	1	1	1	1	1	1
300	1	1	1	1	1	1	1	1	0	0	0	0

Although economically tempting, a common exhaust system for multiple installations is usually not acceptable. Combined exhaust systems with boilers or other engines allow the operating engine to force exhaust gases back into equipment that is not operating.

REFERENCES

[1] NFPA 37, Stationary Combustion Engines and Gas Turbines, National Fire Protection Association, Batterymarch Park, Quincy, MA, 02169.

[2] NFPA 211, Chimneys, Fireplaces, Vents and Solid Fuel Burning Appliances, National Fire Protection Association, Batterymarch Park, Quincy, MA 02169.

[3] Caterpillar Engine Division, Application and Installation Guide, April 1983, pp. 58 and 61.

[4] Cummins Engine Co. Bulletin 3382408, 2/79.

[5] Selkirk Metalbestos Installation Instructions #4, 3/85, p. 15.

Exhaust Silencers

Douglas Bradley

CHAPTER 18

INTRODUCTION

Sound is such a common part of everyday life that we rarely appreciate all of its functions. It provides enjoyable experiences, such as listening to music or spoken communication with family and friends. It can alert us or warn us; for example, a ringing telephone or a wailing siren.

Many sounds are unpleasant or unwanted — these are called noise. The level of annoyance depends not only on the quality of the sound, but also our attitude toward it.

Worst of all, sound can damage and destroy. The most unfortunate case is when sound damages the delicate mechanism designed to receive it, the human ear.

As a result, numerous laws, regulations, and ordinances have been enacted to control excessive noise. Worker exposure to excessive noise over an extended period may result in a permanent loss of hearing. Noise can also be hazardous as well as objectionable to nearby residents. With tighter controls now in effect, every effort should be made to eliminate or reduce noise at its source.

18.1 SOUND DEFINITIONS

Sound is any pressure variation that the ear can detect.

Frequency. The number of variations per second is called the frequency, which is measured in hertz. The lower the frequency, the lower the tone of the sound.

Pressure. The pressure amplitude can be expressed in any typical units of pressure like inches of water or psi. The higher the pressure, the louder the sound.

Decibel. The decibel (dB) is the unit of measurement used in noise control. Because noise is an irritation which causes stress, pain, and fatigue on humans, it is necessary to measure it to understand its effects. The measurement units are decibels, which are measured on various scales (A, B, C, D, and Linear). The A scale is most common since it most accurately reflects what the human ear hears (see Table 18-1).

Table 18-1. Converting Octave Level to an Overall dBA Level

Freq.	Level (dB)	Correction Factor	Correction Level (dBA)
63	89	−24.6	64.4
125	93	−15.1	77.9
250	90	− 8.0	82.0
500	95	− 2.9	92.1
1K	90	0	90
2K	92	+ 1.2	93.2
4K	91	+ 0.9	91.9
8K	87	− 1.4	85.6

18.2 SOUND PRESSURE

The quietest sound pressure detectable by the human ear has an amplitude of about 0.00000008 in. water. The loudest tolerable is about 0.8 in. water. This relates to readings from 0 to 120 dB.

At zero decibels, hearing begins for those with excellent hearing. Fifty to 60 dB is a normal quiet office. Eighty dB is a heavy truck going down the highway. At 120 dB, pain to the normal ear begins. From 1–120 dB is a change in pressure on the ear of one million times. The sound pressure level is over a specified frequency range.

Octaves. Using octaves is a way to summarize frequency (see Table 18-2).

18.3 ADDITION OF LEVELS

If the sound levels from two or more machines have been measured separately and you want to know the total SPL

Table 18-2. Octaves

Frequency Range (Hz)	Octave Center Frequency (Hz)
44.7 to 89.1	63
89.2 to 178	125
179 to 355	250
356 to 708	500
709 to 1410	1000
1411 to 1820	2000
2821 to 5620	4000
5621 to 11200	8000

made by the machines when operating together, the sound levels must be added. However, dBs cannot be added together directly (because of the logarithmic scale). Addition of dBs can be done simply using Figure 18-1 and the following procedure:

a. Measure the SPL of each machine separately.

b. Find the difference between these levels ($L_2 - L_1$).

c. Enter the difference on the horizontal axis of the chart. Go up until you intersect the curve, then go to the vertical axis on the left.

d. Add the value indicated (ΔL) on the vertical axis to the level of the noisier machine (L_2), this gives the sum of the SPLs of the two machines.

e. If 3 machines are present, repeat steps 1 to 4 using the sum obtained for the first two machines and the SPL for machine three.

Example:

1. Machine 1 L_1 = 82 dB
   Machine 2 L_2 = 85 dB

2. Difference $L_2 - L_1$ = 3 dB
3. Correction (from chart) ΔL = 1.7 dB
4. Total Noise =   85 + 1.7 = 86.7 dB

18.4 WHY DISTANCE FROM THE SOUND SOURCE AFFECTS SOUND LEVELS

As the sound radiates from the source, the pressure gets spread over a larger and larger area. Thus, whenever we speak of the sound pressure level, we must mention the distance from the source. Normally the sound is reduced by 6dB, or dBA, by doubling the distance from the source. Actually, it is reduced by only 5 to 5.5 dB due to ground reflections.

All internal combustion engines produce noise, some more than others, while 2- and 4-cycle engines of equivalent hp and rpm produce roughly the same overall noise levels.

The predominant sources that make up engine noise are:

- Exhaust
- Air intake
- Cooling fan
- Mechanical

Our main concern is the control of exhaust noise. The most common noise control method is the installation of silencers.

Typically silencers are divided into three distinct categories; the reactive type, the dissipative absorptive type, and a combination of the two.

In silencing internal combustion engines, the most widely used silencer is the reactive type. It is largely dependent upon an area change to reflect sound energy back to the source and utilizes the attenuation properties of expansion chambers and perforated tubes, etc.

The proper selection and sizing of a silencer is of utmost importance to ensure that pressure drop, acoustical performance, and other specific design criteria are met.

In selecting a silencer, a common approach is to stipulate the maximum allowable noise levels and little more, leaving actual compliance up to the supplier. As the first step in eliminating uncertainties, unsilenced levels for the engine should be obtained from the manufacturer and supplied to the individual making a recommendation. This individual should also be given the exhaust flow of the engine (usually expressed in CFM). If the exhaust flow is unknown, use the equation below to determine the flow.

The silencer manuacturer needs the following information:

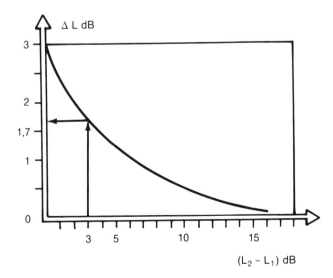

Figure 18-1. Incremental Sound Level, Two Machines

Engine displacement (cubic inches) 638
Rated BHP 250
Full load RPM 1800
Aspiration type .... T
Cycle 4
Exhaust flow rate (CFM)* 1650

*If the exhaust flow rate is not available, it can be approximated by the following equation:

$$(CFM) = \frac{\text{Engine Displ. (cu. in.)} \times \text{Full Load RPM} \times \text{Eff.} \times (\text{Exh. Temp.} \,°F + 460)}{C \times 941760}$$

(Equation 18-1)

Where:

    CFM = Exhaust flow rate in cubic feet per minute
    Efficiency = .85 for naturally aspirated engines
    Efficiency = 1.4 for turbo charged engines
    Efficiency = 1.2 for engines with scavenging blower
    C = 1 for 2-cycle engine
    C = 2 for 4-cycle engine

If exhaust temperarure is not available, use T = 1200° F for a gasoline engine; T = 900° F for a diesel engine.

NOTE: Use this equation only when the exhaust flow rate is not available.

With this information, the silencer manufacturer can, and should, guarantee the performance of a silencer with all exceptions and limitations clarified.

18.5 SYSTEM EVALUATION

It is extremely important to evaluate the total system. For example, while a silencer might theoretically reduce the exhaust noise of an engine to 90 dBA at ten feet without effective silencing or isolation of the engine intake, mechanical and other sources of noise, the measured noise level in the area could actually be considerably higher.

The distance at which the sound levels are to be measured must be included. At the lot line is not sufficient. An actual distance must be given.

The best way to recommend a silencer to do the most effective job is to fit the silencer to the engine based on the exhaust flow, rather than on the pipe size. Although all silencer manufacturers publish an attenuation curve, it will vary from one engine make to another. Figures 18-2 and 18-3 are examples of attenuation curves presently used by two manufacturers.

Two factors affect silencing. First, some engines are much easier to silence than others. Reasons for this are:

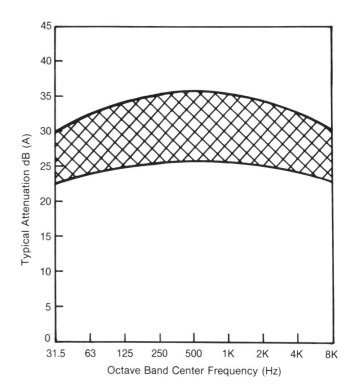

Figure 18-2. Typical Attenuation Curve

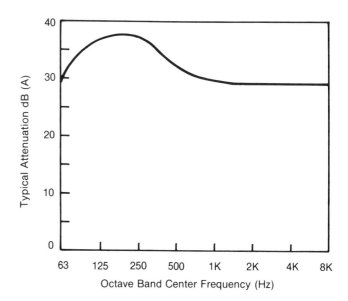

Figure 18-3. Typical Attenuation Curve

number of cycles (2 or 4), number of cylinders, naturally aspirated or turbocharged, inter/after cooling, operating speed, etc. These all affect the sound output and exhaust CFM. The second factor is the amount of air flowing through the silencer. If the silencer is much too large, the exhaust noise simply passes through, using only the initial large expansion for attenuation. Up to a point, all silencers work better with increased back pressure.

18.6 BACK PRESSURE

To determine pressure drop use Equation 18-2.

$$\text{Exhaust Gas Velocity} = \frac{\text{Exhaust Flow Rate (CFM)}}{\text{Silencer Inlet Pipe Area (sq.ft.)}}$$

(Equation 18-2)

The result of using this equation will supply the individual making the recommendation with the exhaust gas velocity in feet per minute.

Refer to Figure 18-4 to determine pressure drop in inches of water for various levels of silencers. To convert inches of water to inches of mercury, divide the calculation of inches of water by 13.6.

If pressure drop requirement is 1 in. of Hg or less and exhaust flow rate (CFM) is known, Table 18-3 can be used to determine silencer size. Find the lowest flow rate on the chart that is equal to or greater than the flow rate of your engine under the appropriate silencing level.

18.7 MATERIAL

Another consideration is material for the silencer. Again, the material must suit the application. In most instances, aluminized steel is preferred for general applications, although a number of silencers are manufactured of mild steel. Aluminized steel is more heat resistant, up to 1250°F, while mild steel is good only to 1100°F. This allows for weight reduction and longer life for the

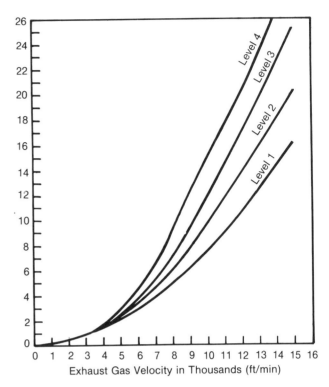

Figure 18-4. Silencer Back Pressure

Table 18-3. Calculated Exhaust Flow Rate for a Silencer with Pressure Drop of 1.0 in. Hg or Less

Inlet Pipe Area (Ft²)	Pipe Dimension	100 Level Silencer	200 Level Silencer	300 Level Silencer	400 Level Silencer
.0031	3/4	50	43	38	—
.0055	1	80	70	62	—
.0085	1-1/4	138	120	106	—
.0123	1-1/2	190	162	145	—
.0218	2	310	270	240	—
.0341	2-1/2	440	380	340	—
.0491	3	680	590	520	—
.0668	3-1/2	900	780	700	—
.0873	4	1160	1000	900	855
.1363	5	1830	1580	1400	1330
.1963	6	2700	2300	2000	1900
.3491	8	4500	3900	3500	3320
.5454	10	7200	6200	5500	5220
.7854	12	10200	8800	7900	7500
1.069	14	14000	12000	10700	10160
1.438	16	18000	15500	14000	13300
1.76	18	23000	19800	17800	16900
2.18	20	28400	24300	21800	20700
2.64	22	34400	29500	26500	25200

NOTE: If pressure drop requirement is 1 in. of Hg or less and exhaust flow rate (CFM) is known, the chart above can be used to determine silencer size without calculations. Find the lowest flow rate on the chart that is equal to or greater than the flow rate of your engine under the appropriate silencing level.

silencer because metal fatigue is reduced. Aluminized steel is more resistant to corrosion than mild steel. For very corrosive environments, 409 stainless steel, or for severe instances, 300 series stainless steel should be used. Since these materials are very expensive, operating conditions should be well understood before specifying either type. Type 409 is most common and will withstand heat up to 1500° F.

A final consideration is the configuration of the silencer. The most widely used styles are end-in end-out, side-in end-out, side-in side-out, and middle side-in end-out. These will accommodate most applications. When using a silencer on top or inside an enclosure, the most popular style is the side-in end-out model. This enables the packager to mount the silencer in applications where space is a problem and where there is a minimum of piping. For horizontal applications where space is again a problem and there is a need for additional piping, the best configuration is a middle side-in end-out silencer. This will solve many mounting problems and give a balanced look to compact applications.

REFERENCES

[1] Bruel & Kjaer Instruments Inc., Measuring Sound, Revision January 1984.

[2] Hutchins, Ed. Nelson Industries, Keep It Quiet-How to Do It Best, September 1980. EGSA Paper 80-F-1100.

Engine Protective Controls

Jack Rogers

CHAPTER 19

INTRODUCTION

Engine protective controls are essential to ensure full time availability of the engine-generator set. The engine can be thoroughly protected and provisions made for immediate shutdown, for warning before shutdown, or for alarm only, without being complicated or expensive.

The alarms or shutdown controls that are necessary to safeguard the engine depend on the circumstances of each installation (Ref. 1). However, certain basic controls are recommended. If an engine loses oil pressure, overheats, or overspeeds, it must be shutdown (in all except life threatening situations), to avoid damage (Ref. 2). The controls discussed here deal with two- and four-stroke cycle engines of all sizes and types. While many recommendations can apply to turbine driven sets, no recommendations are given for these units.

A proper control system not only protects the engine from damage, but is used to identify potential trouble before an emergency shutdown occurs (Ref. 3). Principle considerations in the design of an engine control system are reliability, durability, and simplicity. Controls must be capable of accurately directing the performance of the engine under adverse conditions and, furthermore, they must be easy to install and maintain (Ref. 4). A properly designed system will have an individual guardian for each function (Ref. 5).

A wide variety of types of protective controls are available, ranging from electronic/electrical to mechanical to combination units. Type choice is left to the customer, depending on equipment supplied, available activating sources, and functional requirements. However, fully automatic controls are recommended over indicating-only gauges, or controls that require constant monitoring to detect faults. Dependable controls are far better protection for an engine than a human attendant. Where the attendant may become neglectful, the controls constantly monitor the functions they are set to guard (Ref. 6).

Demanding applications such as hospitals and computer operations often require a higher level of sophistication. Many electronic "supervisory" control systems are available for these applications. Data logging and trending analysis are important features of many of these systems. This logging and analysis provides historical operating data useful in the maintenance program for the generator set. Some systems can interface with computers, which further enhances their offerings.

Engine function gauges are recommended to indicate function status to operating personnel. Through record maintenance they can indicate deteriorating conditions and thus warn of pending problems. Some protective control systems utilize the gauge as an integral part of the system, while others supplement the system with gauges. The gauges should be mounted on an instrument panel located on or in close proximity to the engine and may be included with the generator meters. Accuracy of $\pm 2\%$ is adequate for most all generator set applications. Remote indication for engine function gauges normally is limited to status indication by means of lights, or LEDs, or alarm functions only.

Protective controls should have a means to tell if they are functioning and a means to test their operation and trip set points. Adjustability of the trip set point is desirable to compensate for differences in operating characteristics of various engines, and under various operating conditions. It can also serve as the means to test the trip operation.

A means must be provided to override the shutdown operation during engine startup. This feature is normally included in the automatic start-stop controller for installations requiring this type operation. For manual start-stop systems, override can be as simple as a push button or is available in the control devices of popular protective control systems. The override must be self canceling.

Some protective control applications are mandated by an authority having jurisdiction who states specifically what is and what is not allowed and the sequence of operation allowed under certain situations. The authority may be local, state, national, or others responsible for ensuring public safety. For example, shutdown controls are frequently not allowed for sets involved in life support equipment, fire pumps, or lighting where loss of power might lead to panic, etc. However, recommended practice in these instances is to activate visual and/or audible alarms, warning operating personnel of the fault. Pre-alarms are often used in these applications. They alert operating personnel that a protected function on the generator set is approaching a critical value and that attention to the problem is needed. Pre-alarms also are a desirable feature for generator sets equipped with shutdown devices.

Protective controls are not infallible. They should compliment routine maintenance, not replace it. They should receive regular maintenance and checking on a basis consistent with the type of operation and environment in which they are operating or as required by the authority with jurisdiction.

For the typical engine generator set, the following are the recommended basic points of protection:

a. Lubrication system

b. Cooling system

c. Overspeed

19.1 LUBRICATION

Lubrication is the life blood of an engine (Ref. 7). It reduces friction between moving parts of the engine and assists in removing heat from the combustion process. Without it, the engine would soon seize and stop, generally with major damage to moving parts.

19.1.1 Full pressure lubrication.
For engines with full pressure lubrication, a low oil pressure protective device should be installed, with the trip point set just under the recommended minimum operating pressure range.

For engines without full pressure lubrication, oil temperature is a good indication of lubrication. As oil temperature increases, the oil breaks down and loses its ability to properly lubricate the engine. Oil temperature can also indicate low oil level in the crankcase, since a lesser volume of oil will not carry away as much heat as a larger volume. It is possible to have adequate oil pressure but inadequate lubrication due to low oil level. High oil temperature monitoring is of benefit in this case. Other benefits of monitoring oil temperature will be discussed in the section dealing with cooling.

19.1.2 Oil level.
Optional lubrication protection includes high-low lube level devices. These devices have sight glasses that allow the operator to check lube oil level without shutting down the engine. Low level is normally monitored but monitoring of high level also will indicate overfilling, leaking gaskets and components allowing coolant to enter the crankcase, stuck injectors leaking diesel fuel into the crankcase, etc.

19.1.3 Lube level maintaining devices.
Various lube level maintaining devices are also available to meter lubricant into the crankcase as required to maintain the proper level. Care should be taken to choose a device that will meter only the required quantity of oil and will not overfill the crankcase due to drawdown during normal engine operation. The metering valve design is of major importance in proper operation of these devices. Level maintainers are especially helpful in continuous run operations such as prime power generator sets.

19.2 COOLING

The combustion process produces heat that must be dissipated in order for the engine to function properly. This process is accomplished by circulating liquid coolant throughout the engine and dissipating the heat through a radiator or heat exchanger, or by flowing air across the cylinder and cylinder head of an air cooled engine. In either engine, the operating temperature must be monitored to ensure that the engine does not overheat. Such a condition can cause the engine to seize or even to catch fire. Extreme cases have resulted in various engine parts actually melting. In either type system, the temperature sensor should be installed in an average temperature location in the head or block (Ref. 8).

19.2.1 Liquid cooling.
For a liquid cooled engine, this point is typically located on the engine side of the thermostat. It is important that the sensor be located in a place of circulating coolant. Non-circulating or stale coolant will not register true overall average temperature, thus compromising the effectiveness of this important sensor. Figure 19-1 illustrates the importance of mounting location.

The inlet manifold reading is taken at the rear of the engine where a cab heater connection would be made. This temperature, and temperature measured in the head, rise together until the thermostat opens. During the same

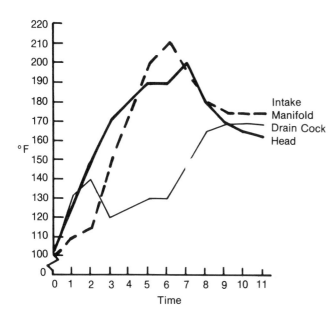

Figure 19-1. Engine Temperatures

time period, temperature in the lower part of the engine, measured at the block drain cock, remains low due to minimal coolant circulation and heat absorption by the block. If the thermostat fails to open and temperature is measured only at the drain cock, the upper parts of the engine would no doubt receive heat damage before alarm or shutdown occurs.

This graph also points out the desirability of having separate gauges and protective sensors for coolant and pressure to take readings at the same location on the engine.

It is important to have an adjustable trip point for temperature devices to compensate for differences in operating temperature. Three prime factors can change the temperature at which coolant will boil. These factors are:

a. The amount and type of antifreeze in the coolant

b. The pressure at which the cooling system operates

c. The altitude at which the cooling system operates (Ref. 9)

An increase in the concentration of ethylene glycol antifreeze raises the boiling point, as does increasing the cooling system pressure. Higher altitudes lower the boiling point of the coolant. The sensor trip point must be adjustable to compensate for these differences.

A sudden loss of coolant such as occurs when a radiator hose breaks, will render most all temperature devices ineffective (Ref. 10). Coolant level protection is, therefore, worthy of consideration as optional equipment. Slow leaks and low coolant level will normally be detected by the high temperature protection device. A coolant level device with a sight glass or dial indicator allows checking coolant level without removing the radiator cap. Removing the radiator cap could allow damaging oxygen to enter the cooling system.

The cooling system should be free of aeration (see paragraph 16.8) since air circulating through it will cause inaccurate readings by temperature sensors. It can also lead to false alarms or shutdowns from coolant level devices.

19.2.2 Air cooling. Air cooled engines require continuous flow of air across the engine and cylinder heads to carry away heat. Most multi-cylinder air cooled engines also incorporate an oil cooler, which assists in cooling the engine overall.

Temperature monitoring on these engines is typically done by placing a suitable sensor in a box located on the side of the cylinder head or into the cylinder head wall itself. Typically only one cylinder head is monitored, although on vee engines it is common to monitor one cylinder of each bank. A cylinder of average temperature, usually the second from the rear on multi-cylinder engines, is chosen to mount the temperature sensor.

A growing trend for these engines is to monitor oil temperature. Points of reasoning include:

a. Oil temperature represents the average temperature of the engine as the oil is dispersed throughout the engine and returns to a common collection point

b. Oil temperature provides an alternative to monitoring all cylinder heads

c. Change in oil temperature occurs more rapidly to conditions that affect engine cooling

Oil temperature sensing alerts operating personnel to potential problems earlier than monitoring cylinder head temperature. This point is borne out by the results of a test conducted on a two-cylinder diesel engine. See Figure 19-2. Similar test results are observed on larger engines, both air cooled and liquid cooled.

In the test (Figure 19-2), oil temperature was monitored in comparison to cylinder head temperature of both heads. Each cylinder head temperature was measured on both the intake and the exhaust sides. Head temperature was taken using a bolt-on heat sink designed for (and which has proved to be reliable) other applications on air cooled engines.

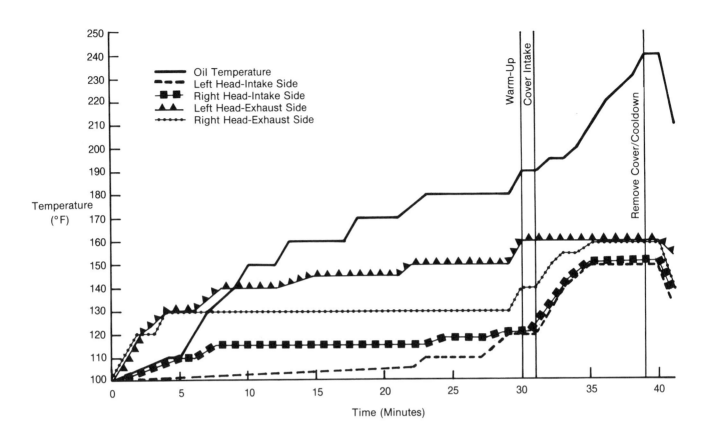

Figure 19-2. Oil Temperature Versus Cylinder Head Temperature

As the graph shows, oil temperature rose at a steady rate as the engine was allowed to warm up, while cylinder head temperatures stabilized.

The cooling air inlet was covered to cause a rise in temperatures. Note that head temperature stabilized at a point where oil temperature continued to rise. If not monitored, these could lead to the false assumption that no overheating had occurred. The engine surely would have been damaged.

Many times loss of cooling ability can be detected before a rise in temperature occurs. Belt break switches activate when v-belts break, thus compromising the cooling ability of the engine. Other sensors react to loss of battery charging alternator output to activate the alarm or shutdown system in belt break situations.

19.3 OVERSPEED

The third recommended protective control is against overspeeding of the engine. Overspeeding can occur because the engine-generator coupling breaks, the speed control governor fails, intake of excessive raw or enriched fuel, etc. Besides destruction of the engine and/or generator, overspeeding may adversely affect the output of the generator and may cause damage to systems connected to the generator output.

Engine shutdown is mandatory when an overspeed condition occurs.

19.3.1 Magnetic sensor. Overspeed protection can be accomplished by mechanical or electronic devices. The most popular method uses a magnetic sensor to detect flywheel speed, and generates an electric signal that is transmitted to a frequency sensitive speed switch. The speed switch is set so that an increase in frequency trips the circuit to activate the engine shutdown system.

19.3.2 Mechanical sensor. Some systems employ a mechanically driven signal generator to send an electric signal to the speed switch. Mechanical speed switches are driven either directly or by cable from the camshaft or other gearing. When using this type of speed switch it is necessary to know the ratio of the gear speed to crankshaft or engine speed and usually the direction of rotation. Except for signal generators, these switches employ flexible shaft drive or flyweight mechanisms that react to speed of the engine and operate contacts of varying types, which operate the engine shutdown mechanism.

Several manufacturers offer combination tachometer and speed switch units. They are available as electronic, mechanical, or a combination of both. All have means of adjusting the switch setting.

19.3.3 Override during startup. Overspeed protection should not be overridden during engine startup since an engine running at unrestrained speed will destroy itself before typical override systems deactivate.

19.3.4 Manifold vacuum. Spark ignition engines normally develop between 4 and 16 in. Hg vacuum from the intake manifold. Measuring this vacuum is an effective and inexpensive method of detecting a runaway engine, since manifold vacuum will increase sharply as the engine overspeeds. However, manifold vacuum is not a good indicator on diesel engines because the level of vacuum is typically very small.

19.4 MISCELLANEOUS

While only lubricating pressure, temperature, and overspeed should be considered essential engine protection, many installations may dictate that additional protective controls be installed. Larger engines generally require additional controls. Many of these controls are helpful on smaller engines also.

Previously mentioned coolant and lube level controls are highly valuable in any protective control system. Coolant pressure controls in liquid cooled engines ensure that coolant circulation is adequate and provide good backup to temperature and level controls.

Monitoring air, oil, and fuel filter restrictions helps the mechanic tune the engine to maximum efficiency. These functions should not be shutdown functions; rather, they should alarm or simply indicate the degree of restriction.

Other points of monitoring could include manifold pressure or vacuum, intake air temperature, exhaust temperature, fuel pressure, excessive vibration, etc.

19.5 ALARMS

The protective controls may activate various alarms alone or in conjunction with an engine shutdown system.

The application will generally determine the type of alarm required. It may include visual or audible alarms, or both. Local or remote alarms, or both, may be required. The alarm(s) may be pre-alarm or shutdown, or both.

19.5.1 Function and general. A properly designed alarm system will designate specifically the function causing the alarm, and may be coupled with a general alarm. All alarms must be clearly identified as to function and to the applicable generator set. If the alarm is used with a shutdown system, it should indicate only the first function to fail, and should inhibit the engine from being restarted until the fault is corrected or the signal is disabled by authorized personnel. The identification feature enables maintenance personnel to quickly locate the fault. The alarm should remain activated until acknowledged by personnel in charge. If visual and audible alarms are used, the visual alarm should remain activated until the fault is corrected. The audible alarm can be silenced as soon as it is practical.

19.5.2 Visual. Visual alarms may include incandescent lamps, LEDs, pop out buttons, rotating magnetic balls, or any suitable means that will alert personnel and serve to indicate the cause of alarm. LEDs are preferred over incandescent lamps because of their long life and low expense. Because they emit cold light, they have no element to fail as do incandescent lamps. They are also far more resistant to shock and vibration.

Because of the high failure potential of incandescent lamps in alarm systems, a push-to-test switch should be installed to facilitate locating inoperative lamps. This test switch is not necessary with LEDs and in fact is discouraged, since a proper test of the alarm system must include the switch contact or trip device at the function sensor. A lamp test switch tests only the lamp without regard to the operability of the sensor, thus creating a false sense of security.

As with the engine protective controls, the alarm system must comply with any requirements set forth by any authority having jurisdiction. For example, NFPA 99 specifically states alarm requirements for engine generator sets used in health care facilities.

Alarm systems with repeater or remote panels should allow for silence and/or reset only from one master panel. Optionally, any single panel may be silenced only if it does not affect the alarm status of other panels.

19.6 SHUTDOWN

Protective or emergency shutdown devices should not be used to stop the engine under non-emergency situations. Normal shutdown procedures recommended by the engine manufacturer should be followed.

19.6.1 Method. Distributor ignition engines are stopped by interrupting the circuit between the ignition coil and the ignition distributor. This operation is performed by the appropriate device or module of the protection system.

Magneto or capacitor discharge ignition engines are typically stopped by grounding the magneto or ignition output. Again, the grounding operation is performed by the appropriate device or module of the protection system.

Larger gas fueled engines typically are shutdown by closing a valve supplying fuel to the engine. The ignition may also be grounded at the same time.

The fuel shutoff valve may be operated electrically, pneumatically, or by any other means that provides certain shutoff. Electrically operated valves may be powered by the ignition system, standby batteries, or line voltage. Pneumatically operated valves may utilize fuel gas pressure, engine oil pressure, engine manifold vacuum or pressure, standby pressure or vacuum, etc. as the operating medium. In any case, provision should be made to ensure that the trip source will be available when called upon. Consideration should also be given to whether the activating force is applied for the valve to open or to close.

19.6.2 Grounding ignition and fuel valve. The practice of closing a fuel valve and grounding the ignition varies among engine manufacturers and packagers. If both options are chosen, the fuel supply should be closed before or concurrent with grounding of the ignition. To do otherwise will allow raw fuel to flow into the engine intake system and be distributed throughout the engine, and possibly throughout the building. A restart with this fuel present could lead to an explosion. It is highly recommended that the fuel valve be equipped to vent to an outside area the fuel trapped between the valve and the engine carburetion when the engine shuts down (Ref. 11).

In all cases, the shutdown circuit components and devices must be properly matched to the trip power source. Devices designed for conventional magneto ignitions typically won't work with capacitor discharge ignitions; systems designed to work from pressure won't work from vacuum; systems designed to work from battery voltage won't work from line power, etc.

19.6.3 Diesel engines. Diesel engines are the prime movers in the 10–1500 HP range of engine gen-sets (Ref. 12). Whether two-stroke or four-stroke cycle, diesel engines can be shutdown by closing off the fuel and/or combustion air.

Fuel Rack Solenoids. The quickest shutdown is achieved by operating the fuel rack or governor control to a no-fuel position. This can be done with solenoids, air cylinders, spring loaded pull devices, etc. Consideration must be given to the amount of force required to operate the shutoff lever, the length of travel required to move the lever from running position to shutoff position, the direction of lever operation, and whether the activating medium is applied to the shutoff device to allow the engine to run or to stop the engine. In the latter case, it may be necessary to supply a means to disconnect the power source after shutdown to prevent battery drain, solenoid burnout, depletion of air supply, etc. This can be done automatically by electronic time delay devices, through fuel pressure switches, or other means that can detect engine shutdown and disconnect the engine and itself from the power circuit.

Some diesel engines have an electric solenoid, which can be used to complete shutdown, built into the injection pump. Typically, these are energized to run, but could be energized to stop. Again, in the latter case the solenoid must be disconnected after shutdown.

Fuel Valves. Solenoid fuel valves can also be used to stop diesel engines. Their application is normally limited to engines under 300 HP and to engines without unit injectors. Some delay in shutdown is inherent in the use of fuel valves. The length of this delay depends on several factors that in total allow an amount of fuel to continue to be injected into the combustion chamber.

The fuel valve must be placed as close to the injection pump as possible to minimize the amount of fuel in the fuel line between the valve and the injection pump. If the injection pump can draw upon this fuel, shutdown will be delayed. Some controls manufacturers offer fittings that allow the valve to connect directly to the injection pump fuel inlet, thus eliminating any supply line between the valve and pump.

All connections must be completely air tight. If the injection pump can draw air from any source, it can continue to pump fuel from any available source. Once pressure in the pump is equalized to atmospheric, the pump can no longer function as a pump; thus fuel cannot continue to be injected.

A third requirement is that functional check valves be placed in all bypass or recirculation lines to prevent fuel and/or air from being re-introduced to the injection system. Again, pressure within the pump must be reduced to atmospheric for shutdown to occur.

An engine operating under load will not necessarily shut down quicker than an engine with no load. As engine speed decreases, the load typically decreases accordingly, thus requiring less fuel for the engine to continue running.

Some engines, most notably those with gear driven fuel transfer pumps, may require pressure relief valves to prevent fuel filter canisters or gaskets from bursting after the shutoff valve closes. As the engine speed decreases after the fuel valve closes, the gear driven transfer pump continues to try to supply fuel to the injection pump.

However, the closed valve prevents that fuel from being delivered and the pressure must be absorbed within the system between the transfer pump and the valve. Additionally, some pressure buildup occurs due to radiated heat from the engine. A pressure relief valve placed after the filter, with a return line to the fuel supply, will open as pressure increases and, thus, relieve that pressure and prevent broken canisters or blown gaskets. The relief valve must not fully open during normal engine operation.

Both the check valve and the pressure relief valve are quite inexpensive and easy to install. Their value to the effectiveness of the engine shutdown system is well worth their expense and may well determine whether or not the engine shuts down.

Engines with unit injectors and those with fuel passages cast into the heads are not good candidates for fuel valve shutdown. Unit injectors, by their nature, will draw fuel from any source. It is not practical to apply check valves to these injectors. Engines with cast-in fuel passages provide a large supply of fuel, much of which will be injected before shutdown can occur. Governor or rack solenoids or air intake systems typically perform much better on these engines.

Air Intake. If the engine is operating in an area where gaseous vapors exist, use of an intake air shutoff device is recommended. Even with the normal fuel supply shut off, many times the engine can continue to operate from the vapors ingested through the air intake system. This applies to both 2-stroke and 4-stroke engines. It applies equally to diesel and to spark ignited engines. Many applications such as in refineries and off-shore platforms are required by various authorities to have air shutoff devices.

If in doubt as to the best or proper system for an engine, consult the engine manufacturer or one of the systems manufacturers.

19.7 SUMMARY

Whether for standby or prime power, whether for convenience or necessity, the engine generator set is expected to be available when called upon. Engine protective controls are a valuable aid in maximizing the availability of the set, by monitoring the engine's critical life signs during operation, alerting operating personnel of problems, and/or shutting down the engine when these life signs reach critical levels. Down time is minimized. Availability is maximized. They watch the engine functions (life signs) when you can't.

Recommended basic controls cover oil pressure or the lubrication system, cooling temperature, and overspeed.

Other engine functions can be monitored according to need or preference. However, the controls must operate as a system and, therefore, must be compatible with each other and with other equipment associated with the entire set. Certainly they must comply with any regulations set forth by any authority having jurisdiction.

All protective controls must have a means of testing their operation through the complete protective circuit. Push-to-test devices do not test the complete circuit and are generally discouraged as a system check. Protective controls can operate alarms or engine shutdown, or both.

Alarms should specifically designate the function causing the alarm, although a general alarm may be included with the specific alarm. Whether visual or audible, the alarm must be readily ascertained by operating personnel so that immediate action can be taken. The alarm may be mounted on the generator set, it may be remote, or both. Only the master panel should have provision for silencing general alarms. For systems incorporating shutdown devices and alarms, only the master panel should allow engine restart.

Required shutdown devices vary according to type and make of engine and available sources of trip power. Spark ignition engines are stopped by interrupting the ignition. Diesel engines are stopped by closing off the fuel or combustion air.

It cannot be stressed too strongly that one must not rely solely on engine protective controls to ensure that the engine never experiences problems associated with the functions being monitored. Controls are a maintenance tool. They must be checked, tested, and maintained periodically. They are the means to an end, not an end by themselves. Choose your controls wisely, take care of them, and they will come through for you with big dividends when you need them.

REFERENCES

[1] EGSA Standard 101S-1988, Engine Driven Generator Sets Guideline Specifications for Emergency or Standby, Electrical Generating Systems Association.

[2] Ibid.

[3] Hood, J. Taylor. Murphy Controls for Engines, Pumps and Compressors, 1981.

[4] Rogers, Jack. Murphy Simplified Protective Controls for Engine-Generator Sets, 1983.

[5] Murphy, Frank W., The Role of Automation in Increasing Engine Sales, 1963.

[6] Ibid.

[7] Ibid.

[8] Rogers, Jack. Murphy Simplified Protective Controls for Engine-Generator Sets, 1983.

[9] Caterpillar Tractor Company, Know Your Cooling System, SEBD0518-REV-01; 1978.

[10] Rogers, Jack. Murphy Simplified Protective Controls for Engine-Generator Sets, 1983.

[11] Hood, J. Taylor. Murphy Controls for Engines, Pumps and Compressors, 1981.

[12] Lehnerer, George J. Major Alternatives Affecting Generator Set Designs, (EGSMA Paper No. 83-F-300), 1983.

Vibration Isolation

Francis J. Andrews

CHAPTER 20

INTRODUCTION

Vibration isolators are used with engine generator sets to attenuate annoying and damaging effects caused by engine generator vibrations. Generator sets create disturbances that can result in a variety of complaints. Installed in occupied buildings, such as hospitals or office buildings, the vibrations may annoy personnel. Such vibrations may also affect the operation and accuracy of equipment when the vibrations are transmitted through the foundation to other machines. In extreme cases, vibrations may result in structural problems, due to cyclic fatigue effects.

There are numerous sources of vibration in engine generators and several methods available to reduce the effects of the vibrations. Properly selected, vibration isolators can attenuate the vibrations over virtually the entire frequency range of concern. Since isolators can reduce the vibrations generated by numerous sources within a generator set (e.g., rotational unbalance, gear frequencies), they are often chosen as a cost effective means of overall vibration control.

In this chapter, the basics in the theory of vibration isolation are reviewed, as well as some practical considerations. A general discussion of isolator characteristics follows, including a review of isolator materials and configurations. Isolation of mechanical shock frequently requires a different approach and is mentioned only briefly in this chapter. Seismic loading of isolators is a different aspect of isolators; this is briefly discussed in Paragraph 20.4.3.1.

Since this chapter is not intended to be a textbook on vibration theory, equations have been kept to a minimum. If the reader is interested in more details, the references include several valuable textbooks on the subject. For the reader not interested in the technical details, Section 20.4, Types of Isolators, provides useful information on the various types of vibration isolators.

20.1 SOURCES OF VIBRATION

Any moving or rotating body is a vibration generator. The vibration generated is frequency dependent, numerous frequencies typically being excited simultaneously over a wide frequency spectra. In some circumstances, even low amplitude source vibrations can be troublesome. For instance, a low amplitude disturbing frequency could be amplified by a coincident structural frequency, resulting in unacceptable vibrations at a location removed from the source.

Rotor unbalance is a primary source of excessive vibration in rotating equipment and is a potential source of metal fatigue and machinery failure. Rotor unbalance creates radial centrifugal forces that occur sinusoidally at rotational frequency and at harmonics of rotational frequency. Sources of unbalance include dissymmetry in design, non-homogeneous material, shifting of component parts due to centrifugal forces or thermal effects and eccentricity of rotors. A unit operating at a rotational speed of 1800 revolutions per minute (RPM) would exhibit a rotational frequency of 30 Hz (1800 RPM/60 seconds per minute = 30 Hz). Harmonics of rotational frequency would be evident at 60 Hz, 90 Hz, 120 Hz, etc.

Additional disturbing frequencies include torsional dynamic pulses due to variations in cylinder gas pressure, and imbalance forces due to reciprocating masses within an engine. These forces can excite critical speeds of a rotating shaft.

Numerous sources of discrete frequencies occur in electric motors and generators, caused by periodic magnetic forces in the air gap between the rotor and stator. Discrete frequencies are also generated in turbines and pumps as blades or impellers rotate past stationary points on the housings. Bearings, gears, and timing belts are additional sources of discrete frequencies. These would occur as multiples of rotational frequency. For instance, a pump with six impellers would generate a discrete frequency at six times rotational frequency — a six impeller pump rotating at 1200 RPM would have a rotational frequency of 20 Hz, an impeller frequency of 120 Hz, and harmonics. A turbine with 100 blades would exhibit a blade passing frequency at 100 times rotational frequency. A gear with 20 teeth would exhibit a gear mesh frequency at 20 times rotational frequency.

From the preceding, it is clear that discrete frequency vibrations can occur over a very broad range of frequencies, depending upon the speed of rotation and the nature of the excitation force (e.g., unbalance, turbine blades, gear meshing, etc.).

Broadband vibration is also generated by high velocity fluids and by restrictions in fluid lines, such as steam valves, or due to high velocity steam exhaust from a steam turbine into a condenser. Such broadband vibration can occur as low as 50 Hz, and can extend as high as 2000–3000 Hz.

With the multitude of sources of discrete frequency and broadband vibration, it is highly likely that structural resonances will be excited by one or more of these sources. The dynamic response of a structure can amplify and aggravate a relatively low amplitude vibration source. The structure can also act as a transmission path for a vibrating source.

20.2 VIBRATION THEORY

Figure 20-1 shows a schematic of a simple mounting system consisting of a mass (M) supported on linear springs having a stiffness (K). The ideal system includes a parallel damping element as shown, having a critical damping ratio (C/C_c). The mass is assumed to be symmetrically supported about its center of gravity and to vibrate only in the vertical direction. The spring and damper are actually mathematical simplifications for a vibration mount. In practice, damping and stiffness are frequently integral to the unit mount.

In the case of generator sets, the purpose of a vibration mount is to reduce the force transmitted from equipment to the foundation. The mount acts as a filter, breaking the transmission path of vibrations from the generator to the foundation and to the surrounding environment.

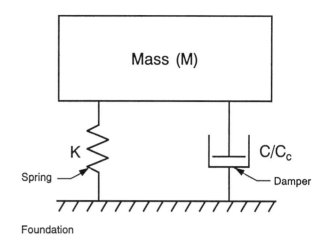

Figure 20-1. Schematic of Simple Mounting System

The equation for the natural frequency of the system of Figure 20-1 can be considered to be the same as for an undamped system for most values of damping. Specifically,

$$f_n = \frac{1}{2\pi} \sqrt{\frac{K}{W}} \qquad \text{(Equation 20-1)}$$

Where:

f_n = isolation system natural frequency, Hz
K = isolation system stiffness, pounds per inch
M = isolated mass, pound-sec per inch

For practical applications, equipment weight can be used in place of equipment mass, as in the following equation:

$$f_n = 3.13 \sqrt{\frac{K}{W}} \qquad \text{(Equation 20-2)}$$

Where:

f_n = isolation system natural frequency, Hz
K = isolation system stiffness, pounds per inch
W = isolated weight, pounds

When more precise estimates of system response are desired, the effect of damping can be included by using:

$$f_{nd} = f_n \sqrt{[1 - (C/C_c)^2]} \qquad \text{(Equation 20-3)}$$

Where:

f_{nd} = isolation system damped natural frequency, Hz
C/C_c = critical damping ratio

If the foundation of Figure 20-1 is vibrated with a sinusoidal input over a frequency range of 1 Hz–500 Hz, and if the response of the mass and the input at the foundation are measured at every frequency, we have a measure of performance for the simple mounting system. The ratio of the output at the mass to the input at the foundation is referred to as **transmissibility**, and is defined by

$$T = \sqrt{\frac{1 + [2(f_d/f_n)(C/C_c)]^2}{[1 - f_d^2/f_n^2]^2 + [2(f_d/f_n)(C/C_c)]^2}} \quad \text{(Equation 20-4)}$$

A transmissibility greater than 1 means the system amplifies the vibration input, while a value less than 1 means that the mounting system reduces, or attenuates, the vibration input. A review of Equation 20-3 shows that transmissibility is dependent upon the frequency ratio (f_d/f_n) and also upon the critical damping ratio (C/C_c). The numerator (f_d) is the excitation frequency and is determined by the dynamic environment (e.g., rotational frequency of the generator). The denominator (f_n) is the natural frequency of the mounting system, and is a function of the isolators as well as the isolated mass. The critical damping ratio is an indication of the degree of damping provided by the mounting system. This is determined by the isolator design and materials.

Figure 20-2 shows five curves relating transmissibility to frequency ratio (forcing frequency/natural frequency, or f_d/f_n) for critical damping ratios varying from 0.01–0.50. At a low frequency ratio, transmissibility is equal to, or slightly greater than 1, meaning that the mass moves with the foundation regardless of the amount of damping present in the system. As frequency ratio increases, transmissibility increases to a maximum at the natural frequency of the mounting system. At the natural frequency of the mounting system, there are large differences in transmissibility due to damping. This is the region where the benefit of damping is seen — damping limits the amplification of forces at the natural frequency of the isolation system and limits the resultant displacements.

At a frequency above the natural frequency of the mounting system, there is a crossover point where the transmissibility equals 1. This crossover frequency is said to occur at 1.414 times the natural frequency and is theoretically independent of damping.

As frequency increases further, the system moves to a region of isolation, in which the vibration input is attenuated, the degree of attenuation increasing with increasing frequency. At the higher frequencies, damping affects transmissibility. In particular, the high frequency attenuation is less for the mounting system with greater damping. As an example, at a frequency ratio of 3.0, the isolator with a critical damping ratio (C/C_c) of 0.01 has a transmissibility of 0.13, while the isolator with a critical damping ratio of 0.50 has a transmissibility of 0.37. This shows the tradeoff required in selecting an isolation system. If control of displacements at resonance by damping is desired, there will be some loss in vibration attenuation at high frequencies. In many cases, these losses are negligible and are frequently masked by other effects.

A properly selected isolation system will provide a natural frequency well below the critical disturbing frequency. In other words, the actual disturbing frequency is well into the region of attenuation. (See the boxed example below.)

Figure 20-2 also shows, on the right hand axis, attenuation as percent isolation. This is simply an alternative way of considering transmissibility. A transmissibility of 0.1 is equivalent to 90% isolation (1 − 0.1 = 0.9), and a transmissibility of 0.02 is equivalent to 98% isolation (1 − 0.02 = 0.98).

For lightly damped systems, Equation 20-3 can be reduced to

$$T = \frac{1}{1 - (f_d/f_n)^2} \quad \text{(Equation 20-5)}$$

Except at resonance (where transmissibility would equal infinity), Equation 20-5 will provide reasonable estimates of transmissibility for critical damping ratios less than about 0.10.

For any isolator with negligible damping, the static deflection of an isolation system is related to the vertical natural frequency by the following:

$$\delta_{st} = \frac{9.8}{f_n^2} \quad \text{(Equation 20-6)}$$

This equation assumes that the dynamic stiffness of the isolation system is equivalent to the static stiffness of the system, which is reasonably correct for isolators with negligible damping, such as steel springs. However, isolators with a significant amount of damping exhibit a higher dynamic stiffness; the resultant natural frequency is higher than for an undamped spring with the same static deflection.

Figure 20-3 shows the curve represented by Equation 20-6, plotted on a logarithmic scale. From Figure 20-3, one can see that a natural frequency of 10 Hz would require a static deflection of 0.1 inch, whereas a natural frequency of 1 Hz would require a static deflection of 10 inches. This simple chart is useful in that it illustrates the large increase in static deflection (and, indirectly, isolator size) which is required to obtain a lower natural frequency, and the consequent improvement in high frequency isolation. Static deflection is frequently a primary concern in selecting an isolation system, especially for spring isolators.

Figure 20-4 combines several aspects of vibration isolation previously discussed. The horizontal scale shows the

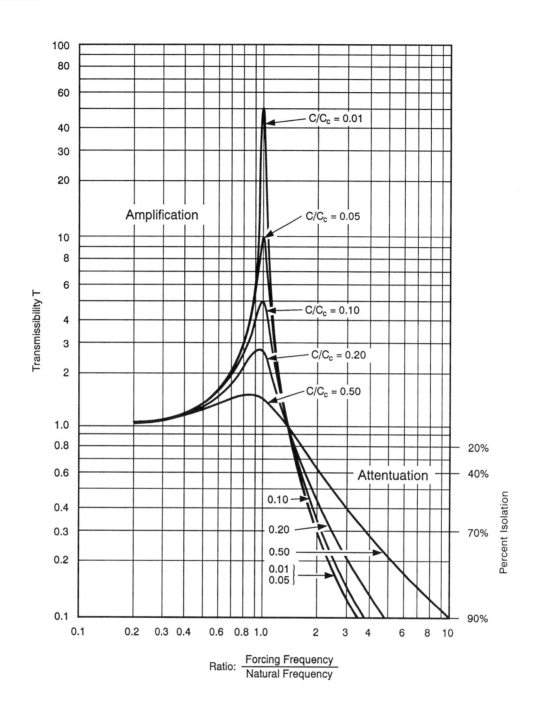

Figure 20-2. Transmissibility Versus Natural Frequency and Critical Damping Ratio

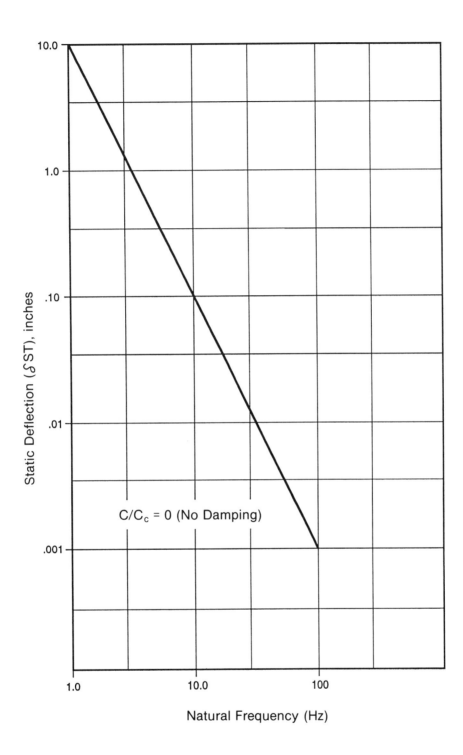

Figure 20-3. Static Deflection Versus Natural Frequency

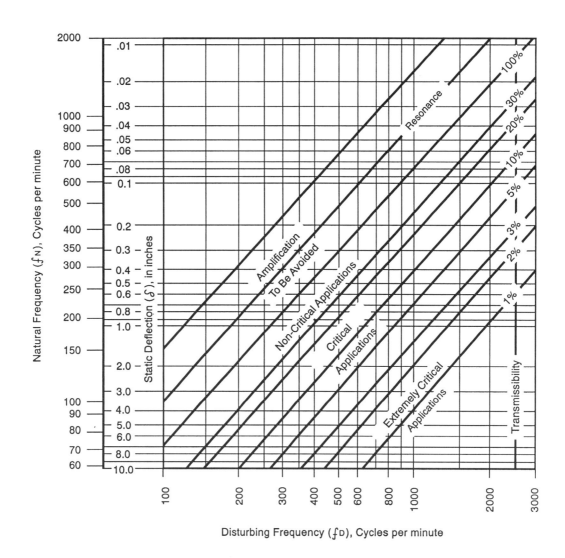

Figure 20-4. Vibration Transmissibility Chart

disturbing frequency in cycles per minute (divide by 60 to obtain cycles per second, or Hz), while the leftmost scale shows the isolation system natural frequency, also in cycles per minute. A series of lines at an angle of 45 degrees shows transmissibility. With this nomenclature, 1% transmissibility = 0.01, etc. These lines assume a critical damping ratio (C/C_c) of zero (no damping). As discussed above, static deflection is related to natural frequency. The second scale from the left, therefore, shows static deflection. From this curve, one could determine the expected transmissibility for any value of static deflection at any given disturbing frequency.

Thus far, the discussion regarding vibration isolation has been limited to a mass symmetrically supported about its center of gravity and vibrated only in the vertical direc-

> **EXAMPLE:**
>
> A diesel generator set rotating at 1800 RPM requires 70% isolation at the rotational frequency (30 Hz). The required transmissibility is 30% or 0.3. From Figure 20-4, on the horizontal scale, Disturbing Frequency, locate 1800 RPM. Draw a vertical line that intersects a transmissibility of 30%. Draw a horizontal line to the left and read the natural frequency as approximately 900 cycles per minute, or 15 Hz. The natural frequency of the isolation system (with negligible damping) should be below 15 Hz. If necessary, damping can be taken into consideration by using Figure 20-2.

tion. In practice, many installations exhibit significant horizontal excitations. In these cases, when the isolation system is free to vibrate horizontally, rocking modes will be set up at two distinct frequencies — a horizontal mode and a pitching mode. These modes are said to be coupled when the natural frequency of one mode can be stimulated by a forcing frequency in another. In other words, an input in one axis can cause a response in another axis. The lower frequency mode acts as a vibration centered about a point well below the center of gravity, while the upper frequency mode acts as a vibration centered closer to the center of gravity.

In the presence of rocking modes, a transmissibility versus frequency curve will show two discernible peaks, one or both showing a frequency range of amplification. Figure 20-5 (Ref. 1) shows a series of response curves exhibiting rocking modes with varying degrees of damping. The horizontal scale is excitation frequency in cps (cycles per second) or Hz, while the vertical scale is transmissibility. Damping has the same effect on the transmissibility curve with rocking modes as it does for the theoretical transmissibility curves shown in Figure 20-2. Specifically, increased damping decreases transmissibility at both resonant peaks, and increased damping also increases transmissibility at higher frequencies.

Factors that influence rocking modes are: spacing between isolators, axial and horizontal stiffness of isolators, mass characteristics (inertia and center of gravity), and damping. Computer analysis is generally required to obtain a detailed characterization of rocking modes, although Ref. 2 presents theoretical curves to estimate the coupled frequencies.

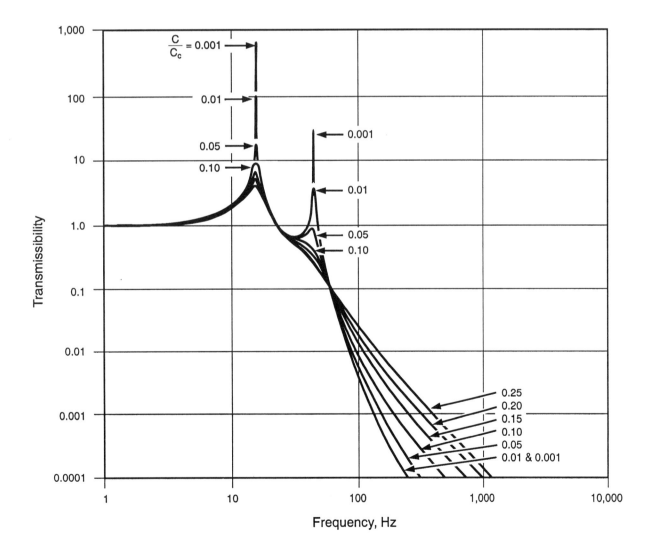

Figure 20-5. Transmissibility Versus Frequency and Critical Damping Ratio, Showing Effect of Rocking Modes

20.3 PRACTICAL CONSIDERATIONS

There are factors which can compromise the effectiveness of a vibration isolation system. A brief discussion of some factors follows:

Sound shorts. Mechanical links that have significant stiffness and bypass the vibration isolators. For example, solid piping that is hard mounted to a generator and also to an adjacent structure provides a ready path for vibrations to bypass the vibration isolators.

Insufficient stiffness in the support structure. Ideally, the structure on both sides of the vibration isolator should have a stiffness at least ten times as great as the stiffness of the vibration isolator. In practice, this may be difficult to achieve, and the supporting structure(s) may need to be analyzed as spring(s) in series with the vibration isolator.

Wave effects. At certain frequencies, decreases in high frequency isolation may be caused by wave effects in isolators. In steel springs, these frequencies are sometimes referred to as surge frequencies. For steel springs with no damping and no noise break material, surge frequencies can compromise vibration isolation at frequencies greater than approximately 100 Hz. For this reason, steel springs are frequently combined with elastomer pads to mitigate these effects. Elastomeric isolators also exhibit wave effects, referred to as standing waves, although damping internal to the elastomer generally limits these effects to the extent that wave effects are seldom conclusively identified. In helical wire rope isolators, a first standing is frequently observed between 500–600 Hz.

Non-linearities. Vibration isolators that provide a significant amount of damping (generally, C/C_c greater than 0.15) exhibit changes in performance with changes in dynamic input. Specifically, for isolators exhibiting such sensitivity, the natural frequency increases with a decreasing dynamic input, and vice versa. For these isolators, damping also changes, although the changes are generally of a secondary nature.

20.4 TYPES OF ISOLATORS

20.4.1 General. The principal characteristics of vibration isolators are (a) resilient load support and (b) energy dissipation. In many isolators, the resilient load support and energy dissipation are provided by a single element, such as a molded elastomeric isolator. Similarly, wire rope isolators provide both load support and damping, the damping being provided by friction effects between strands of interwoven cables. Helical springs, by themselves, provide negligible damping. Consequently, when damping is desired, helical springs are frequently combined with a separate means of introducing friction, viscous or elastomeric (hysteretic) damping. The springs support the load, while the damping element provides energy dissipation. Pneumatic isolators (air mounts), either active or passive, employ gas (usually air) for load support, and are generally lightly damped. Damping can be added to air springs by incorporating capillary flow resistance between the air spring and surge tanks.

The preceding discussion identifies a few of the generic types of isolating media. Within each type, numerous materials are available, each material offering its own advantages and disadvantages. For instance, in the area of elastomeric isolators, there are at least twelve classifications of polymers, with an infinite variety of additives to enhance one or more properties. Similarly, helical springs and wire cables can be manufactured in many materials, the choice depending upon the desired physical characteristics.

20.4.2 Stiffness characteristics. A primary criterion in selecting vibration isolators is stiffness (K). Stiffness is the ratio of the change in force due to a corresponding change in deflection of a resilient element, and is related to natural frequency (f_n) and supported weight (W) (Equation 20-2). The stiffness (K) term used in Equation 20-2 is dynamic stiffness, frequently estimated from a static load deflection curve.

As discussed in Paragraph 20.3, many isolators exhibit a sensitivity in performance to dynamic input. This means that the dynamic stiffness can be greater than the static stiffness, or that the dynamic-to-static stiffness ratio is greater than 1.0. Isolators with negligible damping (e.g., helical springs and lightly damped elastomers) would exhibit a ratio of approximately 1.0, while highly damped isolators may exhibit a dynamic-to-static stiffness ratio as high as 3.0, depending upon the dynamic input and specific materials.

Static load deflection curves are frequently used to estimate dynamic performance. For helical spring isolators, the stiffness calculated from a static load deflection curve can be used as dynamic stiffness and inserted in Equation 20-1 to accurately determine the natural frequency of the isolator. For isolators with damping, a dynamic multiplier would typically be applied to the static load deflection curves. Since this multiplier would vary with materials and dynamic input, no single value can be reported. Consult the isolator manufacturer for recommendations.

For purposes of estimating vibration isolation performance, the static load deflection characteristics of isolators are generally taken to be linear. This assumption is valid over a limited range of deflection. However, since any isolator may potentially be loaded beyond its linear range, especially for severe dynamic loadings, the load deflection characteristics of the isolator need to be considered at the design stage. Also, different load deflection characteristics may be advantageous for different applications, as discussed below.

Figure 20-6 shows some generic static load deflection curves, briefly discussed below:

— Curve (a) shows a linear curve, such as might be obtained with a helical spring isolator. At a certain deflection, the spring bottoms out, becoming solid, and there is a dramatic increase in force.

— Curve (b) shows a curve that is initially linear, but which gradually becomes stiffening. Curves such as this are provided by elastomeric mounts in compression or by air springs with elastomeric snubbing capability. At normal loads, the isolator functions in the linear range, but for increased loads, such as shock, the isolator limits deflection without the solid contact of a bottomed out spring.

— Curve (c) shows a load deflection curve that is approximately linear over a fairly large deflection. Elastomeric mounts in shear provide a relatively large linear deflection. Wire rope isolators also exhibit relatively large linear deflections in the shear and roll axes.

— Curve (d) shows a buckling curve, indicating a high energy storage capacity. Isolators with this characteristic are frequently used for shock applications since they are capable of absorbing the increased energy of a severe mechanical shock input. This characteristic can be attained by wire rope isolators and with elastomeric isolators of special configurations.

20.4.3 Isolator types. Numerous types of isolators may be applied in supporting generator sets and support equipment. Many factors may need to be considered in selecting an isolator. First, the desired performance characteristics of the isolation system must be determined. If a 600 RPM (10 Hz) disturbing frequency needs to be isolated, one would select a 3 Hz isolation system over a 20 Hz isolation system (refer to Paragraph 20.2).

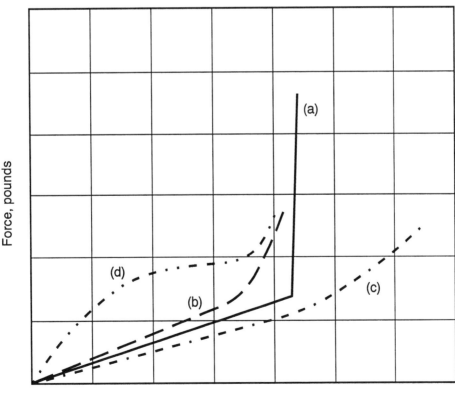

Figure 20-6. Static Load Deflection Curves for Several Types of Vibration Isolators

Obviously, the weight to be supported by the isolator is also an important factor. Operating temperature and fluid exposure are two primary environmental factors that should be considered for isolating generator sets.

Ultimately, the cost of a vibration isolation system is the driving factor — the cost must be commensurate to the value derived from the isolation system. Generally, compromises are required in selecting any isolator but there are too many variables in terms of performance, materials and size to provide any realistic across-the-board guidelines regarding cost comparisons. The recommended approach is to work with a knowledgeable manufacturer of several types of isolators to determine which isolator is best suited for your application.

Housed helical spring isolators are frequently used in stationary applications because they offer a wide load range in an efficient package and provide exceptional vibration isolation. Spring isolators are frequently rated by their linear deflection capability in the vertical direction (e.g., 1-inch deflection or 2-inch deflection). These values are related to Equation 20-5 and Figure 20-3 in that a 1-inch deflection provides a natural frequency of 3.1 Hz and a 2-inch deflection provides a 2.2 Hz natural frequency.

For mobile applications, elastomer and wire rope isolators offer all-directional isolation with captive features. For reasons of drift and stability, vertical natural frequencies for both isolator types are most frequently greater than 5 Hz, usually in the 10 Hz to 15 Hz range. Spring isolators used on mobile applications would generally require an external lock out feature during transport to avoid uncontrolled motion of the unit and isolator.

Auxiliary equipment attached to the gen-set, such as control boxes, is frequently supported by low profile elastomeric isolators. Isolator natural frequencies for these applications are generally in the 10 Hz to 30 Hz range, so isolation of generator rotational frequency would not be as great as for the lower frequency spring isolators. Isolators with a natural frequency in this range may actually amplify the generator rotational frequency.

20.4.3.1 Spring isolators. Helical (coil) spring isolators are typically used in installations where very low natural frequencies are desired — lower than can be provided by most passive isolators. Vertical natural frequencies as low as 2 Hz are not uncommon with helical springs. Helical springs provide great freedom in obtaining a desired stiffness and avoid the drift and creep problems sometimes associated with low frequency isolators. Springs can be fabricated in a variety of metals. Selection is based upon factors such as the desired load on the spring, operating stresses, supported mass, available space, fatigue life, and environmental factors such as temperature and corrosion.

As with any material, the elastic modulus of typical spring materials exhibits some temperature sensitivity, although the sensitivity is less than for elastomeric materials. Over the temperature range of $-100°F - +300°F$, the elastic modulus is within $\pm 5\%$ of the room temperature elastic modulus. At elevated temperatures, changes in yield strength may dictate the choice of materials. Consult with the spring manufacturer in these cases.

Figure 20-7(A) shows a typical housed spring mount. This design incorporates an adjustable damping mechanism to control amplification at resonance. Damping is obtained via frictional forces between an elastomer element and the top housing and loading plate. An adjustable snubber bolt varies preload to the elastomer, varying the amount of damping. This design frequently includes a neoprene base pad that serves two purposes: provides non-skid feature and noise break to mitigate high frequency structure-borne noise.

With additional springs or with larger springs, and with different housings, load bearing capacity can be increased significantly. Mounts of this construction are available for loads in excess of 25,000 pounds.

Housed spring isolators provide restraint in the horizontal directions, resulting in a captive, stable isolation system. However, this configuration results in a horizontal stiffness greater than the vertical stiffness and reduces vibration isolation in the horizontal directions.

Figure 20-7(B) shows an open spring isolator. This design can provide a low resonant frequency in all directions, but has no damping and is not captive. Load bearing capacity in this configuration is much less than for a similarly sized housed spring isolator. With no restraint and little damping, special care should be taken with regard to vibration input at the isolator natural frequency.

In many installations, isolators must be rated for seismic activity. When seismic loading is specified, the isolator and attachments are required to remain structurally intact when subjected to vertical and horizontal loads of specified magnitudes. The magnitude is primarily controlled by the geographical location, but is also influenced by other considerations, such as type of building, floor height above grade, and installation. Spring isolators designed for seismic applications are typically housed and have special restraints designed to limit motion in all directions to approximately 1/4-inch. It should be noted that seismic isolators are not designed to isolate, or attenuate, seismic disturbances. Rather, seismic isolators are intended to function as a conventional vibration iso-

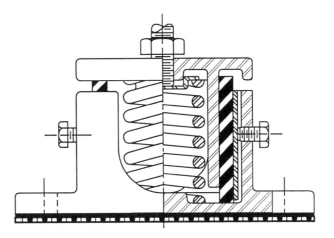

(A) Housed Isolator

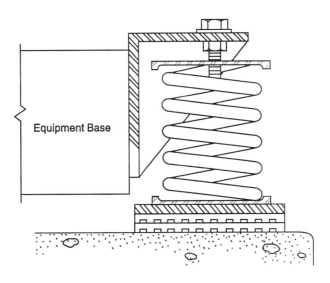

(B) Open Spring Isolator

Figure 20-7. Helical Coil Spring Isolators

lator during normal operation, but have the required strength to remain intact during seismic disturbances, thereby preventing attached equipment from overturning and creating additional hazards.

Figure 20-8 shows a partial cross section view of a seismic spring isolator which has built-in travel limiters. This design operates as a normal vibration isolator, but deflection is limited during a seismic event. When subjected to the large displacements of a seismic event, neoprene pads act as snubbers, limiting motion and preventing metal-to-metal contact within the isolator.

20.4.3.2 Elastomeric isolators. Elastomeric isolators provide a flexibility in design unavailable with other materials used for vibration isolation. They may be molded into many configurations and sizes using many materials, and stiffness can be varied over a wide range. Elastomers are unique in that they may sustain relatively large deformations and return to their original condition with minor changes in dimensions. With the proper combination of elastomer bonded to metal components, there is virtually no limit to the range of vertical and horizontal stiffnesses that can be provided with elastomeric isolators.

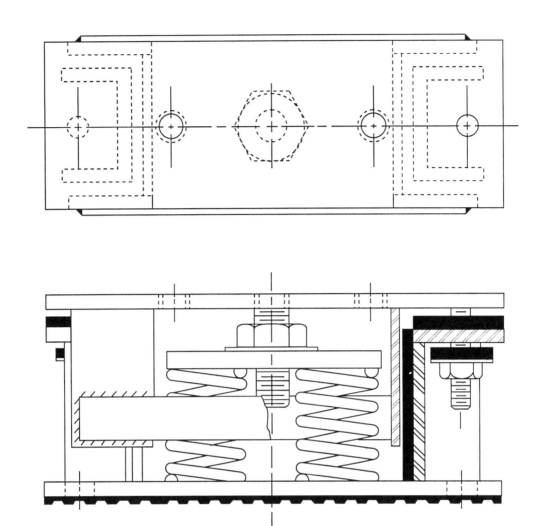

Figure 20-8. Seismic Helical Coil Spring Isolator

At least twelve families of elastomers can be selected for any application. Some of the more frequently used elastomers in generator set applications are briefly discussed below:

— **Neoprene:** good general purpose material, moderate fluid resistance, excellent aging and ozone resistance, operational temperature range of $-20°F - +180°F$.

— **Natural rubber:** low cost, excellent strength and fatigue characteristics, poor fluid resistance, better low temperature characteristics than neoprene, but poorer aging resistance.

— **Nitrile:** superior resistance to fluids, good heat aging.

— **Silicone:** relatively high cost, useful over broad temperature range $-65°F - +300°F$ (higher temperatures for occasional exposure). Excellent aging and moderate oil resistance.

Additional materials that could be used for specific characteristics include, but are not limited to, SBR (styrene butadiene), butyl, EPDM (ethylene propylene), polybutadiene, and fluorosilicone. Most elastomers can be formulated to provide a hardness ranging from 40–70 durometers. This provides the capability of producing an isolator with a fairly large range in stiffness within a given configuration. Damping varies with elastomer types, although there are ingredients that can be added to vary damping within each elastomer family. For the four elastomers discussed above, typical critical damping ratios vary from 0.05 to 0.15. Higher and lower values of damping can generally be obtained with special formulations.

The most basic elastomeric isolator consists of a rubber pad, such as that shown in Figure 20-9(A). These are most frequently molded in neoprene, although they can be produced in alternate materials, including SBR, nitrile, silicone, and cork. Pads of this style frequently have a deflection capability on the order of 0.15 inch, but pads may be combined in series to provide a softer system. Pads load rubber in compression, which is the most common mode of loading. Ribs or other cutouts are frequently included in the pad to provide a skid resistant surface.

Two alternative means of loading rubber in compression are shown in Figures 20-9(B) and (C). These designs bond rubber to metal sections and include the capability of mechanically attaching the mount to the equipment. Design (B) includes a leveling feature. Design (C) can also be used to load the rubber in shear, which can result in a lower stiffness and natural frequency, and improved high frequency isolation.

Figure 20-10(A) shows a cylindrical shear mount, with steel threaded studs bonded to the rubber at both ends. Loaded in shear, this style can provide a natural frequency as low as 5 Hz with a relatively large deflection capability. In compression, this design would provide a natural frequency greater than 10 Hz.

Figure 20-10(B) shows an unbonded cupmount. In this design, rubber is pre-loaded in compression in all directions. Consequently, the mount can be loaded in compression, tension, and shear. A compact design, this style provides elastomeric snubbing in all directions, an advantage where high transient loads may be expected. Natural frequency is in the 20 Hz–30 Hz range.

Figure 20-11(A) shows a fully bonded general purpose isolator. With a snubbing washer installed at the bottom, this design is fail-safe, and provides snubbing in all directions. These mounts can be loaded in either the axial or radial axis. They are available in many sizes, with axial load capability to 2700 pounds. Natural frequency ranges from 10 Hz–20 Hz.

Figure 20-11(B) shows a semi-bonded general purpose isolator with all attitude capability. Natural frequency is as low as 10 Hz and axial to radial stiffness ratio is approximately 1:1.

Figure 20-12 shows a fully bonded semispherical isolator. This design has a relatively high deflection capability and exhibits non-linear buckling characteristics in compression. This mount can be used for vibration protection, but is superior where shock protection is required.

The preceding discussion of commercially available isolators presents a small portion of the available configurations and sizes. Consult a manufacturer for assistance in evaluating performance, materials, and cost.

20.4.3.3 Wire rope isolators. Wire rope isolators combine some of the advantages of elastomeric isolators with those of helical coil springs. Friction effects between the wire strands provide damping values comparable to those of fairly highly damped elastomers, yet the metal components provide the superior environmental resistance of a helical coil spring.

Performance characteristics of the wire rope isolators are determined by the diameter of the wire rope, the number of strands within the cable, the cable length, configuration (i.e., height and major diameter) and the cable twist (or lay). The wire rope materials can be varied to obtain certain characteristics such as elevated temperature performance, electrical conductivity, or heat transfer.

Figure 20-13(A) shows a typical wire rope isolator. Designs of this style are available for loads up to 15,000 pounds per mount. This design provides a buckling characteristic in the axial loading direction, which is desirable for shock loading applications. Horizontal stiffness is less than the vertical stiffness, resulting in improved vibration isolation in the horizontal directions. These isolators are stable in all directions and housings are not required.

Figure 20-13(B) shows a circular wire rope isolator configuration, available in static loads up to 200 pounds.

20.4.3.4 Pneumatic springs (air springs). A pneumatic spring (including air mounts) is a column of gas confined in a container designed to utilize the pressure of the gas as the force medium of the spring.

Air springs can provide a natural frequency lower than can be achieved with elastomeric isolators, and with special designs, lower than helical springs. They provide leveling capability by adjusting the gas pressure within the spring. Air springs require more maintenance and temperature limitations are more restrictive than for helical springs.

Stiffness of air springs varies with gas pressure and is not constant as is the stiffness of other isolators. As a result, the natural frequency does not vary with load to the same degree as other methods of isolation.

Damping in air springs is generally low, with a critical damping ratio (C/C_c) on the order of 0.05 or less. This damping is provided by flexure in a diaphragm or sidewall, by friction, or by damping in the gas. Damping may be increased by incorporating capillary flow resistance (adding an orifice to the flow line) between the cylinder of the air spring and the connecting surge tanks. (See Ref. 7.)

Figure 20-14 (from Ref. 7) shows four commonly used types of air springs. Design (A) is a bellows style air

Vibration Isolation

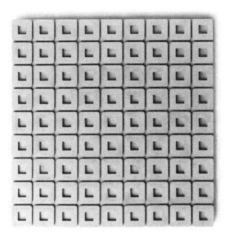

(A) Pad

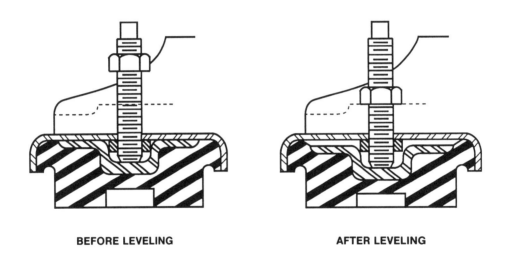

(B) Low Profile Leveling Machinery Mount

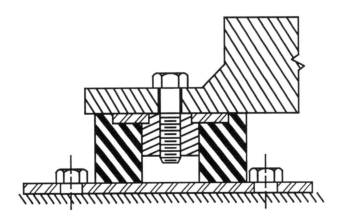

(C) General Purpose Compression Mount

Figure 20-9. Elastomeric Isolators

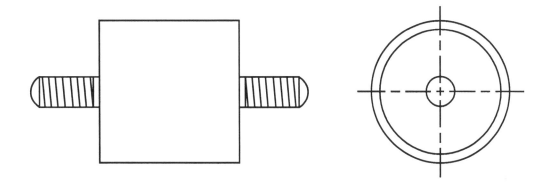

(A) Cylindrical Shear Mount

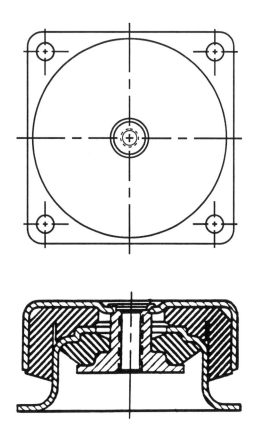

(B) Unbonded Cupmount

Figure 20-10. Elastomeric Isolators

Vibration Isolation

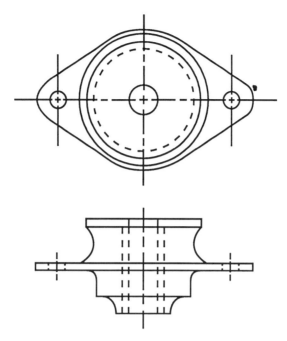

(A) Fully Bonded General Purpose Mount

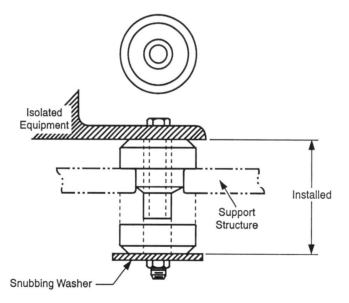

(B) Semi Bonded General Purpose Mount

Figure 20-11. Elastomeric Isolators

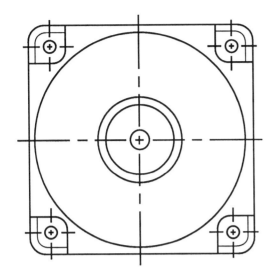

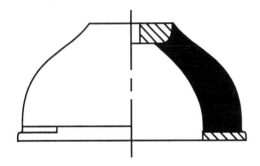

Figure 20-12. Fully Bonded Buckling Mount

spring, which is available in one, two or three convolutions, with girdle hoops (rigid rings) to control expansion of the bellows. Design (B) is a piston type rolling lobe air spring. The flexible member is guided by rigid structures attached to both ends of the flexible member. Designs (C) and (D) both use elastomeric diaphragms to control stiffness. Design (D) has elastomeric sidewalls to provide gradual snubbing in the event of severe dynamic loading.

REFERENCES

[1] Shock & Vibration Handbook, Third Edition, Chapter 3, Vibration of a Resiliently Supported Rigid Body, Harry Himelblau, Jr. and Sheldon Rubin.

[2] Shock & Vibration Handbook, Third Edition, Chapter 30, Theory of Vibration Isolation, Charles Crede and Jerome Ruzicka.

[3] Engineering Vibrations, L. Jacobsen and R. Ayre, McGraw-Hill Book Company, Inc.

[4] Fundamentals of Vibration, R. Anderson, The Macmillan Company.

[5] Theory of Vibration with Applications, W. Thomson, Prentice-Hall, Inc.

[6] Spring Design Manual AE-11, Society of Automotive Engineers, Inc.

[7] Shock & Vibration Handbook, Third Edition, Chapter 33, Air Suspension and Active-Isolation Systems, R. W. Horning and D. W. Schubert.

Vibration Isolation

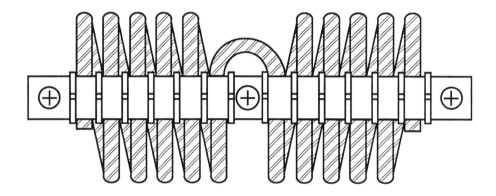

(A) Helical Design

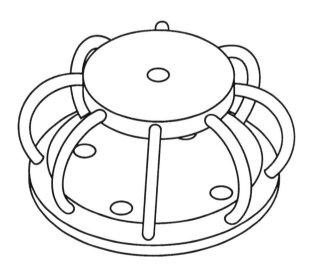

(B) Circular Arch

Figure 20-13. Wire Rope Isolators

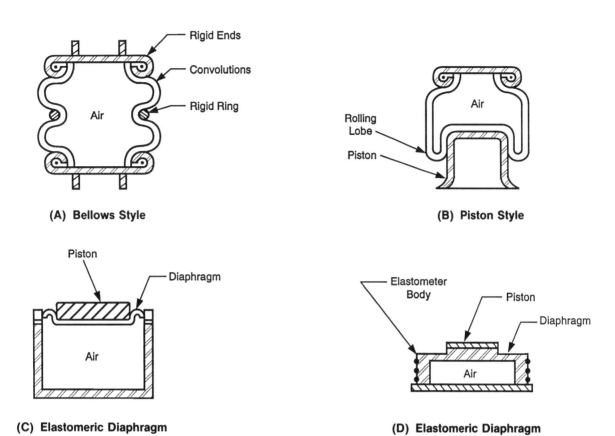

Figure 20-14. Pneumatic Springs

Vibration Analysis for a Sound Generator Set Design

Dr. Simon Chen

CHAPTER 21

21.1 TYPES OF GENERATOR SETS

EGSA members produce a wide power range of generator sets, from a portable 1 kW set to a continuous running 10 MW utility power generator. Prime movers commonly used include: four-cycle diesels, two-cycle diesels, gasoline engines, natural gas engines, and gas turbines. Steam turbines, water turbines, and windmills are also used. Generators must be designed to match these drivers and their rotational speeds. Variations also include voltage requirements and duty specifications. The generator frame, mountings, shaft, and bearings must match structurally the overall gen-set design in order to produce electric power reliably and efficiently. In this chapter, the structural dynamic characteristics of gen-sets are studied. The causes and some solutions of torsional as well as linear vibrations are reviewed.

Figure 21-1 shows different types of generator sets produced today. Take note of their structural differences. For easy reference, different types of gen-sets are classified as follows:

a. Overhung

b. Four-point mounting

c. Six-point mounting

d. Eight-point with two-bearing generator

e. Tandem set

f. Turbine set

21.2 GENERATOR ANATOMY

First let us review the basic anatomy of a generator set. The simplified generator set illustrated in Figure 21-2 is composed of an engine, a flange mounted generator, a four-point mounting system, and a base.

21.2.1 Mass-elastic shaft system. The generator is connected to the engine shaft at the flywheel end, as shown in Figure 21-3.

We can calculate the torsional natural frequency of the system if we know the moment of inertia of the different rotating members and the torsional elasticity of the shaft.

From the cylinder indicator diagram of each cylinder (Figure 21-4) and the shaft geometry (Figure 21-5), the torsional excitation input to the shaft at different harmonics can be obtained. The excitation from cylinder gas pressure causes significant shaft vibration amplitude at certain critical speeds, depending on the firing order and shaft configuration.

The complete shaft system is supported by several engine main bearings and one or two generator bearings. For a single-bearing generator, the engine main bearing must share part of the generator rotor weight and all the extra burden due to imbalance and misalignment. The bearing support of this drive end main bearing must be designed to carry this extra load when the generator is directly coupled to the engine. For a two-bearing generator set, generator and engine weights are carried by their own bearings. The engine crankcase and the generator frame structure take the overall engine torque reaction (not the instantaneous torque fluctuations) transmitted from the engine to the generator. The alignment of the generator along the crankshaft bearing support centerline (see Figure 21-6) is crucial for long life and vibration free operation.

21.2.2 Skid and mountings. The engine and generator are supported by the skid through engine and generator base mountings. The sub-base or skid and its mountings will be subjected to the static load of the engine and the generator. It will also be subjected to the

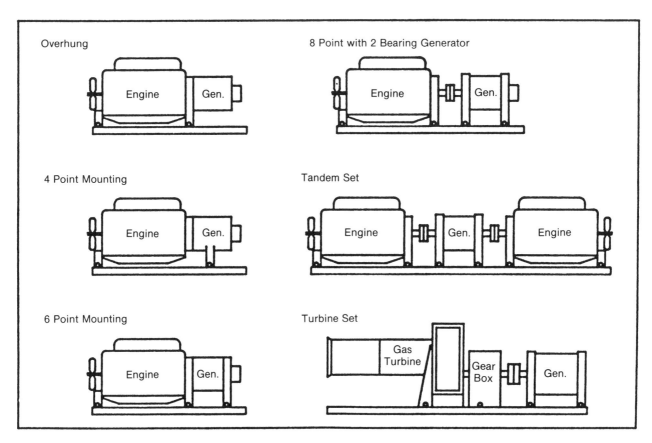

Figure 21-1. Different Types of Generator Sets

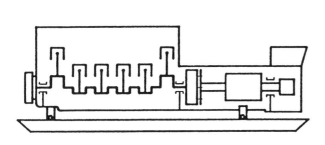

Figure 21-2. A Gen-set Exposed

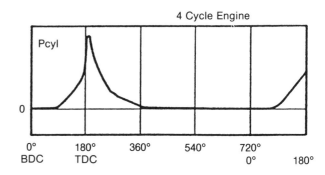

Figure 21-4. Cylinder Gas Pressure

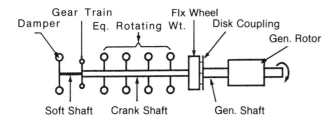

Figure 21-3. Shaft System

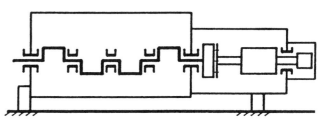

Figure 21-5. Bearing System

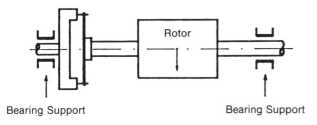

Figure 21-6. Crankline Alignment

dynamic loads including uneven firing, imbalances, and torque reactions. See Figure 21-7.

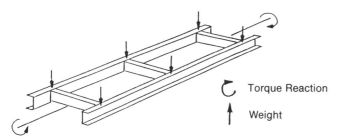

Figure 21-7. Force and Torque Reaction

The skid should be stiff enough so that there is no structural vibration in resonance with some engine vibration orders. If the skid is not properly designed, it could also cause abnormal noise. The type and location of mountings are important for the smooth running of a gen-set. Good mountings make the sub-base an integral part of the whole structural design. The skid also serves as a base for accessories, controls, cooling, oil and starting systems and sometimes, a complete housing. Further details of skid design are discussed in Section 21.8.

21.2.3 Responsibility. Our major goal is to build a trouble free gen-set for our customer. The interrelationship of the engine, generator, and skid requires complete cooperation between generator set builder, generator manufacturer, and engine manufacturer. Table 21-1 shows the normal responsibilities of the engine manufacturer, generator manufacturer, and set builder.

21.3 CALCULATIONS TO AVOID LATERAL VIBRATION FAILURES

21.3.1 Natural frequency of a six-point mount gen-set. When a generator set is on the drawing board, the lateral natural frequency (whipping speed) of the generator shaft must be evaluated. The simplest method is to assume a simple supported beam with generator rotor and flywheel as two major concentrated loads, as in Figure 21-8. For illustration, the last engine bearing is considered as one support and the generator bearing the other. Illustrated is a single bearing generator design with its shaft rigidly connected to the engine crankshaft, like many large generator sets used in municipalities. The shaft is considered as divided into sections, with various diameters and weights of these sections taken into account. This calculation assumes rigid bearing support. Some engine builders request that the lateral natural frequency of the generator shaft be higher than three times the synchronous speed. The purpose is to avoid the possibility of causing a critical lateral vibration or of giving the engine shaft some unexpected bending stress.

Table 21-1. Design and Performance Responsibilities

Engineering Items	Responsibility
1. System design and controls	Set builder
2. Lateral and torsional vibration	Set builder, checked by engine Mfr. as well as generator Mfr.
3. Linear vibration and unbalance	Set builder and engine Mfr.
4. Alignment	Set builder
5. Vibration and sound isolation	Set builder
6. Performance	Set builder and engine Mfr.
Fuel cost $/kW hr	Both
Reliability	Both
Load recovery	Both
Exhaust emission	Engine Mfr.
Sound level	Both
7. Damper design	Engine Mfr.
8. Generator overheat, motor starting capability	Generator Mfr.

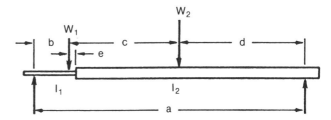

Figure 21-8. Simple Model for Lateral Natural Frequency

Figure 21-9 shows a more complicated model considering the elasticity of bearing and bearing support. This configuration gives a lateral resonance lower than the first, simplified method.

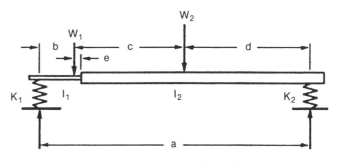

Figure 21-9. Bearing Flexibility

Vibration Analysis For a Sound Generator Set Design

Many generator sets use a flexible disc coupling. The simulation of this torsionally stiff, but laterally flexible joint, becomes more complicated, as shown in Figure 21-10.

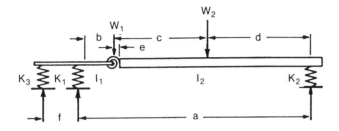

Figure 21-10. Flexible Disc Coupled Gen-set

A comparison of calculated results of these three methods is shown in Table 21-2. The calculation is based on a V-8 engine driving a four-pole generator. The design is quite safe.

Table 21-2. Calculated Lateral Vibration Frequencies

Simple model	6994 cpm
Consider bearing flexibility	5095 cpm
Consider coupling flexibility	4551 cpm

Only the critical speed is analyzed in this chapter. For high speed rotating machinery, studies on bearing stability and rotor whirl orbits are needed for trouble-free operation.

21.3.2 Turbine generator shaft. For a two-bearing generator, used in a gas turbine gen-set, the shaft is sometimes quite long. See Figure 21-11. In this case, the natural frequency will be lower and closer to the operating RPM. Since this long shaft is supported by its own generator bearing and coupled to the gas turbine shaft with a flexible coupling, little bending or flexure will be transmitted to the engine shaft and its bearing. If the coupling design is properly specified, the situation is less critical than for the single-bearing generator. In any case, the generator shaft should be designed so that the actual lateral vibration natural frequency is at least 30% higher than the operating speed. When the calculated result is judged too low, the shaft diameter or stiffness must be increased. Bearing support lateral stiffness should also be scrutinized to provide a smooth-running system.

21.3.3 Shaft axial vibration. As illustrated in Figure 21-12, firing in a cylinder will cause axial movement of the crank. For some high BMEP diesels, this axial movement may be 10 mil each time there is a firing in a cylinder.

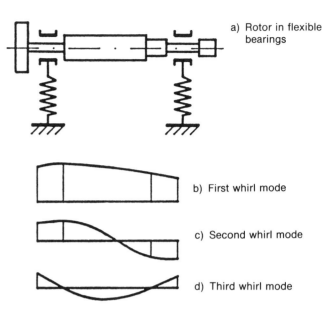

Figure 21-11. Long Turbine Generator Shaft

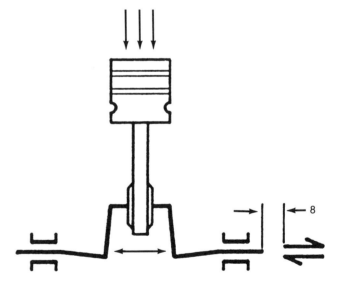

Figure 21-12. Axial Vibration

For a single-cylinder four-cycle engine, the firing causes a half-order frequency (once every 2 RPM). This axial vibration of the shaft, relative to the bearing, could cause failure for a spherical roller bearing commonly used on a single-bearing generator. When this axial movement is coupled with some torsional or bending resonance, the situation worsens. When the measurement shows excessive shaft vibration, and the cause is due to engine crank deflection, the generator bearing must be replaced with a design that is capable of coping with this axial movement. A spherical-backed roller bearing has been used successfully.

21.4 CALCULATIONS TO AVOID TORSIONAL FAILURES

When there is a first mode torsional problem, the shaft could fail close to the first mode nodal point (which is close to the flywheel). In practice, the shaft could fail on any middle crank section between the fan and the flywheel, since more than one mode and many orders are involved.

A torsional failure on a gen-set is a serious engineering design error. Excessive gear train chattering is an indication of a torsional problem. When gear train chattering is detected, the generator set should be given a complete torsional checkup. The weakest torsional link in the shaft system is the engine crankshaft and, sometimes, the coupling between the engine and generator.

A torsional crack generally starts at a high stress concentration corner (where section is changed) or around an oil hole (see Figure 21-13). Sometimes cracks initiated by torsional overload and finished off by bending load (across a cheek section, for example) are observed.

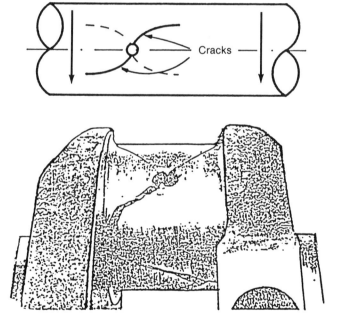

Figure 21-13. Cracks from Radial Hole

21.4.1 Torsional excitation input.
Three major factors affect the torsional vibration of a shaft system:

a. *The mass elasticity system*, which determines the natural frequencies

b. *The torque excitations*, which affect the torsional amplitudes

c. *The damping in the system*. In classical solutions, the natural frequency is determined first and then the torque excitation inputs and the many harmonic components are determined. When the engine is operated at a speed near a major order of the natural frequencies, the shaft will vibrate (oscillate) at this frequency. The amplitude of the vibration depends upon the magnitude of the excitation input, how close the engine speed is from the resonance, and the amount of damping available.

There are two major sources of torque excitation input:

1. *Inertia torque harmonics*. T_{ine} is caused by the reciprocating weight, W_{rec}. Rotating weight causes a uniform inertia torque and no vibration under the stable RPM case. The reciprocating motion of piston and connecting rod causes torque harmonics due to the mechanism. T_{ine} has first order (same as engine RPM), second order (twice RPM), and some third order harmonic components. Higher order ($n>4$) may be neglected. This T_{ine} is not as important for low speed engines, but could become a factor for very high RPM engines.

2. *Gas pressure torque harmonics*. Gas pressure torque, T_{gas}, is derived from the cylinder pressure. It is composed of mean torque, T_{mea} (could be expressed as I_{mep}), and gas pressure torque harmonic components, superimposed on T_{mea}. The harmonic components of gas pressure tangential effort are called TE_{gas}.

For a multi-cylinder engine, the overall torque excitation put to the shaft is the vector sum of T_{ine} and TE_{gas}. The vector summation is needed because of the phase angle between cylinders and the firing order.

(a) *Inertia torque harmonic excitation*. For clarity, we will first look at the T_{ine} derived for cylinder (mass) number m as a function of shaft angle. See Equation 21-1.

$$T_{ine} = \sum_{n=1}^{24} TI_n \cdot \sin[n(\overline{\omega} \cdot t + \alpha_m)]$$

(Equation 21-1)

$$= \frac{W_{rec} \cdot \omega^2 \cdot r^2}{2 \cdot g \cdot A \cdot r} \{ [1/2(r/L) + 1/8(r/L)^3] \sin(\omega \cdot t + \alpha_m)$$

$$- [1.0 + 1/16(r/L)^4] \sin[2(\omega \cdot t + \alpha_m)]$$

$$- [3/2(r/L) + 0.56 \cdot (r/L)^3] \cdot \sin[3(\omega \cdot t + \alpha_m)]$$

$$- [0.5(r/L)^2 + 0.25(r/L)^4] \cdot \sin[4(\omega \cdot t + \alpha_m)]$$

$$+ \ldots \ldots \}$$

Vibration Analysis For a Sound Generator Set Design

TI_n = harmonic component of nth order for inertia torque, a power series of r/L, $(r/L)^n$

n = harmonic order, use 1, 2, 3, and maybe 4

ω = shaft angular velocity

r = crank radius

L = crank length

W_{rec} = reciprocating weight

α_m = crank phase angle between cylinders

For some textbooks (Ref. 1) the inertia torque harmonics are normalized to become a function of T_{mea}. See Equation 21-2.

$$Y_n = \frac{t_n}{T_{mea}} \cdot \sin \cdot n(\omega \cdot t + \alpha_m) \quad \text{(Equation 21-2)}$$

b. *Gas pressure torque harmonic excitation.* The total time-varying gas pressure torque, T_{gas}, is composed of the mean torque output, T_{mea}, and its superimposed harmonic components, expressed in a Fourier series of sine and cosine terms.

$$T_{gas} = T_{mea} + T_{mea} \sum_{n=0.5,1}^{n} [U_n \cdot \sin n(\omega \cdot t + \alpha_m)$$
$$+ V_n \cdot \cos n(\omega \cdot t + \alpha_m)] \quad T_{mea} = I_{mep}$$

(Equation 21-3)

U_n and V_n are the coefficients of sine and cosine terms normalized by T_{mea}.

The torque exerted on the shaft caused by gas pressure is pronounced in the combustion/expansion stroke. Since a four-cycle engine fires once every two revolutions (Figure 21-14), it has a pronounced half order. During the intake and exhaust strokes of a four-cycle engine, the pressures on the piston are rather constant. From the $T_{gas} d\theta$ – = P · dV relationship, the gas pressure torque or tangential

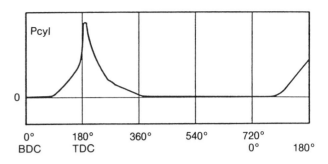

Figure 21-14. Four-Cycle Engine Cylinder Gas Pressure

effort, T_{gas}, can be plotted as shown in Figures 21-15 and 21-16. The mean torque, T_{mea}, is also shown. The T_{mea} is the time-averaged mean of the T_{gas} and is the torque output.

When the engine is operated at part loads, the torque excitation T_{gas} is reduced in the firing-expansion stroke, as shown in Figure 21-17. The dotted curves mark half-load and idle. Harmonic components are correspondingly reduced.

The determination of T_{gas} and its resultant harmonic components (both sine and cosine terms) was complex until the use of computerized Fourier Series techniques. For each order considered, both harmonic amplitude and phase angle are involved. Ker Wilson (Ref. 2), Porter (Ref. 3), and many others have calculated T_{gas} or tangential effort based on similar engine concepts. For example, Lloyd (Ref. 4) provides typical gas pressure tangential effort, T_m, and amplitude data up to 240 I_{mep} (indicated mean effective pressure). See Figure 21-18. Note that the symbol used for this T_m is T_{gas}.

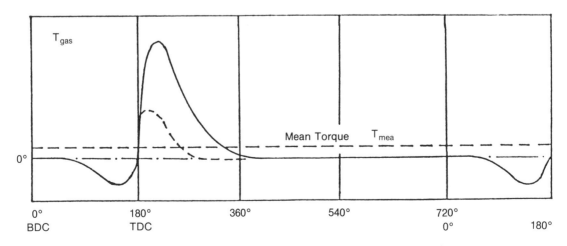

Figure 21-15. Single Cylinder Gas Torque on the Shaft

Vibration Analysis For a Sound Generator Set Design

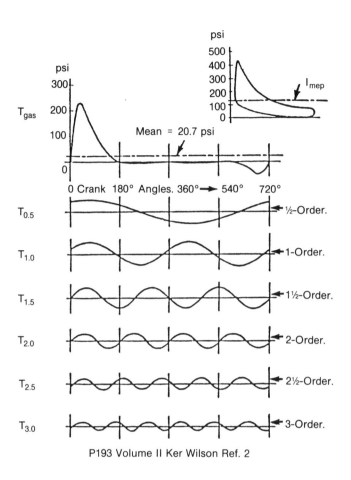

Figure 21-16. Rated Load T_{gas} and Its Harmonics

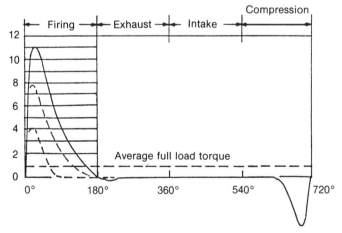

Figure 21-17. Gas Torque T_{gas} at Part Load

Today, this gas pressure tangential effort can be directly obtained from the pressure time diagram of a particular engine by using a data acquisition system and an engine cycle analyzer. This method was described in Diesel Progress, North American, December, 1986, page 26 (Ref.

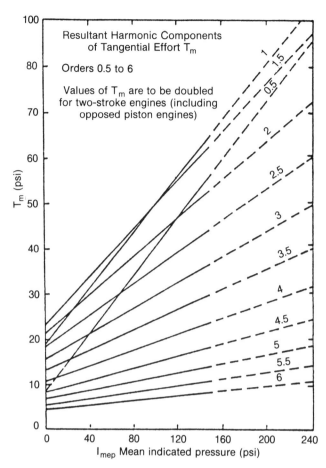

Figure 21-18. Lloyd Tangential Effort Table (Ref. 3)

5). These tangential effort tables are for an average engine of a certain type; the readings could vary as much as 30% for a specific engine.

c. *The combined inertia and torque tangential effort* (*TE*). The combined harmonics is comprised of the sine terms of the T_{ine} and T_{gas}, as well as the cosine term of T_{gas}.

$$TE_m = \sum_{n=0.5,1}^{n} (T_{ine} + T_{gas}) = \sum_{n=0.5,1}^{n} TE_{m,n}$$

(Equation 21-3)

At cylinder m for nth order

$$TE_{m,n} = T_{mea}[(U_n + Y_n) \cdot \sin n(\omega \cdot t + \alpha_m) + V_n \cdot \cos n(\omega \cdot t + \alpha_m)] \text{ for } n = 1,2,3\ldots$$

$$TE_{m,n} = T_{mea}[U_n \cdot \sin n(\omega \cdot t + \alpha_m) + V_n \cdot \cos n(\omega \cdot t + \alpha_m)] \text{ for } n = 0.5, 1.5, 2.5\ldots$$

Where $Y_n = TI_n/T_{mea}$

d. *Instantaneous torque and number of cylinders.* An instantaneous output torque diagram of four-cylinder and six-cylinder in-line engines is shown in Figure 21-19. The output torque fluctuation is clearly evident. This output was obtained by having a rigid shaft and equally exerting all gas pressure pulses at the output shaft. Note that torsional vibration and an elastic shaft will distort the output illustrated in the graph.

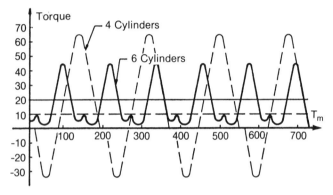

Figure 21-19. Output Torque

21.4.2 The determination of tangential effort. The single order tangential effort, $TE_{m,n}$ caused by any one cylinder with mass number m, is the vector sum of T_{ine} and T_{gas} for that particular single order n being studied.

$$TE_{m,n} = T_{mea}[(U_n + Y_n) \cdot \sin n(\omega \cdot t + \alpha_m)$$
$$+ V_n \cdot \cos n(\omega \cdot t + \alpha_m)] \text{ for } n = 1,2,3\ldots$$

$$TE_{m,n} = T_{mea}[U_n \cdot \sin n(\omega \cdot t + \alpha_m)$$
$$+ V_n \cdot \cos n(\omega \cdot t + \alpha_m)] \text{ for } n = 0.5, 1.5, 2.5\ldots$$

(Equation 21-5)

The total tangential effort TE_m for cylinder number m is the vector sum of orders of all those single order $TE_{m,n}$.

$$TE_m = \sum_{n=0.5,1}^{n} (T_{ine} + T_{gas}) = \sum_{n=0.5,1}^{n} TE_{m,n}$$

(Equation 21-6)

$$T_{ine} = \sum_{n=1}^{n} TI_n \cdot \sin n(\omega \cdot t + \alpha_m)$$
$$= (W_{rec}/2g) \cdot w^2 \cdot r^2 \cdot \{0.50 \cdot (r/L) \cdot \sin(\omega \cdot t + \alpha_m)$$
$$- 1.0 \cdot \sin 2(\omega \cdot t + \alpha_m)$$
$$- [1.5 \cdot (r/L) + 0.56 \cdot (r/L)^3] \cdot \sin 3(\omega \cdot t + \alpha_m)$$
$$- 0.50 \cdot (r/L)^2 \cdot \sin 4(\omega \cdot t + \alpha_m)\ldots\}/(A \cdot r)$$

$$T_{gas} = T_{mea} + T_{mea} \sum_{n=0.5,1}^{n} [U_n \cdot \sin n(\omega \cdot t + \alpha_m)$$
$$+ V_n \cdot \cos n(\omega \cdot t + \alpha_m)]$$

$$T_{mea} = I_{mep}$$

For calculating inertia torque T_{ine}, whole order numbers n(1, 2, 3, etc.) are used for either two- or four-cycle engines. For gas pressure torque T_{gas}, half-order numbers n(0.5, 1.5, etc.) are required for four-cycle but not for two-cycle engines.

The classical method is to treat T_{ine} and T_{gas} harmonic components separately and then to add them vectorially. Both harmonic amplitudes (t_n, V_n, U_n) and vector angle n($\omega t + \alpha m$) must be considered for a particular cylinder. For parametric study t_n, V_n, and U_n are normalized by T_{mea}. This was referred to in Section 21.4.1.c.

For a particular engine, T_{ine} can be computed by knowing the reciprocating weight, the t_n harmonic components, the crank radius, crank radius to connecting rod length ratio, rotational speed, and crank phase angle. Only sine components are involved.

The resultant T_{gas} is the vector sum of both sine and cosine components. Not until recently has the instantaneous cylinder pressure diagram been readily available. A tangential effort table was used extensively for the sake of estimating torsional amplitudes of an engine under study. The use of such a tangential effort table is based upon the engine similarity rule. For example, all gasoline engines have the same indicator diagram at the same I_{mep}, all diesels have another indicator diagram, etc. This is not sufficiently accurate for present day analysis.

Many of the classical methods handle the complex vector analysis and sum of order calculation by breaking the T_{gas} into sine and cosine terms harmonic components, U_{nm} and V_{nm}, respectively. U_{nm} can then be combined with Y_n, which is also sine term harmonics. In addition, these U and V harmonic components are normalized to mean torque, T_{mea}. In this way, similarity rules can be applied so that various tabulations can be used for approximate torsional solutions, without actually solving the complex equations governing the torsional vibration of an engine.

Today this complex problem is solved directly by using advanced mathematics and a computer. Many of the classical steps outlined so far are bypassed. However, these methods including the Holzer, the forced vibration, the flank speed calculation, etc., have their educational value. It helps in understanding the physical implication of crankshaft torsional vibration. For example:

- The major and significant orders are caused by inphase vector summation of the torsional amplitudes from several cylinders. Firing order is involved here.
- A vee engine can be treated as an in-line engine by using a bank angle multiplier factor.
- A star diagram is used to explain the cylinder summation factor.
- The natural frequency of a crank system can be determined by trial and error, i.e., the Holzer method.
- The modal node is formed by using the concept of the normal elastic curve and relative amplitude obtainable by Holzer tabulation.
- The forced vibration solution is treated by energy balance, by the normal elastic curve, and by use of the equilibrium amplitude concept.
- The flank speed resonance amplitudes are determined by using empirical dynamic magnifying factors, and by introducing some simple damping correlations.

These concepts are discussed below.

a. *Number of cylinders and major torsional orders.* The combined gas pressure excitation torque at a particular mass number is a vector sum of each T_{gas} harmonic of all the cylinders, based on their firing interval angle and elasticity between cylinders. For a four-cycle engine, each cylinder fires every two revolutions; for a two-cycle engine, each cylinder fires every revolution. For a six-cylinder in-line four-cycle engine or a three-cylinder two-cycle engine, the major firing impulse is six times every two revolutions, or three times per revolution or third order. For a four-cylinder, four-cycle, or a two-cylinder two-cycle in-line engine, the major order is two. Some illustrations are shown in Figures 21-20 and 21-21. $\Sigma \beta$ is the cylinder torque vector sum multiplier. Now you can see why second order is the major for four-cylinder engines and third and sixth orders are majors for six-cylinder engines.

Multi-Cylinder Gas Torque Vector Analysis

Four-Cycle Four-Cylinder In-line Engine
Firing Interval $\alpha = 180°$
Firing Sequence   1, 3, 4, 2

n Order	$n\alpha$				$\overrightarrow{\Sigma\beta}$
	1	3	4	2	
1/2	0	90	180	270	0
1	0	180	360	180	0
1-1/2	0	270	180	90	0
2	0	360	360	360	4

Four-Cycle Six-Cylinder In-line Engine
Firing Interval $\alpha = 120°$
Firing Sequence   1, 5, 3, 6, 2, 4

n Order	$n\alpha$						$\overrightarrow{\Sigma\beta}$
	1	5	3	6	2	4	
1/2	0	60	120	180	240	300	0
1	0	120	240	360	120	240	0
1-1/2	0	180	360	180	360	180	0
2	0	240	120	0	240	120	0
3	0	360	360	360	360	360	6
6	0	360	360	360	360	360	6

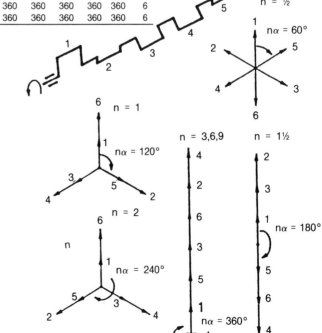

Figure 21-20. Four-Cycle In-line Engine Vector Sum

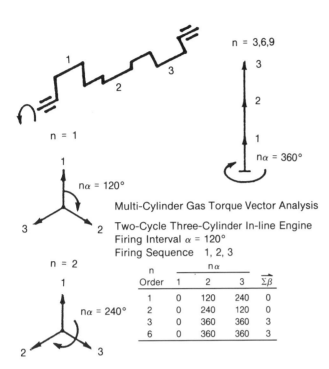

Figure 21-21. Two-Cycle Three-Cylinder In-line Engine Vector Sum

In summary, for a one-cylinder two-cycle engine the significant orders are 1, 2, 3, 4, etc., and for a four-cycle engine they are 0.5, 1, 1.5, 2, etc. For a three-cylinder two-cycle engine the significant orders are 4, 8, 12, etc., and for a four-cycle engine they are 2, 4, 6, etc. See Table 21-3.

Table 21-3. Significant Orders of Gas Pressure Torque

Number of Cylinders	Cycle	Orders Appearing, n
1	2	1, 2, 3, 4, etc.
1	4	0.5, 1, 1.5, 2, 2.5, etc.
2	2	2, 4, 6, 8, etc.
2	4	1, 2, 3, 4, etc.
3	2	3, 6, 9, etc.
3	4	1.5, 3, 4.5, 6, 7.5, etc.
4	2	4, 8, 12, etc.
4	4	2, 4, 6, 8, etc.
5	2	5, 10, 15, etc.
5	4	2.5, 5, 7.5, etc.
6	2	6, 12, 18, etc.
6	4	3, 6, 9, 12, etc.

b. *Vee engine bank angle multiplier and firing order.* When a bank angle of a vee engine is involved, another multiplier, BAM, is needed to complete the vector sum of the torque excitation of the whole engine. Each bank is treated like an in-line engine first and the BAM multiplier is then applied as in Equation 21-7.

$$BAM = 2 \cos n \cdot \sigma / 2 \quad \text{(Equation 21-7)}$$

σ = firing interval between banks

n = harmonic order number

For four-cycle engines with an even number of cylinders in each row, "alternate firing" is commonly used.

σ = vee angle + 360°

For four-cycle engines with an odd number of cylinders in each row, and for all two-cycle engines, "consecutive firing" is used.

σ = vee angle

With this rather simplified explanation, the BAM for vee angle = 45° can be tabulated as in Table 21-4 for the two firing cases discussed.

Table 21-4. Multiplier BAM, vee angle 45 degrees

| | BAM | |
n	Alternate Firing	Consecutive Firing
0.5	0.390	1.962
1	1.848	1.848
1.5	1.112	1.662
2	1.414	1.414
2.5	1.662	1.112
3	0.765	0.765
3.5	1.962	0.390
4	0	0
4.5	1.962	0.390
5	0.765	0.765
5.5	1.662	1.112
6	1.414	1.414

c. *Cylinder summation factor (CS) and star diagram.* The torque coefficients plotted above are for one cylinder only. Torque harmonics from other cylinders will have a phase relationship with the gas torque from the reference cylinder. This phase angle is n times α m, or $n \cdot \alpha$ m. α m is the angle between firing. This was used in a star diagram where a rigid crank was assumed. In that case, for a four-cycle six-cylinder in-line engine, the cylinder summation factor ($\Sigma \beta_m$) is six at the third order, as shown in Figure 21-22.

This third order torsional is the major order for a six-cylinder four-cycle engine. Fourth order is the major order for an eight-cylinder engine. There could be other lesser, but significant, orders in the operating range also.

For an elastic crank, considering realistic stiffness, the amplitude by this torque input is reduced toward the drive end corresponding to the relative amplitude diagram (or normal elastic curve). Forced vibration will have the same relative amplitude shape as a free vibration situation at or near the resonant speed, but with a different reference amplitude depending on the excitation torque input. The first mode star diagram and $\Sigma\beta_m$ will be as shown in Figure 21-23. The cylinder summation is reduced from the rigid shaft case.

21.4.3 A review of the classical Holzer forced vibration torsional analysis.

a. *Mass-elasticity data.* A mass-elasticity model is needed with correctly evaluated inertia masses (I), shaft

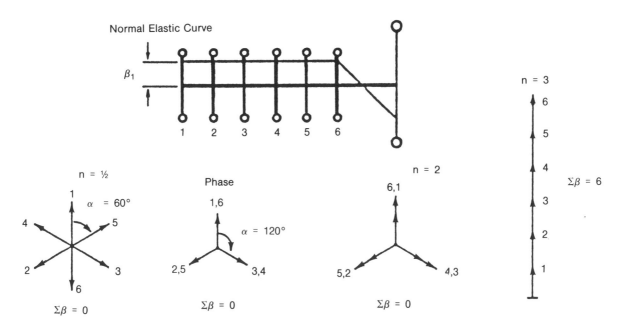

Figure 21-22. Non-Elastic Shaft

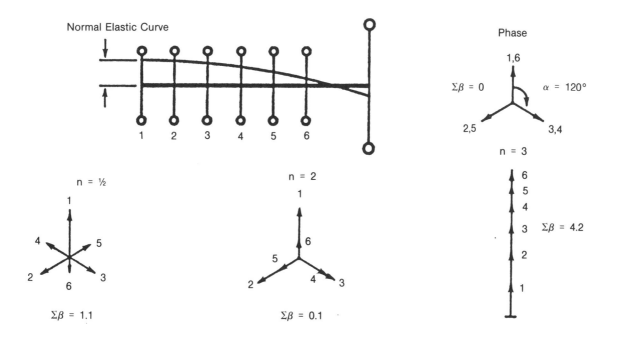

Figure 21-23. Elastic Shaft

Vibration Analysis For a Sound Generator Set Design

stiffness (K) of different segments of the crankshaft and its associated inertia members, couplings and generator shaft. Adding a generator rotor lowers system natural frequency. When accurate prediction of torsional behavior and natural frequency is required, the mass-elasticity inputs (Figure 21-24) must be precisely determined. The stiffness calculation of crank web, etc. can now be more precisely done by finite element methods (FEM). The calculation of mass inertia and center of gravity of complex shapes can be easily achieved by using CAD or FEM techniques.

b. *Holzer Table and natural frequencies.* Natural frequency of the shaft and the mode shape analysis (normal elastic curve) are based on the free vibration principle. The number of natural frequencies is equal to the number of masses, minus one. For example, there is only one major resonance in a two mass system. The most established natural frequency determination used is the Holzer Table. This was established 60 years ago and improved by Timoshenko, Den Hartog, Ker Wilson, Porter, and many others. First, the mass-elasticity system is simulated. Using the Holzer method, the inertia torque summation ($\Sigma I \omega_0^2 \cdot \beta$) with an assumed natural frequency (ω_0) is calculated. From the principle of free vibration, the sum of inertia torques must equal zero when resonance occurs. This means that no excitation or external torque is required to sustain free vibration at resonance (Ref. 1.)

$$ST_m = \sum_{m=1}^{m} I_m \cdot \omega_0^2 \cdot \beta_m = 0 \qquad \text{(Equation 21-8)}$$

The angular vibration amplitude or $\Delta\beta$ between any station is:

$$\Delta\beta = \beta_m = \beta_m + 1 = T_m/K_m = (1/K_m) \cdot \Sigma I_m \cdot \omega_0^2 \cdot \beta_m$$
$$\text{(Equation 21-9)}$$

The Holzer Table is designed to facilitate this free vibration calculation. It is a reiterative procedure, with successive assumptions of vibration frequencies until $T_m = \Sigma I_m \cdot \omega_0^2 \cdot \beta_m$ is equal to zero at the free end of the shaft. When boundary condition is satisfied, the frequency assumed is the correct system natural frequency ω_0. The first mode natural frequency of a V-8 engine and the relative amplitudes at different mass stations are shown in Table 21-5. This iteration can be quickly done on a computer.

Using this procedure, higher mode natural frequencies can also be calculated. For most engines, it is not necessary to go beyond second mode. The disturbing orders caused by higher mode resonances will fall beyond the speed range. For a slow speed engine, assuming the first mode resonance cpm = 255 and the second mode resonance cpm = 1380, we can calculate:

- First mode third order rpm = 255/3 = 85 rpm
- Second mode third order rpm = 1380/3 = 460 rpm

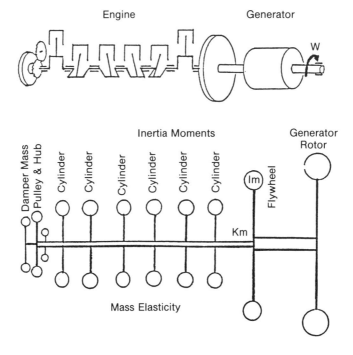

Figure 21-24. Generator Crankline and Mass Elasticity Determination

Table 21-5. First Mode Amplitudes and Natural Frequencies

			First Mode ω_o= 791.7 rad/sec (or 7560.7 cpm)				
m	Im	$\mathrm{Im}\,\omega_o^2$	β	$\mathrm{Im}\cdot\omega_o^2\cdot\beta$	$\Sigma\,\mathrm{Im}\,\omega_o^2$ $Tm\omega_o^2$	Km	$\Delta\beta=\mathrm{Tm/Km}$
1	.8269	518306	1.000	518306	518306	.74 10E6	.736
2	1.0810	67679.9	.999	202972	721278	12.07 10E6	.060
...	...	...	...	...	...	...	...
m	2.61793	16420849	−0.0424	−696244	0		
	(lb-in^2−sec^2)		(rad.)			(lb-in/rad)	

c. *Mode-by-mode nodes and relative amplitudes.* The Holzer Table also gives the relative amplitudes $\Delta\beta$ and the torsional stress levels between each station considered. Since this table is based on an assumed unity amplitude ($\beta_1 = 1$ radian) at the free, only the relative amplitudes and their corresponding relative stress levels are determined for all the shaft sections involved. If the actual free end amplitude is known by measurement, then actual amplitudes at various stations and the associated stress levels can be estimated and plotted. The highest shaft stress or the largest torque excitation (at the nodal point) of a particular mode can thus be determined. The first and second mode free vibrations of a V-8 shaft are shown in Figure 21-25. This relative amplitude curve at natural frequency and a particular mode is called the **normal elastic curve** of the shaft system.

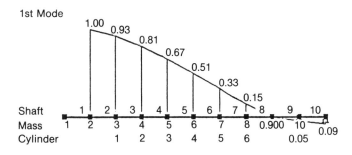

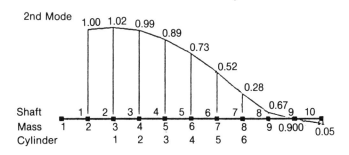

Figure 21-25. Normal Elastic Curve

d. *Forced vibration.* So far we have determined the natural frequency of the shaft section and the excitation torque harmonics of a free vibration mass-elasticity system. From these results, the amplitude of forced vibration is calculated (meaning having some gas pressure and inertia excitation inputs to the vibration system).

There are three basic assumptions in calculating forced vibration amplitudes in the traditional Holzer method:

- Energy balance between vibration and excitation input for determining static amplitude
- The use of static amplitude and dynamic magnifier for the determination of amplitude at resonance
- Torsional amplitude away from resonance is determined by an empirical relationship between dynamic magnifier and speed ratio.

Assumption 1. Energy balance. The vibrational energy (kinetic) of the masses is equal to the total excitation work caused by gas pressure and inertia torque. The energy balance produces a static amplitude θ_s, as shown below. This was advanced by Ker Wilson (Ref. 1). Let us base our discussion on an n order vibration for a specific mass m. The natural frequency of vibration is ω_o.

Kinetic (vibrational) energy is:

$$KE = 1/2 \cdot \omega_o^2 \cdot \sum_{m=1}^{m} I_m \cdot \theta_{sm}^2 \quad \text{(Equation 21-10)}$$

Excitation energy input for each order is:

$$EI = (1/2 \cdot \sum_{m=1}^{m} TE_{m,n} \cdot \theta_{sm}) \cdot A \cdot r$$

(Equation 21-11)

Energy balance is:

$$\omega_o^2 \cdot \sum_{m=1}^{m} I_m \cdot \theta_{sm}^2 = \sum_{m=1}^{m} TE_{m,n} \cdot \theta_{sm} \cdot A \cdot r$$

(Equation 21-12)

The static amplitude θ_s thus denotes that imaginary equilibrium amplitude of a forced vibration system when engine speed ω is zero or insignificant when related to ω_o. The solution is actually based upon the free vibration output tabulations of the Holzer Table. From the Holzer tabulations, there is an equilibrium amplitude factor (EAF) for a specific mode and order n. This EAF is used to construct the normal elastic curve of the free vibration case. For in-line engines:

$$\text{EAF} = \frac{\Sigma \vec{\beta}_m}{\omega_o^2 \cdot \Sigma I_m \cdot \beta_m^2} \quad \text{(Equation 21-13)}$$

Both $\Sigma \beta m$ and $\omega_o^2 \Sigma I m \beta_m^2$ values are from the Holzer Table. Note that $\Sigma \beta_m$ is a vector sum including firing order and crank phase angle. It is a function of order n. The static amplitudes of each vibrating mass of the forced vibration case will have the same shape of normal elastic curve as the free vibration case. When the free end static amplitude θ_{sf} is determined, you can find the static amplitude of all the masses along the normal elastic curve and the equilibrium amplification factor (EAF) of that mass system. For free end mass no. 1, the static amplitude of n order vibration is:

$$\theta_{sf} = \text{EAF} \cdot \text{TE}_n \quad \text{(Equation 21-14)}$$

For a vee engine, a bank angle multipler (BAM) must be included:

$$\text{EAF} = \frac{\vec{\text{BAM}} \cdot \Sigma \vec{\beta}_m}{\omega_o^2 \cdot \Sigma I_m \cdot \beta_m^2} \quad \text{(Equation 21-15)}$$

EAF = equilibrium amplification factor
TE$_n$ = combined torque excitation amplitude for even in-line firing, n order harmonic component
ω_o = natural frequency n order for the mode
$\Sigma \vec{\beta}_m$ = vectorial summarized from normal elastic curve, firing order, and crank phase angle, also illustrated in the vector "star" diagram
$\Sigma I_m \cdot \beta_m^2$ = effective inertia of the system, a Holzer output
$\vec{\text{BAM}}$ = bank angle multiplier for vee engine
$\vec{\text{CS}}$ = cylinder summation factor
= BAM $\times \Sigma \beta_m$

For mass m along the normal elastic curve of a specific order n:

$$\theta_{sm} = \text{EAF} \cdot \text{TE}_n \cdot \beta_m / \beta_1 \quad \text{(Equation 21-16)}$$

This is portrayed graphically in Figure 21-26.

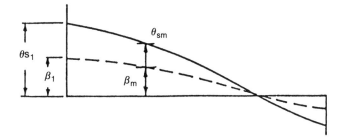

Figure 21-26. Amplitude for Mass m and Order n

Assumption 2. Dynamic amplifier at resonance. The forced vibration determines first the static amplitude θ_s (when ω is zero or insignificant). Then a dynamic amplifier is used to determine the dynamic amplitude at resonant speed ω_o is as follows:

$$\theta_{om} = \theta_{sm} \cdot \text{DM}_o, \text{ (for n order)} \quad \text{(Equation 21-17)}$$

For an undamped forced vibration system, DM$_o$ and m are infinite at resonant speed ω_o. However, most engine shaft systems and cylinders have a certain amount of friction, which provides some amount of damping (Figure 21-27) even if an external damper is not used. The DM$_o$ is an inverse function of the damping.

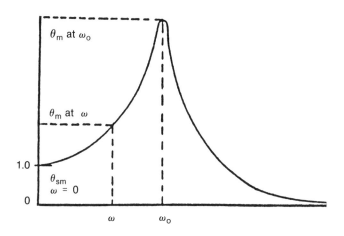

Figure 21-27. Amplitude at Resonance

Assumption 3. Torsional amplitude away from resonance. At all other speeds, the actual amplitude for a specific vibration order n is determined by using the empirical relationship between DM and speed ratio ω/ω_o as follows:

$$\text{DM} = \frac{\theta_m}{\theta_{sm}} = \frac{1}{((1 - \omega^2/\omega_o^2)^2 + \omega^2/\omega_o^2 \cdot 1/\text{DM}_o^2)^{1/2}}$$

Equation 21-18

e. *Dynamic multiplier and damping factor.* The dynamic multiplier at resonance (DM_o) is expressed as $Cc/2C$, to denote the inverse ratio of damping factor (C) relative to critical damping (Cc) used in the force torsional vibration differential equation. See Figure 21-28. The damping factor is experimentally determined by adding one extra mass to the engine system. DM_o in the range of 20–50 is used by some engine crankshaft designers. This empirical value could be verified by taking torsional data close to resonance.

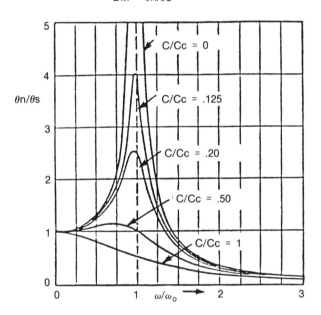

Figure 21-28. Dynamic Multiplier

Several damping factors are involved, as follows:

- *Cylinder or mass damping.* The damping relative to ground, caused mostly by friction between piston and liner, bearing friction and attached pumps, etc.
- *Shaft damping.* The damping between adjacent cylinders caused by shaft material hysteresis.
- *Damper damping.* This is an external damping introduced to act as a detuner, or an absorber, for the suppression of harmful torsionals. This is a special case of shaft damping.

Typical damping factors are necessary inputs for the crankshaft designer.

f. *Actual amplitude* (θ_n). Based upon this dynamic magnifier DM concept, the actual normal elastic curve of a forced vibration case can be constructed based upon the static normal curve. This in turn is based upon the normal elastic curve of a free vibration case of the same shaft system. See Figure 21-29.

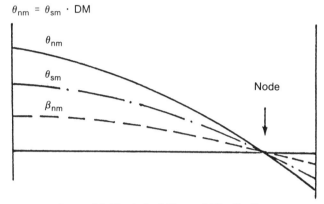

Figure 21-29. Actual Normal Elastic Curve

For a gen-set using a new engine, experimental determination of torsional characteristics of the new shaft is recommended, if for no other reason than to verify the mass-elasticity system simulation, its natural frequency, and the projected damping factor.

g. *Stress level* ($S\tau$). The shear stress, Equation 21-10, at any station at a particular order and a particular mode can be obtained when the forced vibration torsional amplitude is known. The stress level is proportional to the relative amplitude between the two cylinders considered. The highest single mode stress level will be found at the nodal point of each mode of vibration analyzed. The torsional stress level is proportional to the relative amplitude $\Delta\theta$ between the two shaft sections considered.

$$S\tau = \frac{\Delta\theta_m \cdot K_m \cdot SCF}{Z_m} \qquad \text{(Equation 21-19)}$$

K = stiffness of shaft section

Z_m = section modulus = $\pi \dfrac{d^3}{16}$ at station m

$\Delta\theta_m$ = relative angular amplitude between two shaft sections considered

SCF = stress concentration factor

Combined vectorial sum torsional amplitudes caused by more than one mode and more than one order must be vectorially added, since both shaft phase angle and amplitude for each order are involved. When classical Holzer-forced vibration tabulations are used, the vector sum solutions for more than six orders become quite time consuming. A modern harmonic synthesis method will be discussed later in this chapter.

To avoid torsional overload, it is important to observe the following rules:

- Alleviate stress concentration factor (SCF) by providing ample fillet, polished oil hole, shot peening, etc.
- Limit free end amplitudes to a safe level at the normal running conditions (certainly not more than ± 0.5°).
- Avoid prolonged running at ± 10% of the major critical speeds.
- Apply a damper when required.

h. *Example using Holzer-forced vibration method.* The result of a typical six-cylinder engine driving a dynamometer is shown in Figure 21-30. This information is obtained by computing the mass-elasticity data and the cylinder gas pressure data.

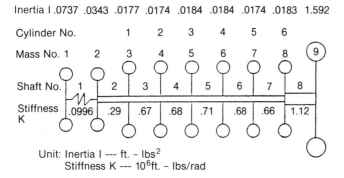

Figure 21-30. Mass-Elasticity System

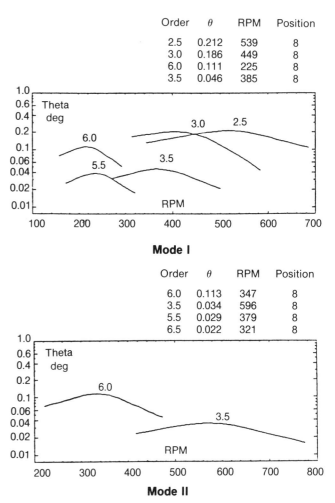

Figure 21-31. Resonance Curve

The Holzer method calculates the first and second mode (and higher if necessary) natural frequencies quite accurately. Using a pressure-time diagram (or TE table) as input, the resonance amplitudes θ for those ordal resonances are calculated (see Figure 21-31).

The nominal torsional severity is a function of vector sum torsional amplitude, while most classical methods only provide torsional solutions mode by mode for simplification.

21.4.4 Gen-set crankshaft evaluation criteria.
Each engineering department has its own criteria in evaluating the torsional and the shafting adequacy of a gen-set system. If the criteria are not met, changes might be necessary in the following areas:

- Generator shaft stiffeners
- Engine counterweight
- Flywheel size
- Modifying damper
- Adding a damping coupling to a two bearing generator system

For large stationary and ship engines, guidelines have been established by DEMA (Diesel Engine Manufacturing Association). Crankshaft design criteria is defined by ABS (American Bureau of Ships). Recently a more thorough analysis based upon European practice was made by CIMAC (Counsel International des Machines a' Combustion).

a. *DEMA.* Stationary and Marine Standards (Ref. 6) allows a maximum nominal torsional stress of 5000 psi due to any single order of torsional stress operating within the speed range, but not more than 7000 psi for all orders combined. Nominal stress does not consider stress concentration factor (SCF).

b. *ABS.* For ship installation of engines over 135 HP, ABS (Ref. 7) requires the submission of the crankshaft arrangement for approval. A torsional analysis report is also required. Quoting ABS, "...calculations are to be submitted including tables of natural frequencies, vector summations for critical speeds of all significant orders up to 120% of rated speed and stress estimates for criticals whose severity approaches or exceeds the limits indicated in (Section) 34.47 and Table 34.3."

Section 34.47 states that the allowable stresses of a single harmonic exciting factor at the resonant peak not exceed the value determined by an empirical equation, which considers the minimum tensile strength of the shaft material and some shaft design factor. It stipulates that the total vibratory stress in the interval from 90% to 105% of rated speed due to resonant harmonics, and the dynamically magnified parts of significant nonresonant harmonics, not exceed 150% of the allowable stress for a single harmonic exciting factor. It stipulates also that torsiograph tests may be required to verify the calculations and to assist in determining the ranges of restricted operation, etc.

c. *CIMAC and IACS*. Counsiel International des Machines a' Combustion (an affiliate society of ASME) has drafted a recommendation on crankshaft stress calculation for a standardization of design requirements of all classification societies (Ref. 8). This draft is being reviewed by the International Association of Classification Societies (IACS). The calculation considers nominal torsional stress, nominal bending and shear stresses, stress concentration factor, combined alternating stress, fatigue strength of material, and a design safety factor. PEI Consultants has prepared a software program to compute this calculation quickly (Ref. 9). This calculation should be properly carried out by the engine manufacturer to assure the reliability and durability of a crankshaft under load.

d. *Gen-set crankshaft adequacy*. For the gen-set application, we should check the torsional stress first and then recheck the lateral frequency evaluation with the latest shafting, coupling, and flywheel data. For some long gen-sets, such as a 2500 kW gas turbine driving a two bearing generator, some final adjustment and reiteration might be required so that torsional stress, lateral frequency, and stress level requirements can all be met satisfactorily.

The total inertia of the system is also related to load recovery capability of the gen-set. For a no-break system, an auxiliary flywheel might be needed. See Figure 21-32. A large flywheel is used to provide additional inertia for coping with sudden load increases without the engine bogging down or smoking beyond the limit allowed. However, a large flywheel lowers the natural frequency of the system and additional overhang load to the main bearing.

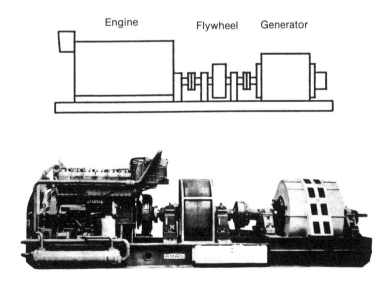

No-break alternator set 1500 kVA (alternator coupled with the mains and working as synchronous motor)

Figure 21-32. No-Break Gen-set

21.5 HARMONIC SYNTHESIS METHOD

Torsional analysis methods described thus far are classical methods used by many engine companies. However, with the development of the computer, complex torsional problems can now be quickly and accurately solved.

Two harmonic synthesis computation methods were described recently in an SAE paper published at the Government/Industry Meeting and Exposition in Washington, D.C. in May of 1987 (Ref. 9). PEI Consultants described an "Ordal Superposition" Torsional Simulation Code called TSC (SAE 861226) (Ref. 8) while FaAA developed a "Modal Superposition" called SHAMS (Shaft Harmonic Analysis by Modal Superposition, SAE 870870) for the analysis of torsional vibration under steady-state. Additional methods have been developed for transient analysis. SHAMS, for example (SAE 870870), is capable of calculating the instantaneous torsional amplitudes and nominal stresses based on the vector sum of (24) orders for all modes. Both methods are precise computations of a set of mass-elasticity equations with torsional excitation. For the PEI method, the torsional solutions of a single order for all harmonic modes are performed first. These single order solutions are then synthesized for "sum of order" solutions, and it is feasible to incorporate damper parameters related to order or engine RPM.

The FaAA STAMS code is an extension of the single mode Holzer forced vibration method discussed earlier and is, therefore, called a **modal superposition method**. The results of these two methods were compared with a carefully taken dynamic torsiograph. See Figure 21-33. Both methods checked the dynamic free end amplitude test data very closely. The eight-cylinder in-line engine was not equipped with a damper.

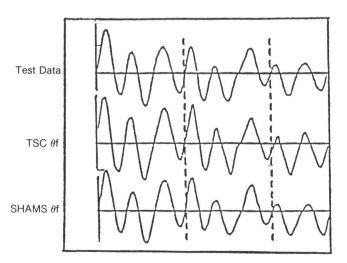

Figure 21-33. Comparison of Data

21.5.1 Torsional simulation code.
The modeling (Figure 21-34) of an in-line eight-cylinder crankshaft system is used as an example.

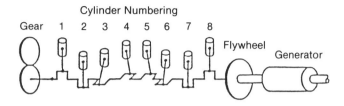

Figure 21-34. Model of Eight-Cylinder System

a. *The mass-elasticity simulation.* The shaft section-to-section simulation is shown in Figure 21-35.

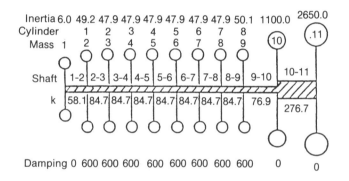

Figure 21-35. Mass-Elastic Simulation, Eight-Cylinder System

The torque excitation T_i on mass number i is illustrated in Figure 21-36.

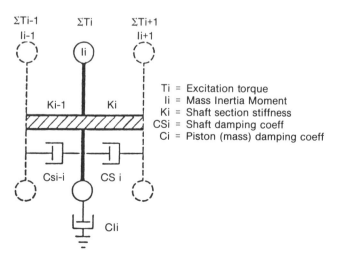

T_i = Excitation torque
I_i = Mass Inertia Moment
K_i = Shaft section stiffness
CS_i = Shaft damping coeff
C_i = Piston (mass) damping coeff

Figure 21-36. Torque Excitation, Mass i

b. *Mathematics.* The exciting torque expressed as T_{mn} is the summation of inertia and gas torques for all orders, and for mass m. The mathematical model of the mass-elasticity system for (m=1 to n) mass number m is:

$$\begin{cases} I_1 \cdot \ddot{\theta}_{1,n} + CI_1 \cdot \dot{\theta}_{1,n} + CS_1 \cdot \Delta\dot{\theta}_{1,n} + K_1 \cdot \Delta\theta_{1,n} \\ = TE_{1,n} \cdot A \cdot r \\ \qquad\qquad\qquad\qquad\qquad\qquad \text{(Equation 21-20)} \\ I_2 \cdot \ddot{\theta}_{2,n} + CI_2 \cdot \dot{\theta}_{2,n} - CS_1 \cdot \Delta\dot{\theta}_{1,n} + CS_2 \cdot \Delta\dot{\theta}_{2,n} - K_1 \cdot \\ \Delta\theta_{1,n} + K_2 \cdot \Delta\theta_{2,n} = TE_{2,n} \cdot A \cdot r \\ \cdots\cdots\cdots\cdots\cdots\cdots\cdots\cdots\cdots\cdots \\ \cdots\cdots\cdots\cdots\cdots\cdots\cdots\cdots\cdots\cdots \\ \cdots\cdots\cdots\cdots\cdots\cdots\cdots\cdots\cdots\cdots \\ I_m \cdot \ddot{\theta}_{m,n} + CI_m \cdot \dot{\theta}_{m,n} - CS_{m-1} \cdot \Delta\dot{\theta}_{m-1,n} - K_{m1}\Delta\theta_{m1,n} \\ = TE_{m,n} \cdot A \cdot r \end{cases}$$

For a system with m number of masses, m-1 equations are needed. PEI Consultants developed a torsional simulation code (TSC) to solve these equations. For details see Ref. 10 (also contained in Ref. 9).

c. *Outputs.* Computer outputs include graphs and tables to describe the torsional characteristics and to provide the simulation results. The outputs provide quasi-transient solutions for start-up and coastdown. the software provides vector sum of (24) orders and all modes on both torsional amplitude and nominal torsional stress at any shaft section being studied.

In Table 21-6, amplitudes for each order and vector sum of (24) orders solved by the TSC code are compared with the classical methods of Holzer and TORVAP C (Ref. 11). TORVAP C, a computer program capable of producing sum of (six) orders results, was developed in Great Britain during the 1970s.

21.6 TORSIONAL DAMPERS

After the prediction of shaft critical speeds too close to synchronous speeds, the engine manufacturer will suggest changes in the mass-elasticity system to satisfy the required conditions. Changes can be introduced to generator shaft stiffness (cross section or material), or engine counterweight, or flywheel, or in many cases, by adding a damper. The dampers used are elastomeric (rubber), viscous, and, infrequently, pendulum absorber (used today on the FM 38 — 8-1/8 OP engines).

21.6.1 Elastomeric damper. A rubber mounted pulley is used in many automotive engines. It is, basically, a two mass system with a spring (rubber) connection tuned to counteract the original resonance. It is, therefore, a detuner.

The best reference is that of a pure elastometric damper of Bremer (Ref. 12). A schematic of the damper and its performance is shown in Figure 21-37. The tuning characteristic of this type detuner is to replace a high amplitude first mode resonance order with two lesser peaks caused by both the first and second modes of the same resonance order.

For rubber dampers, there are three important parameters: the inertia mass, the spring constant of the connecting elastomer, and the damper magnifier. The damper magnifier of the elastomer material used today ranges from two to four.

Table 21-6. Torsional Stress Calculations Versus Test

Order	Holzer Lloyd T_n	TORVAP C Lloyd T_n	TSC, Test T_n	θf TEST	Order	Holzer Lloyd T_n Shaft 9	TORVAP C Lloyd T_n Shaft 6	TSC, Test T_n Shaft 6
0.5	0.05	0.07	0.07	0.06	0.5	520	265	271
1.0	0.02	—	0.00	0.01	1.0	221	—	1147
1.5	0.12	0.14	0.18	0.17	1.5	1136	1637	2012
2.0	0.04	—	0.00	0.00	2.0	379	—	106
2.5	0.10	0.11	0.14	0.13	2.5	934	1257	1540
3.0	0.01	—	0.00	0.00	3.0	44	—	174
3.5	0.05	—	0.06	0.06	3.5	444	—	330
4.0	0.36	0.31	0.34	0.33	4.0	3424	2540	2754
4.5	0.07	0.07	0.07	0.06	4.5	621	450	488
5.0	0.04	0.02	0.03	0.03	5.0	357	—	251
5.5	0.13	—	0.12	0.13	5.5	1206	840	875
6.0	0.01	—	0.01	0.01	6.0	99	—	32
Sum	0.42	0.56	0.64	0.69	Sum	4065 psi	5029 psi	6812 psi
Orders	(12)SRSS	$\Sigma(6)$	$\Sigma(24)$	$\Sigma(24)$	Orders	(12)SRSS	$\Sigma(6)$	$\Sigma(24)$

End Amplitude for 12″ Shaft

Vibration Analysis For a Sound Generator Set Design

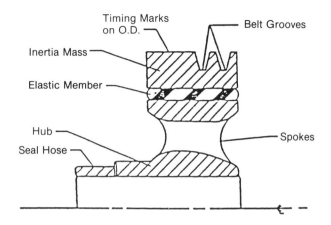

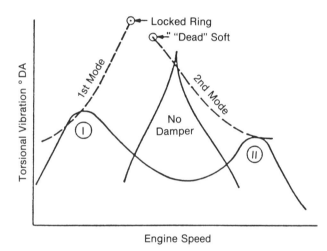

Figure 21-37. Rubber Damper/Pulley

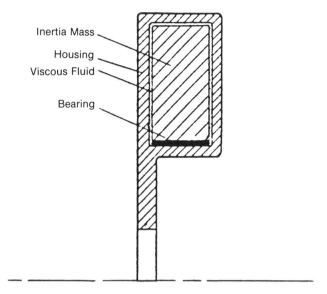

Figure 21-38. Pure Viscous Type Torsional Vibration Damper

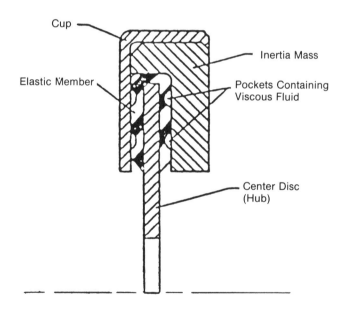

Figure 21-39. NuLastic Rubber-viscous Torsional Vibration Damper

Much more fundamental work is needed to simulate this type of damper more completely because of the nonlinear relationship between the stress and strain of the elastomeric material used in the rubber damper.

21.6.2 Viscous damper. The viscous design dissipates vibratory torque energy through fluid friction between driving and driven plates and, sometimes, through small orifices. The examples for high speed engines shown in Figures 21-38 and 21-39 are taken from Bremer, Ref 12.

Due to the viscous friction, the viscous damper reduces the peak resonance amplitude, as shown in Figure 21-40.

Viscous fluid introduces damping resistance to vibration and it works at a wide range of speeds. On the other hand, the Geislinger type, (marketed by Eaton), with both elastic member and viscous fluid, functions both as a detuner and a viscous damper. The Geislinger type, due to its complexity and cost, is only used in large engines.

21.6.3 Pendulum absorber. The pendulum absorber is a form of tuned absorber whose natural frequency varies in direct proportion to its rotational speed. See Figure 21-41.

$$W_o = \text{rpm} \sqrt{R/L} \qquad \text{(Equation 21-21)}$$

R, L = pendulum design parameters

As in the case of the simple tuned absorber, the pendulum absorber must have sufficient mass to keep the amplitude small. Furthermore, the effectiveness of the

Vibration Analysis For a Sound Generator Set Design

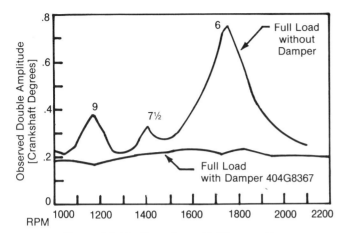

Figure 21-40. Damping with Viscous Damper

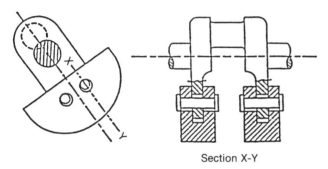

Figure 21-41. Pendulum Absorber

absorber is very dependent on the dimensions of the holes and pins. This requires precise machining.

Fairbanks Morse OP engines have used this feature since World War II. Some old data shows how effective this damper is for a wide range of speeds. See Figure 21-42.

21.7 PROCEDURES TO AVOID SHAFT BENDING FAILURE AND ABNORMAL VIBRATION

Crankshaft failures can be caused by torsional or bending stresses, or by combined stress. For most modern high bmep engines, main bearings are strategically placed between every crank throw. Therefore, single throw specimens are used in simple analysis of shaft bending stress. Most of the time, the shaft is designed to resist firing gas load and imbalance. However, additional consideration must be given to gen-set shaft misalignment and additional imbalance due to the generator rotor, etc. Bending of the shaft, when excessive, causes an additional axial vibration mode as shown in Figure 21-43.

21.7.1 Bending failures. The crankshaft bending stress must be considered in crankshaft stress calculation (CIMAC, Ref. 8). Figure 21-44 illustrates some bending failures. The engine manufacturer generally specifies how much web deflection is safe. Listed below are some of the

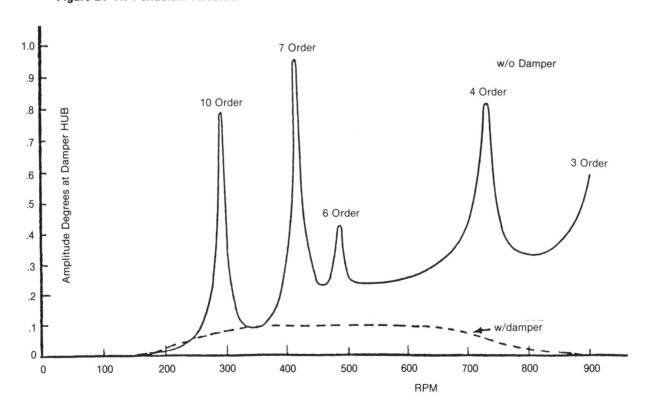

Figure 21-42. 10 Cylinder 8.125 × 10 OP Diesel

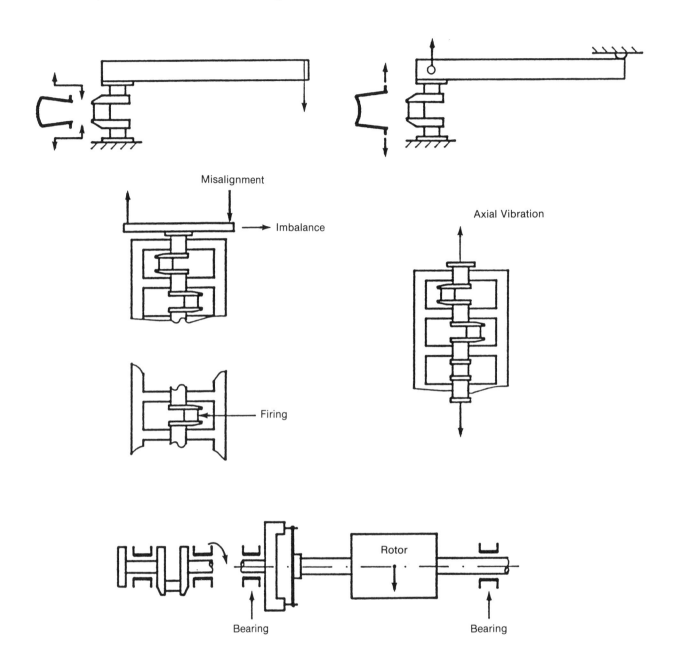

Figure 21-43. Crank Shaft Bending

factors that cause bending overstress. To safeguard the reliability and durability of a large generator set, these must be corrected when the gen-set is installed on the final foundation.

a. Combustion pressure or hydraulic lock of the power cylinder

b. Misalignment during installation, foundation sinking; the symptom is excessive web deflection

c. Imbalance

d. Bearing failure or excessive wear

e. Excessive overhang load

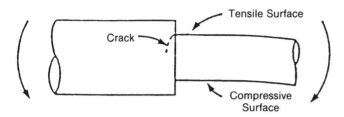

Filleted Shaft, Crack Due to Bending

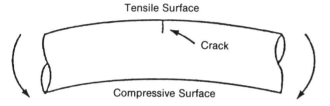

Cylindrical Shaft, Crack Due to Bending

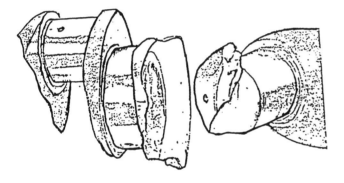

Figure 21-44. Filleted Shaft

21.7.2 Shaft misalignment. Poor shaft alignment is the most common and serious problem for a properly designed engine. The misalignment could be either angular (when the engine shaft axis does not line up 100% with the generator shaft axis), or it could be caused by a lateral displacement of the two axes due to some concentricity problem on either the flywheel or the flanges. Flywheel concentricity can be checked by a dial indicator; only .002 – .004 in. is generally tolerated. The same thing holds true for the concentricity deficiencies on the generator shaft, flange, or the drive disc. See Figure 21-45.

The gen-set builder should refer to the engine manual for the procedure to check crank line alignment.

The importance of alignment is illustrated by comparing the amount of imbalance tolerated by a generator manufacturer and the potential imbalance caused by a misalignment of .005 in. of a 2000 lb. rotor:

$$\text{imbalance due to misalignment} = 2000 \times 16 \times .005$$
$$= 160 \text{ oz-in.}$$
$$\text{imbalance limit by gen. manufacturer} = 2 - 6 \text{ oz-in.}$$

The misalignment, like imbalance, causes a one order vibration. The vibration increases with RPM. The misalignment could be within limits during installation and become excessive when the engine crankcase reaches running temperature. The use of flex-disc coupling minimizes somewhat the angular displacement problem.

Uneven air gap of the generator could also cause some imbalance due to uneven magnetic pull. This is usually quite small at low load and intensifies with increasing load.

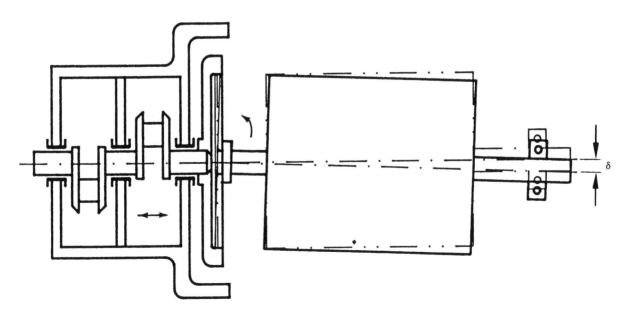

Figure 21-45. Misaligned Generator Shaft

Alignment of the gen-set shaft can usually be checked by laser scoping, dial gauge, etc., but the most effective method is checking crank strain (commonly known as **crank web deflection**) while rotating the shaft. See Figure 21-46. This is a mandatory procedure for larger diesel gen-sets. The crank strain change, sometimes indicated by a dial indicator, should be kept within a certain limit when the shaft is rotating 360°. The engine builder must be consulted for maximum allowable limit. This procedure should be carried out when the engine is warm.

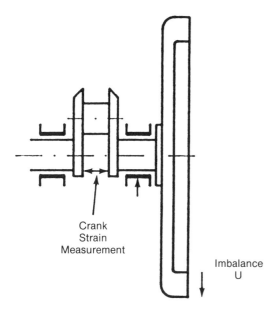

Figure 21-46. Crank Web Deflection

The alignment can also change after the engine is hot, since the engine block and base expand from the base 13 mil for every 100°F increase in temperature, assuming the engine shaft center line is 20 in. above the base. This thermal expansion must be compensated in the alignment program to achieve a smooth running engine. Due to this thermal expansion the vibration levels could change during warmup.

21.7.3 Imbalance. The imbalance of the generator is normally limited to a few oz-in., depending on the weight of the rotor (Wt), at the manufacturer's test floor. Due to minor mismatching and misalignment, etc. during the installation, trim balance might be needed to minimize the gen-set allowable linear vibration limit to 1-2 mils.

MIL-STD-167 specifies maximum allowable residual imbalance (U) as follows:

For 1000 RPM and below

$$U = 4000 \frac{Wt}{RPM^2} \text{ oz-in.}$$

For above 1000 RPM

$$U = \frac{4Wt}{RPM} \text{ oz-in.}$$

Generator manufacturers are capable of balancing the rotor better than these limits. The generator rotor contains steel lamination, copper wire, and insulation. It is conceivable that the insulation might yield somewhat when the rotor is overheated or run over the speed limit, thus changing the balance. This does not occur often, but when it does, the generator manufacturer must be alerted at once, since the condition could worsen.

21.7.4 Overhang and web deflection. The engine manufacturer generally specifies how much overhang load can be exerted on the shaft and how much web deflection is safe. This is necessary to safeguard the reliability of a large generator set. The web deflection test must be done when the gen-set is installed on the final foundation.

21.7.5 Main bearing distress. The main bearing distress is usually caused by:

a. Lack of lubrication, especially during startup

b. Excessive vibration

c. Brinnelling during transportation

d. Failure of the oil filter, allowing debris to come through

e. Failure of the air filter, allowing airborne dust and hard particles into the crankcase

During investigations (Ref. 9) of why many nuclear standby gen-sets fail prematurely, the major cause has been proven to be insufficient lubrication reaching some critical bearing (turbocharger and upper crank of an OP engine, for example). The beginning of a main bearing failure is indicated by a sudden temperature rise signal when a bearing temperature indicator is used. The other indications are increased vibration and abnormal noise. At this point it is too late to avoid failure of the gen-set.

21.7.6 Abnormal vibration. Abnormal vibration could be either shaft induced or bearing support induced. The misalignment and imbalance that give additional bending load to the shafts also give additional fatigue load to the bearings.

Axial shaft vibration (relative to bearing support) is a major cause of failure for some spherical bearing applications when the generator shaft is rigidly mounted to the engine shaft. This axial vibration is caused by high BMEP combustion (high firing pressure) and an axially over-flexible crankshaft. When generator bearing failure occurs, there is a good chance of causing engine main bearing or thrust bearing failure. The generator bearing support could also cause resonance vibration if the support is not sufficiently stiff in the lateral or axial directions. This is readily detected by field measurement. For a medium speed engine, axial vibration over 10 mils displacement is dangerous.

When a set is not running, the ball or roller of an anti-friction bearing could be damaged due to pounding and brinnelling caused by floor excitation. This is one of the reasons why vibration isolators are used for standby sets. Damage can occur during transportation. There have been several cases where bearings were damaged during rail transit. This is why the rotor must be braced and specially supported when the generator or the gen-set is being packaged for long haul freight.

Sleeve bearings, properly designed, are more tolerant of vibration problems. When pressure lubricated, they also tolerate more heat build-up. However, they are more prone to suffer oil starvation during emergency starting when the gen-set must be raced to full speed in just a few seconds, especially when the oil is cold and the initial supply pressure is deficient.

The bearing will fail readily when shaft current, coupled with high vibration is present. Many large generators that use segmented stator lamination design, have shaft current leakage problems. Some combinations of segments and poles give more trouble than others. An insulated bearing support is used to prevent bearing pitting. For 1800 RPM generators with single piece stator lamination design, no insulated bearing is needed. For eight-pole and six-pole generators, check with the generator manufacturer for assurance.

21.8 SOUND SKID DESIGN AND FOUNDATION REQUIREMENTS

The skid design, Figure 21-47, is an important part of the gen-set structure and must be treated accordingly. The skid design is not complete unless the engine and generator structural capability are considered and mounting system, vibration isolation, and foundation and vibration characteristics are fully determined.

To avoid later dispute, the gen-set packager should include a set of vibration and noise data in the first article test record. This could alleviate future complaints about vibration and excessive noise.

21.8.1 Skid design.
The skid or sub-base, depending on its destination and application, must satisfy the following situations:

a. Provide a self-contained base for packaging a complete transportable power package. The skid must, therefore, be stiff enough to maintain alignment between engine and generator during lifting, transportation, and operation. No overhang member should exist between the engine mounting and the skid.

FM 012
12-Cylinder
BPS Generator

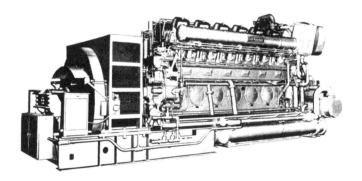

Colt PC 2.3
12-Cylinder
BPS Generator

Figure 21-47. Generator Sets on Skids

b. Limit static deflection during lifting or transporting to a prescribed figure in order to avoid damaging connections, flexible couplings, and/or distorting engine structures beyond their limits. See Figure 21-48.

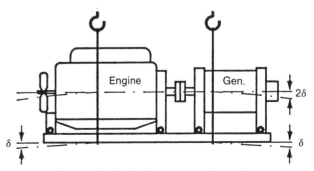

Figure 21-48. Method of Limiting Static Deflection During Lifting

Vibration Analysis For a Sound Generator Set Design

c. Satisfy certain minimum seismic and vibration requirements:

1. A minimum seismic requirement of 4G to resist humping during transportation.

2. No lateral natural frequency of the siderail (longest member) should fall within three times the synchronous speed.

3. Stiff enough for vibration isolator, since many standby sets are installed in high-rises, hospitals, office buildings, small boats, vehicles, etc.

d. Resist torque reaction without distortion. For a two bearing generator design or long tandem set, the torque reactions from the prime movers must be considered. The cross member must be stiff enough to resist torque reactions. See Figure 21-49. Excessive skid distortion may cause vibration and undue fatigue stress on the engine block or generator frame.

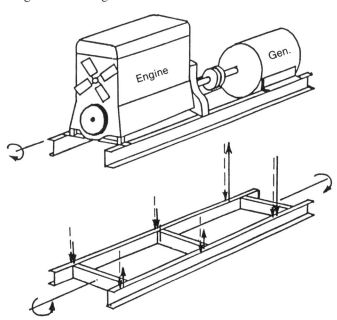

Figure 21-49. Couples Acting on the Skid

For a one-bearing generator design, the engine block and generator frame provide the structure resisting the engine torque outputs, and the skid design can be simplified.

e. Satisfy special seismic requirements.

1. *Earthquake*. Seismic design (G loading) will depend on the location where the set is to be installed.

2. *Nuclear standby*. Seismic calculation must be performed for the specified G loading and the natural frequency of the sub-base must be sufficiently high; one manufacturer specifies "a minimum of 3 × RPM" and another states "preferably above 33 Hz."

3. *Navy shock requirement*. For some severe applications, loading is specified as 50 G. The gen-set must perform after a mine is exploded within a few feet of the test barge where the gen-set is installed. The sub-base stiffness and natural frequency must be designed for this requirement.

4. *Air drop requirement*. This requires special seismic treatment and is an expensive skid.

It can be seen from the different requirements described above that the skid must be designed to suit a specific application. A standard skid will be overdesigned for some applications and underdesigned for others. Some gen-set builders stock modules of parts and can produce skids quickly for various end user applications.

21.8.2 Engine flange load and SAE flywheel designs. For a flange mounted generator, especially for overhung designs, the engine rear mounting flange or flywheel housing flange must be strong enough to carry extra generator static loads, dynamic loads, and torque reaction. By the same token, the generator frame must be sufficiently stiff so that any vibration problem can be contained. For the overhung design (Figure 21-50), the support of the generator bearing is basically cantilevered from the engine frame. The engine flywheel housing, as well as the rear engine bearing, must be properly designed for the overhung load. There are SAE standards governing the generator adapters with the engine flywheel housings. There are also generator flexible discs matching the flywheels for the generator-to-engine final connection.

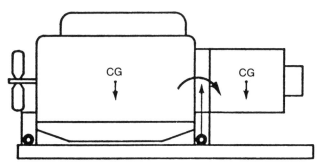

Figure 21-50. Overhung Load

For a flange mounted generator design with four-point mounting, the flange load can be minimized by proper location of generator feet. See Figure 21-51. To minimize the flange loading, the CG (center of gravity) of the engine and generator and their weights must be considered.

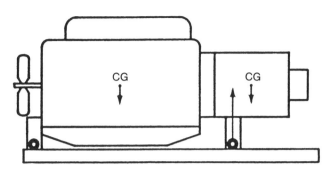

Figure 21-51. Four-Point Mounting and Minimal Flange Load

A generator that mounts at the rear increases engine flange load. See Figure 21-52.

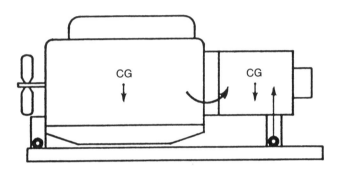

Figure 21-52. Four-Point Mounting and Increased Flange Load

When the engine flange is not sufficiently strong, a cradle mount can be used. A generator with an adapter directly mounted to the base is also helpful. See Figures 21-53 and 21-54 for examples of generator mountings.

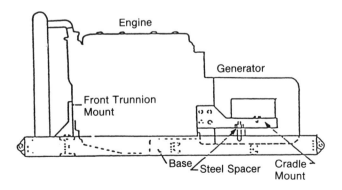

Figure 21-53. Simple Base with Cradle Mount

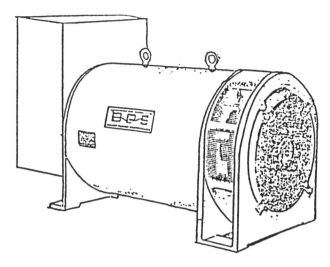

Figure 21-54. CEBRA Generator with Front Foot Mount to Support the Engine Weight

21.8.3 Front engine mounts. The front engine mounts can be either the rigid type or the trunion type (like an automobile engine).

a. *Rigid type.* All force from the engine will be transmitted to the skid. Proper alignment and shimming is mandatory when the engine is installed on the skid.

b. *Trunion type.* The trunion type uses a rubber sandwich design that permits some limited fore and aft movement and some rotational movement of the engine at one axis of the crankshaft, thus relieving part of the residual torque reaction (if any) before it is transmitted to the skid. This elastic mount also provides some limited protection to the engine frame when there is some distortion of the skid (caused by uneven ground, for example).

Figure 21-55 is taken from an application guide prepared by Cummins Engine (Ref. 13). It shows two variations of front engine mount design using a rubber sandwich for isolation. Most larger engines use solid front and rear engine mounts.

21.8.4 The selection of mounting system. For most larger gen-sets, rear engine mounts are of the rigid type. The location of rear mount(s), however, can be quite diverse, as shown previously. Factors affecting the selection of a mounting system are:

1. Engine flange load capability and generator feet design.

2. Engine shaft and generator shaft torsional and lateral vibrational compatibility.

3. Skid design and cost depend upon the rigidity of the engine frame itself.

4. Gear box requirement.

Vibration Analysis For a Sound Generator Set Design

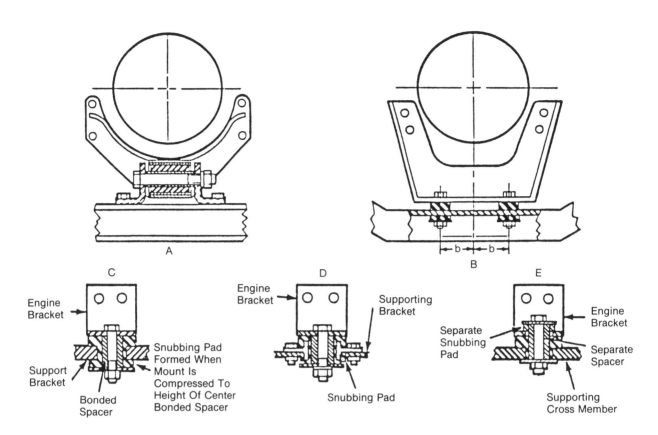

Figure 21-55. Rubber Sandwich Front Mount

a. The four-point mount is generally for smaller kW sets, and when the generator frame and mounting feet are sufficiently stiff to support half the engine weight and torque reaction. Otherwise, a six-point mount is commonly used to provide the smoothest installation. See Figure 21-56.

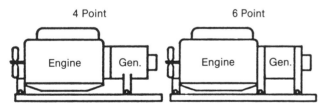

Figure 21-56. Four- and Six-Point Mounting

b. The eight-point mount, Figure 21-57, is used when it is more desirable to isolate the engine shaft from the generator shaft for torsional disturbance or lateral vibration. This requires a two-bearing generator. In this case a flexible coupling is used to transmit the torque. Additional mounts will be required when a gear box is involved, such as in a turbine gen-set. When an eight-point mount design is used without a flexible coupling, the alignment and thermal growth become extremely difficult to control.

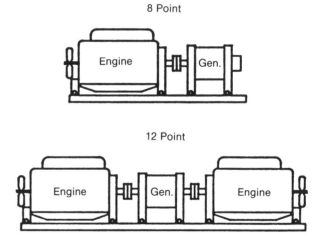

Figure 21-57. Eight- and Twelve-Point Mounting

c. For a long skid, such as a turbine skid (Figure 21-58), the type of mounting used and its location become even more critical. The thermal growth of the engine frame both vertically and longitudinally (laterally) with respect to the shaft system and to skid must be carefully evaluated to avoid undue stress, hot misalignment, and vibration.

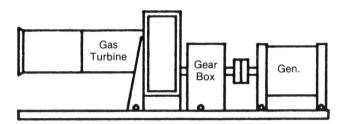

Figure 21-58. Gas Turbine Driver and Two-Bearing Generator

21.8.5 Vibration isolator. For a more demanding application, it is necessary to isolate any vibration or torque reaction of the gen-set from the surrounding structure. This calls for either a strong foundation isolated from the surroundings, or the use of vibration isolators under the sub-base.

Many spring (Figure 21-59) or rubber type isolators are commercially available. The selection is based on the load per mount and the major orders of existing vibration. When the natural frequency of the isolator, with the known skid load, is many times less than that of the vibrating skid (first or second order of the synchronous speed), very little vibrational energy will be transmitted through the mounting to the ground. This is quite important for a ship application for both vibration and noise isolation. The ship structure is generally very prone to vibration caused by engine operation.

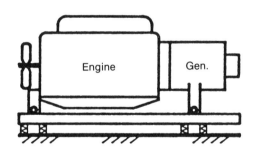

Figure 21-59. Shock Mount Between the Skid and the Foundation

The vibration isolator also protects an idling gen-set from vibration transmitted through the foundation. There are definitely cases when the generator anti-friction bearing is damaged by pounding vibration even when the set is not running at all. When the skid or sub-base is not sufficiently strong, it is important to test the vibration level of the gen-set with isolators installed the same way as in the final installation.

The isolators must be positioned so that they all carry the same weight. To do this, the weight and center of gravity of the gen-set must be calculated. The computation required for vibration isolation could get quite involved by using Modal Analysis and finite element methods. For more information on vibration isolators see Chapter 20.

21.8.6 Foundation considerations and bearing capacity of the soil. The foundation of a gen-set must be able to support the engine skid and its load without yielding or cracking through many years of operation.

The foundation is less critical for a gen-set equipped with a strong sub-base that is designed for frequent transportation and for rough terrain.

For a gen-set with a weak skid, or without a skid, the foundation must be properly designed and built. However, the settling of the foundation can be a problem. Major engine frame cracks and excessive vibration problems have occurred due to improper foundation or faulty installation.

When the gen-set is destined for a marine or a railroad application, a sub-base with sufficient rigidity is necessary to protect the gen-set from the expected "hull" distortion.

The skid or sub-base is generally set on the foundation by foundation bolts and chocks, or by vibration isolators. When these are not properly treated, abnormal vibration will happen. This is critical for large, medium, and slow speed engines.

Since the skid will be resting on the foundation, the load carrying capacity is based on the projected bearing area of the skid and the total load.

When a vibration isolator is used, the projected bearing area is quite limited and the foundation must be designed accordingly. A permanent concrete foundation should be able to bear 150,000 lb/sq. ft. For other semipermanent or temporary foundations, the figures in Table 21-7 are commonly quoted as permissible bearing loads:

Table 21-7. Soil Load Characteristics

Solid ledge of hard rock, granite, etc.	50,000 lb/sq. ft.
Shale, medium rock, requiring blasting for removal	20,000 lb/sq. ft.
Hardpan, cemented sand, gravel or hard clay, requiring picking for removal	10,000 lb/sq. ft.
Loose coarse sand or gravel	5,000 lb/sq. ft.
Loose fine compact sand, stiff clay	3,000 lb/sq. ft.
Soft clay	2,000 lb/sq. ft.

When medium or low speed diesels are used, soil loading should be based on natural frequency evaluation, to prevent resonance with operating speed. See Figure 21-60 for two kinds of soils.

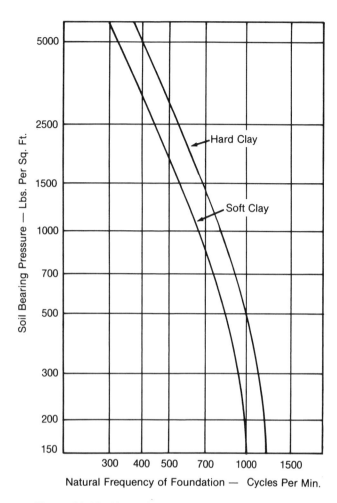

Figure 21-60. Natural Frequency of Foundations of Two Types of Soil

21.9 TROUBLESHOOTING FOR VIBRATION

Some guides for vibration troubleshooting are summarized below.

a. *Alignment check.* Misalignment causes first order vibration at no load or at loads. Alignment at the hot and running condition is most important.

b. *Imbalance of rotating parts.* Imbalance causes first order vibration at no load or loaded. Vibration will increase with speed at no load. If trim balance does not help, the problem could be caused by:

1. Bent shaft
2. Unseated drive disc
3. Oil dipping by crankshaft

c. *Imbalance of reciprocating parts.* Vibration increase with speed at no load, second order vibration:

1. Due to mismatched pistons or rods
2. Four-cylinder engine secondary order balancer malfunctions

d. *Resonance.* When the speed is increased at no load, extreme vibration and noise will appear at some resonant speeds. The system suddenly goes through a narrow range of speed with a large amount of vibration:

1. The natural frequency of some sheet metal panels, some brackets, piping, air cleaner, exhaust support, etc. fall in the operating speed range.
2. Skid's siderail not stiff enough
3. Generator's bearing arm is not stiff enough

e. *Torque reaction not secured.* Vibration increases as load is applied.

1. Weak base, or unsecured mounting on a two-bearing generator
2. Improper foundation, or unsecured skid to the foundation

f. *Misfiring.* For a four-cycle engine, misfiring causes a half-order vibration and for a two-cycle engine, one order vibration.

1. Check injection/ignition
2. Check valves
3. Check compression

g. *Bearing failures.* Same as imbalance but becomes progressively noisier.

h. *Torsional problems.* Look for gear chattering, coupling overheating, or misfiring. Use a torsiograph to determine the amplitude and order of vibration. It is necessary to check this test result against the computed torsional analysis. Note that some damper designs, especially those using viscous fluids, do not live forever and torsional amplitudes should be checked again when there are signs of excessive stress.

i. *Vibration test log.* Figure 21-61 is a suggested gen-set vibration test log.

21.10 SUMMARY

In a diesel generator set, the mass elastic system typically consists of the engine having several inertial masses along the crankline, the flywheel, and the generator rotor. The structure components to be considered for sound generator set design are: shaft system, bearing and bearing

Vibration Analysis For a Sound Generator Set Design

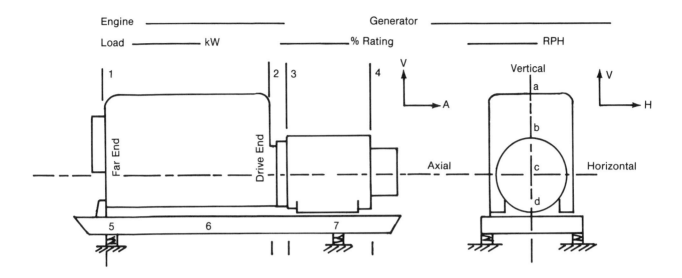

Pickup		Overall		Order											
				1/2		1		1 1/2		2		2 1/2		3	
No.	Axis	mil	in./sec	mil	in./sec	mil	in./sec	mil	in./sec	mil	in./sec	mil	in./sec	mil	in./sec
1	H														
	V														
	A														
2	H														
	V														
	A														
3	H														
	V														
	A														
4	H														
	V														
	A														
5	H														
	V														
6	H														
	V														
7	H														
	V														

Instrument Used _____   Tested By _____
Comments _____   Date _____

Figure 21-61. Generator-Set Linear Vibration Test Log

support, mountings, skid, and foundation. These components have to be properly matched to satisfy the following criteria:

[1] The total WR of the system must be sufficient to satisfy the inertia requirement for specified response characteristics.

[2] The torsional natural frequencies of the system must be such that no torsional criticals occur near or at the synchronous speed of the generator set. The stress level should be under control.

[3] The crankshaft stress should be calculated using up-to-date calculation methods. Many classical methods, such as the Holzer Forced Vibration Method, do not normally provide torsional amplitudes of all the shaft sections based upon vector sum of orders for all modes.

[4] Viscous or tuned damper must live according to the specified life. A failed damper is dangerous to the crankshaft. Accurate damping inputs are a difficult part of the torsional calculation.

[5] The lateral natural frequency of the generator shaft and the skid must be sufficiently high. It should be at least 2.5 times the rated speed.

[6] Linear vibration level must be below an irritating and potentially dangerous level, or 5-8 mils, depending on the size and speed of the power package.

[7] The gen-set package should not produce "beyond the limit" vibration or noise transmission to its environment. A first article test is important. A generator set vibrational test log is included for the user's convenience. See Figure 21-61.

REFERENCES

[1] Wilson, W. Ker. Practical Solution of Torsional Vibration Problems, Volume I, Frequency Calculations, 1956.

[2] Wilson, W. Ker. Practical Solution of Torsional Vibration Problems, Volume II, Frequency Calculations, 1956.

[3] Porter, F. P. Evaluation of Effects of Torsional Vibration, SAE War Engineering Board, October, 1945.

[4] Lloyd's Register of Shipping: Guidance Notes on Torsional Vibration Characteristics of Main and Auxiliary Oil Engines, 1976.

[5] Data Acquisition system for Engine Research, Diesel Progress, North American, December 1986.

[6] Standard Practices for Low and Medium Speed Stationary Diesel and Gas Engines, Sixth Edition, Diesel Engine Manufacturers Association (DEMA), 1972.

[7] Rules for Building and Classing Steel Vessels, American Bureau of Shipping (ABS), Section 34.13, 1984.

[8] 1983 CIMAC Recommended Rules for the Calculation of Crankshaft for Diesel Engines (For IACS Adaptations).

[9] Crankshaft and Component Adequacy; Update of Analysis and Testing Developed for Nuclear Standby Engines, SAE SP-714, 1987.

[10] Chen, Dr. Simon K. and Chang, Thomas. Crankshaft Torsional and Damping Simulation, An Update and Correlation with Test Results, SAE 861226, 1986.

[11] Torsional Vibration Analysis Program (TORVAP), Computer Aided Design Centre, UK, June 1975.

[12] Bremer, Jr., Robert C. A Practical Treatise on Engine Crankshaft Torsional Vibration Control, SAE SP-79/445, 1979

[13] Walter, John C. A Guide for Power Plant Installations in Trucks, Cummins Engine Co., Inc., SAE 810001.

ACKNOWLEDGMENTS

Technical material for this chapter is from my own files. Some historical aspects of this subject come from Ker Wilson's Practical Solution of Torsional Vibration Problems, Den Hartog's Mechanical Vibrations, SAE War Engineering Board's Evaluation of Effects of Torsional Vibration, and from Taylor's Internal Combustion Engine in Theory and Practice.

The 1978 edition included contributions from Roland Yang (retired) of DeLaval Enterprises, Chuck Newton (deceased), Bob Maddock (retired) of Colt/Fairbanks Morse, and Dr. Micheal Wen of Beloit Power Systems (now at Westinghouse-Round Rock, Texas). These are greatly appreciated.

The revised 1987 edition includes main contributions from Thomas Chang of the PEI Consulting staff and additional references identified in the report. Rocky Lin, a former graduate student at UW-Madison (now at Cummins Engine) assisted in reading and correcting both of these editions.

Dr. Simon K. Chen

August, 1989

Enclosure Design

Charles Gears

CHAPTER 22

INTRODUCTION

In the last decade of the 20th century, electric power generated by something other than the local utility is increasing at a rate that few people could imagine just a few short years ago. Where generator sets were once used mainly by hospitals and as backup to airports and military installations, now nursing homes, computer centers, shopping malls, waste water treatment plants, and offshore oil and gas rigs are installing emergency power for either an existing need or as required by state and local codes.

The proliferation of systems for transmitting information via satellite has forced communication companies to install prime power generating sets that run totally unattended in remote locations. Add to this the companies either cogenerating or peak shaving to reduce utility demand charges, and the requirements for generator sets and switchgear seem endless.

22.1 DESIGN CONSIDERATIONS

When generator sets are an integral part of a new building design, provisions are made to house a supplemental generator system either within the building structure or directly adjacent to it. Often designers and engineers are not afforded this luxury and must install a generator set near an existing building where no provisions for location have been made. Due to lack of adequate space or to requirements of local codes, engineers have to consider alternative means of housing a standby or emergency power source. In some cases this could be outside the building.

What are the design parameters? To what degree does the user desire protection from the local environment for his generator set? Weather protection, contaminated air from sand and dirt, inside temperature control, and enclosure appearance are just a few important considerations. In addition, a significant reduction in noise level may be required by federal, state, or local codes, particularly if the system is to be located in a residential neighborhood. Does the system have to be transported to the job site and positioned by a crane or some other method? Is the user trying to minimize expenses and simplify coordination of the installation by having one source responsible for complete system design and integration for direct job site shipment? Is the unit indoors but located in an area where it needs to be protected for either safety or security purposes?

With these very broad parameters in mind, a good working knowledge of electrical generating systems is required to properly design and specify a cost effective and operationally efficient enclosure system. More often than not, a generator system is installed in a pre-engineered and prepackaged housing to maximize cost efficiencies and to minimize delivery lead times. Typically, a portable gen-set enclosure is made out of light-gauged metals for the roof and wall assemblies. The underframe, which must support the entire weight during handling and delivery, is usually made out of structural shapes and heavier gauge steel. Lift rings must be carefully selected to ensure safe handling.

22.2 MATERIALS

Environmental conditions at the job site *must be* considered when selecting the type of metal to be used for enclosure construction. For example, if the site is located in an area that has a great deal of salt-laden air, such as coastal areas, and corrosive air from industrial waste, highly non-corrosive metals such as aluminum and stainless steel should be considered.

Today, commercially produced, prepainted aluminum sheets that display excellent weathering qualities in highly corrosive atmospheres are readily available. They exceed a life cycle of twenty years with very little deterioration of appearance to the factory-applied painted finishes. For general applications in areas where corrosion is of less concern, prepainted aluminum and mild steel are commonly used for the enclosure construction; the steel sheets are prime coated and finish paint is applied in a variety of ways.

Along with selecting the basic enclosure material, careful consideration should be given to fasteners and all hardware such as hinges, door handles, and louvers as to the degree of corrosion resistance desired. A number of different plating methods are available, such as zinc, cadmium, zinc with gold chromate, or powder coating. Stainless steel or aluminum is generally the most corrosion resistant selection. Fasteners and hardware should be selected on the basis of corrosion resistance, strength requirements, and finished appearance.

22.3 OPERATING ENVIRONMENT CONSIDERATIONS

In cold climates, it is often essential to include a minimum interior heat level to protect equipment, ensure automatic startup capability, and provide a comfortable environment for maintenance personnel. Insulating the enclosure, as well as selecting properly sized heaters, motor operated inlet louvers, and exhaust dampers will usually result in an acceptably controlled interior climate.

To design a system that is weatherproof, climate controlled and/or sound attenuated, careful consideration should be given to both engine combustion and radiator cooling air. Since a large volume of air is required for both combustion and cooling, special consideration should be given to the type of air flow control devices. The simplest means of getting air into an enclosure is by punching openings in either the aluminum or steel side panels. However, due to limited surface area, this process is practical only on smaller units. On larger systems, the most common method for moving air is the fixed blade louver. Screening is recommended to keep out birds and rodents.

For radiator exhaust air, gravity-type exhaust dampers are generally used in conjunction with fixed louvers. However, open expanded metal is not uncommon. A more elaborate method is the motor operated louver and damper combination, which is operational when the unit is running and closed when it is not. This arrangement is necessary for climate control, of course, and also makes the housing more secure.

Intake and exhaust air hoods are often used in combination with louvers and dampers; Figures 22-1 and 22-2. They ensure that water does not enter the enclosure, prevent snow and ice buildup on louvers, and reduce engine noise. Environmental factors affect selection of methods of regulating combustion and the flow of cooling air into the enclosure.

A diesel generator that is to be located in a dusty or sandy area may need additional filtration, such as fine screens or filters. Intake louvers and dampers may have to be oversized to overcome the reduction in efficiency due to these filtration methods.

The most important consideration in any air handling scheme is that engine inlet and exhaust air restriction limits are not exceeded. These restriction limits are generally set at 0.5 inches of water, and although the inlet restriction is expressed as a negative value (-) and the exhaust restriction as a positive value (+), they are additive and combine without regard to sign to form the total restriction of the air handling system.

If the acceptable limits are exceeded, serious operational problems may result, most notably air starvation to the radiator, which often causes engine overheating, with subsequent engine shutdown or damage.

Other problems associated with exceeded limits include generator overheating, difficulty in opening enclosure doors while the gen-set is operating, and combustion air starvation. High face velocities at the enclosure air inlet openings cause excess amounts of moisture, ice, snow, and debris to enter the enclosure.

22.4 NOISE CONTROL

Another factor to be considered is the possible need for reducing system noise due to locations near hospitals, city and suburban neighborhoods, or anywhere that humans might be exposed to noise levels exceeding OSHA maximum standards. Because of the large openings required in enclosure walls for cooling and combustion air, reducing noise requires a very specialized design. The size of the electrical generating system, job-site conditions, the volume of air required, and the amount of noise reduction desired will generally dictate both exterior and interior enclosure design. It is recommended that generator enclosure manufacturers experienced in noise control be consulted to develop a functional system that will result in the desired noise level.

22.5 TEMPERATURE CONTROL

A number of insulating materials are commonly used to insulate the enclosure for temperature control. The more

Figure 22-1. Enclosure with Intake Hood

common materials used are polyurethane foam and semi-rigid unsupported fiberglass. The use of semi-rigid fiberglass is preferred over batt-type unsupported fiberglass because it will not settle into the wall cavities as a result of vibration. For the coldest climates, polyurethane foam is generally selected because it possesses the highest insulating qualities of any other commercially available material. Insulating material should not be left exposed on the interior walls but should be covered with some form of interior lining. The most practical lining materials used are light-gauge, perforated aluminum or steel panels, particularly when non-combustibility is of primary concern. Other materials are used however, such as plywood, fiberglass, and prefinished wood panelling, and all are dictated by the preference of the designer and the intended use of the enclosure.

22.6 ANCILLARY EQUIPMENT

Portable enclosures can be plumbed to a large fuel tank system, which is usually done when the application is prime power, or the enclosure can incorporate its own fuel tank base, which is often utilized for standby applications. Other design considerations should be given to removable wall panels or end walls to allow for equipment installation and replacement. Roof hatches are frequently provided to allow the installation or removal of switchgear equipment. Often with very large gen-sets,

Figure 22-2. Intake Louvers

an integral roof crane/rail system is desired to remove such parts as cylinder heads, turbochargers, and other machinery items that usually weigh less than two tons.

22.7 FUTURE CONSIDERATIONS

At the initial design phase, consideration should be given to additional electrical needs in future years. If this is done in the original planning stages, it is relatively simple and economical to expand the original enclosure or install additional enclosures by connecting them to the original system.

22.8 SUMMARY

Generator enclosures that are properly designed, using high-quality non-corrosive materials, can be expected to provide many years of satisfactory performance. When a portable gen-set enclosure is placed next to, or in the vicinity of an existing building, there is no reason why the new building should not be aesthetically pleasing or blend in with the landscape.

Sound Attenuation

Susan Brown

CHAPTER 23

INTRODUCTION

Why noise control? In the 70s, as people became increasingly concerned with the quality of our environment, the desirability for quiet was added to the desire for clean air and pure water.

Any unwanted sound is called noise. It can annoy, distract, or interrupt. It can damage and destroy, as when a sonic boom shatters windows. Excessive noise can destroy the ability to hear, and may also put stress on the heart, the circulatory system, and other parts of the body.

There is no cure for physical damage done by excessive noise. A hearing aid can make speech louder but cannot make it clearer. A worker who has had long term exposure to loud noise will first lose the ability to hear high frequency sounds. The hearing impaired individual typically says "I can hear you, but I can't understand you."

23.1 REGULATIONS, THE NUMBERS GAME

It often seems that the maximum permissible noise level for an electrical generating system is different for each project. Part of the reason for this variety lies in the increasing number of regulating agencies, laws, and standards.

In the US, under the Occupational Safety and Health Act of 1970, every employer is legally responsible for providing a work place free of hazards, including excessive noise.

The OSHA standard limits workers' noise exposure to 90 dBA averaged over an eight hour period. There are shorter time limits for higher noise levels. The relation used by OSHA allows 5 dB increase in level for the reduction of 2 to 1 in exposure time. This relation is often called a 5 dB exchange rate. In other countries, a 3 dB allowance or exchange rate is used.

In the United States, if noise exposure rises above these levels, the employer may use engineering controls. Engineering controls are changes in the physical work environment to reduce noise levels, such as enclosing machines or installing acoustic materials. If engineering controls are not practical, administrative controls can also be used for compliance. This would limit an individual employee's exposure time. At the bottom level of acceptable compliance measures, personal protective devices may be adopted. This is the least desirable control measure. Ear plugs, for example, can interfere with the ability to hear instructions, and are disliked by most employees.

In many industries, meeting the 90 dBA limit is expensive. Still, studies have shown that even at the 90 dBA level, exposure over a long period will lead to measurable hearing loss in some people, and an 85 dBA limit would be more protective. The 85 dBA limit has been discussed, but not adopted.

Under the Noise Control Act of 1972, the Environmental Protection Agency was mandated to establish noise limits protective of public health and welfare, with an adequate margin of safety. These recommended levels include 75 dBA for an eight hour exposure, and 55 dBA outdoors.

In Europe, there are even more variations. The USA considers noise a public health hazard, such that Congress has made a commitment to provide a tranquil environment. In the United Kingdom (UK), noise is a nuisance, and the remedy is through common law. This means taking your neighbor to court. Each case is considered on its own merits, and the judgments are not always consistent.

In France, the Ministry of the Quality of Life decreed on Oct. 17, 1975 that products "must not in any way cause excessive inconvenience," and "their noise levels must be brought to the knowledge of users and purchasers." It is not the intention of the Ministry to establish upper limits for noise. Instead, the philosophy is that by requiring manufacturers to publish noise levels of their products, competition will force product noise levels down.

In Germany, the Ministry of Labor and Social Affairs has defined maximum noise levels at the work place. These levels are lower than necessary to protect against hearing loss. They have been chosen to provide for an "acceptable work place," in this instance through lack of annoying noise. Legal control rests with state authorities.

Scandinavia has no quantitative limits in law. In industry, the noise levels must be satisfactory to the workers. Sweden has a Foundation for the Protection of Workers. Each employer pays a 0.1% tax on his annual payroll. The resulting Swedish Work Environment Fund, operated jointly by employers and labor unions, conducts research and education for improvement of the work environment.

23.2 BASICS OF SOUND

To apply the principles of noise control, one should be familiar with the behavior of sound. Sound is defined as any **pressure variation** (in air, water, or other medium) that the human ear can detect. Sound is produced when a sound source sets the air nearest to it in wave motion. Like ripples in a pond, sound waves move away from the source at a constant speed. The speed of sound in air varies slightly with temperature and humidity, but in general the speed is about 340 meters per second. In water, the waves travel at about 1500 meters per second, and at 5000 meters per second in steel.

The barometer, which measures air pressure variations that occur with changing weather patterns, cannot respond fast enough to measure sound. Pressure variations must occur at least 20 times a second to be heard and, therefore, to be called sound.

The number of pressure variations per second is called the **frequency of sound**, which is measured in cycles per second, called hertz (Hz). The range of human hearing extends from approximately 20 Hz to 20,000 Hz (20 kHz) while the range from the lowest to the highest note of a piano is 27.5 Hz to 4186 Hz. By dividing the speed (340 m/s) by the frequency of a sound, we can find the wavelength; that is, the physical distance in air from one wave top to the next. At 20 Hz this gives a wavelength of 17 meters while at 20 kHz, one wavelength is very short,

only 1.7 cm. In designing for noise control, the engineer will select techniques and materials that are appropriate for the wavelengths of the particular sound. For example, high frequency (short wave) sound is strongly directional and more easily reflected. It does not travel around corners easily. A barrier may be quite effective in shielding a worker near a machine emitting high frequency shrieks or whistles. At 8000 Hz, the wavelength is 4.25 centimeters. At 125 Hz, the wavelength is 2.72 meters. This long wave easily bends around corners, travels through holes, and continues to travel in all directions. A shield or barrier has little effect unless it is very large.

23.2.1 Sound power level. Noise control engineering, like any other discipline, has a specialized vocabulary. You may see references to sound power level in a manufacturers' published test data. The action of giving off or sending out sound is **emission**. The preferred description for this action is the **sound power level**, often abbreviated as PWL, LW, or 10^{-12} watt. Describing the total acoustic power of a source, or sound power level, is analogous to describing a light bulb by its wattage.

Small sound powers can have a large effect on the ear. It has been estimated that the roar of a football crowd produces only enough energy to heat a cup of coffee. A full symphony orchestra only produces about one acoustic watt.

23.2.2 Sound pressure level. Immission is the opposite of emission. The OSHA level of 90 dBA refers to **immission** as the noise from all sources coming to the ear. The preferred description is **sound pressure level**, SPL, or Lp, where the reference is to atmospheric pressure. The sound pressure level is analogous to the brightness of a light bulb, which is dependent upon the room in which it is placed. In a bright, daylight room the bulb will seem to have a different brightness than it will in a dark room at night. Therefore, when using the sound pressure level to describe the noise of an engine, it is not enough to say 90 dB. One must say the sound pressure level is 90 dB at 10 feet to convey meaningful information.

23.2.3 The decibel. The weakest sound that a healthy human ear can detect is 20 millionth of a pascal (or 20 μPa – 20 micropascals), which is a factor of 5 billion less than normal atmospheric pressure (one Kg per square cm). This change of 20 μPa is so small that it causes the membrane in a human ear to deflect a distance less than the diameter of a single atom. Amazingly, the ear can tolerate sound pressures up to more than one million times higher. Thus, if we had to measure sound in Pa, we would end up with large and unmanageable numbers. To avoid this, another scale has been devised — the decibel (dB) scale.

The decibel scale uses the hearing threshold of 20 μPa as its starting point or reference pressure. This is defined as zero dB. Each time we then multiply the sound pressure in Pa by 10, we add 20 dB to the dB level, thus 200 μPa corresponds to 20 dB, 2000 μPa to 40 dB, and so on. The dB scale compresses the million to 1 range into a 120 dB range.

Another useful aspect of the dB scale is that it gives a much better approximation to the human perception of relative loudness than the Pascal scale since the ear reacts to the percentage change in level, which corresponds to the decibel scale, when 1 dB is the same relative change everywhere on the scale. One dB is the smallest change we can hear. A 6 dB increase is a doubling of the sound pressure level, although a 10 dB increase is required to make it sound twice as loud. Similarly, a reduction in sound of 10 dB makes it seem as if the intensity has been reduced by half.

The important thing to remember about the decibel is that it represents a relative measurement, or ratio. With decibel we also use the word level, sound power level, or sound pressure level, etc. to remind us that we are referring to a reference quantity and that our measurement is relative, not absolute.

23.2.4 Response to sound.
To achieve a reduction in noise, you must start with accurate information about the source. Humans hear loudness and pitch. The human ear is not equally sensitive at all frequencies. It is most sensitive in the 500 Hz to 6000 Hz range, and least sensitive at extremely high and low frequencies. The same sound pressure level at 2000 Hz sounds much louder to the human ear than does that same sound pressure level at 500 Hz.

Electronic circuits have been designed so that sensitivity varies with frequency the same way as in the human ear. There are three different standardized characteristics called the "A", "B", and "C" weighing networks, each one centered on different sound pressure level curves. A specialized characteristic, the "D" weighing, has been standardized for aircraft noise measurements.

The "A" network is most widely used since it gives the best correlation to subjective tests. Legislation is most often written with reference to the "A" scale, such as the 90 dBA OSHA standard. This means the noise coming to the ear across all frequencies is combined and read as a single numerical value on the "A weighted" sound level.

It should be noted that the single number dBA sound level provides practically no information that can be used in analysis leading to a reduction in noise levels.

23.2.5 Octave bands.
For more detailed analysis, it is common practice to divide the range of frequencies we can hear into eight octave bands. The top frequency in an octave band is always twice the bottom one and the band is referred to by its center frequency. For example, 500 Hz is the center frequency for the octave band whose lowest frequency is 354 Hz and whose highest frequency is 708 Hz.

23.2.6 Adding noise levels.
Noise is a disorderly mixture of tones at many frequencies, but the decibel levels for two or more sounds cannot be added directly since the values are logarithmic. The combined effect of sounds depends on the difference in their decibel levels.

To calculate the estimated overall dBA (Figure 23-1), you must first correct the octave band readings to allow for the way the human ear hears. To correct for the A-scale these numbers are applied to the readings (-26, -16, -9, -3, 0, $+1$, $+1$, -1). The noise of a tractor had these readings: 91, 98, 91, 85, 86, 82, 80, 65. After correcting for the A-scale, the corrected sound pressure levels are 65, 82, 82, 86, 83, 81, 64 for the various frequencies.

Next we arrange the dB value from the highest to lowest without regard to band sequence. The highest value is 86, the lowest is 64. Because we are working with logarithms, we cannot simply add the numbers directly. The logarithmic addition of two equal sound pressure levels equals 3 dB more. For example, 82 dB + 82 dB = 85 dB. Refer to Figure 23-2 to determine what number to add after calculating the difference between two levels.

Because of rounding at various stages, the numerical end value is 91 (Table 23-1). It reads 90 on a meter.

We must understand the addition of noise levels in many parts of a noise control problem. If the overall noise level of a particular gen-set measures 90 dBA at the lot line, installing an identical set next to the first would increase the noise level to 93 dBA. (When the difference between noise levels is 0, add 3). If a third identical set were added, the measured result would be 95. (When the difference is 3 (90, 93) add 2 to the higher number).

23.2.7 Ambient noise.
The noise that is present in the environment before the installation of a gen-set must be accounted for. In this example, the requirement for the completed installation is 65 dBA, measured at point X. Because of transformers and fans, the pre-installation noise level is already 63 dBA at point X. Therefore, in order to meet the 65 dBA spec, the gen-set package should not exceed 60 dBA. (The difference between levels is 3, so add 2 to the higher number.)

OCTAVE BAND	1	2	3	4	5	6	7	8
FREQUENCY	63	125	250	500	1000	2000	4000	8000
BAND CORRECTION	−26	−16	−9	−3	0	+1	+1	−1

EXAMPLE (TRACTOR)

BAND SPL	91	98	91	85	86	82	80	65
BAND CORRECTION	−26	−16	−9	−3	0	+1	+1	−1
CORRECTED SPL	65	82	82	82	86	83	81	64

Figure 23-1. Calculating dBA From an Octave-Band Analysis

dB Difference Between Two Levels	0	1	2	3	4	5	6	7	8	9	10
No. to Add to Higher Level	3	2.5	2	2	1.5	1.5	1	1	.5	.5	0

Figure 23-2. Calculating dBA From an Octave-Band Analysis

Table 23-1. Example of Logarithmic Addition of L_wA Octave Band Value Using Table 1 Factors

List dB Values from Highest to Lowest without Regard to Band Sequence	Diff Bet. (1) & (2)	Factor from Table 1	Factor + (1)	Diff Bet. (3) & (4)	Factor from Table 1	Factor + (3)	Diff Bet. (5) & (6)	Factor from Table 1	Factor + (5)
Highest									
(1) 86	3	2	(3) 88						
(2) 83				3	2	(5) 90			
(1) 82	0	3	(4) 85						
(2) 82							5.5	1	*(7) 91
(1) 82	1	2.5	(3) 84.5						
(2) 81				17	0	(6) 84.5			
(1) 65	1	2.5	(4) 67.5						
(2) 64									

23.2.8 The mass law. The mass law, which relates to the transmission loss of walls, says that in a part of frequency range, the magnitude of the loss is controlled entirely by the mass per unit area of the wall. The law also says that the transmission loss increases 6 decibels for each doubling of frequency, or each doubling of the wall mass per unit area. For example, lead sheet which has a transmission loss of 13 dB at 63 Hz, has 13 + 6, or 19 dB at 125 Hz. At 250 Hz it is 19 + 6, or 25 dB. If you double the thickness of the mass from 1/16 in. to 1/8 in. thick, the transmission loss at 63 Hz becomes 13 + 6, or 19 decibels.

In relation to the mass law, the lighter engines that have been a response to energy conservation have made noise control engineering more critical. If you study the noise levels of a series of engines, you will see that in many cases the smaller engines are noisier than the largest models. In general, the casing is less massive, allowing more transmission of mechanical noise. However, addition of mass usually is not a practical solution for noise control. Today we look for combinations of lightweight materials that can do an even better job than sheer solid mass.

23.2.9 Resonance.
Each material has its own resonant frequency, that is, its natural mode of vibration. The particular frequencies are dependent upon many characteristics, one of which is mass. Less massive skid bases sometimes result in a higher noise level because the forcing frequency (the engine) excites the base natural frequency, strengthening the sound pressure level at that frequency. A number of damping materials that can be applied to change the transmitted force amplitude are on the market. The selection of vibration isolators also becomes very critical, since the wrong isolator supports may not only fail to give any improvement, they can make the situation worse. Spring isolators with good internal damping may be required to prevent a reinforcement of vibrations at the fundamental frequency. The reduction of noise by vibration isolation could be a lengthy subject by itself. We will make further references to the importance of vibration isolation, but it is covered in detail in Chapter 20.

A noise control problem can be divided into three parts. There is a source, a path (or paths), and a receiver. Often it is desirable to treat the noise at the source.

23.3 SOUND TRANSMISSION (PROPAGATION)

The word *sound* usually means sound waves travelling in air. However, sound waves also travel in solids and liquids. These sound waves may be transmitted to air to make sound we hear. Airborne sound is usually caused by vibration in solids or turbulence in fluids.

An example of vibration in solids is the vibration of strings in a violin. The vibrating strings rest on a bridge, which transmits the vibration to the sound box or body of the instrument. The vibration causes the wood to vibrate, and sound is then transmitted to the air. If damping material was applied to the bridge so that string movement was isolated from the sound box, the amount of airborne sound would be cut drastically. If strips of damping material were applied inside the sound box, the airborne sound could be decreased even further. Knowing where to place the damping material is important. Sometimes the major point of vibration transmission is very obvious. In other cases, it may be necessary to use instrumentation to identify the offending source. By applying damping material we can reduce the response of the solid, because the damping material will actually absorb or dissipate some of the energy.

Turbulent fluid flow within pipes produces sound that can be radiated from the pipes, and even transmitted to the building or enclosure structure. If an enclosure has been selected as the noise control solution, rigid wall couplings for pipes and electrical connectors can drastically affect the attenuation. Connections should be flexible, or isolated to prevent transmission of vibration to the enclosure walls.

Vibration in solids and liquids can travel great distances before producing airborne sound. One example is the vibration from a train, which can be heard in the rails a long distance away. Sometimes people are disappointed with the results of an enclosure that is assembled around a gen-set, and which rests on top of the gen-set base. Without adequate vibration isolation at the skid base, vibration can travel through the skid, so that much of the noise is not treated by the enclosure.

Ideally, the gen-set should be on isolators, totally within an enclosure that is a complete box. Or the set should be vibration isolated on a concrete pad, with the enclosure completely surrounding the base, with gasketing and tie downs to prevent any noise leaks through gaps in an uneven concrete surface.

23.4 Sound absorption.
Sound is transmitted or propagated in a number of ways. Some sound waves are reflected. They hit a hard surface and bounce off. We can reduce some of the reflected sound by providing an absorbent surface. In a "hard" room, soft materials such as absorbent ceiling panels, carpeting on the floor, and drapes or special absorbent wall coverings, will reduce noise by reducing the reflected sound. Keep in mind that only reflected sound can be treated in this way. Direct sound will not be affected. Many times the effectiveness of acoustic ceiling panels lasts only until the first spring cleaning. If a coat of paint is applied to the acoustic panels, the pores become clogged and the absorptive properties are reduced.

In evaluating materials for their ability to absorb acoustic energy, that is, the ability to absorb reflected sound, the data published is usually the absorption coefficient. This is a number between 0 and 1, and the closer it is to 1, the more acoustic power is absorbed.

The absorption coefficient is dependent on the frequency. A material can have a very high value in a high frequency and a very different value at another frequency. For example, ordinary window glass is better at the 125 octave band (0.35) than at 4000 Hz (0.04). Fiberglass is better at 4000 Hz (0.99) than at 125 Hz (0.23) (for 1 in. thick, 3 lb density).

The major mechanisms of sound absorption thought to exist in absorbing materials are through viscous losses in air cavities and mechanical friction losses caused by fibers rubbing together. This helps to explain why the short wavelengths of high frequency sound are so much easier to treat than the very long waves of low frequency sound. The short waves are more easily trapped in the air cavities.

In order to maintain the best absorption values of the chosen materials, the air channels should all be open to the surface so that sound waves can propagate into the material. If pores are sealed, as in a closed cell foam, the material is generally a poor absorber. Pores should not be sealed by painting, and any protective shielding for the absorbing material must be perforated.

23.5 TRANSDUCTION

Transduction refers to a change in form of sound energy. Very high sound pressure levels will cause an absorbent material to get warm. However, it is unlikely that you would be able to actually feel that warmth. The example that I used before was that of the cheering football crowd, where the total sound energy, changed to heat, would only heat one cup of coffee.

Transduction is also the change that occurs when sound is changed to vibration. The sound energy strikes a rigid surface and makes it vibrate.

23.6 REFRACTION

Refraction refers to the bending of sound. The speed of sound, which we have said is 340 meters per second (m/sec.) as a rule of thumb, is actually proportional to temperature. Sound moves slower in cold air. The normal temperature gradient, where temperature decreases with increasing height, tends to bend sound waves upward and adds to the attenuation. But this situation is changed when there is a temperature inversion at the ground (temperature increasing with height near the ground). This inversion often occurs on a clear night, and the sound wave is bent downward. This can cause noise complaints about an installation that is normally acceptable. In other words, the noise level can change appreciably from time to time, even with a constant source level.

Wind conditions also affect propagation of sound. Near the ground, winds tend to increase with height. This gradient causes sound traveling with the wind to be bent toward the ground. But a sound wave traveling opposite to the wind will be bent away from the ground, and there will be increased attenuation. It is best to avoid unusual conditions of wind, temperature, and humidity when taking sound level measurements.

23.7 DIFFRACTION

Diffraction refers to the bending of sound waves around an obstacle. Sound bends very readily. This is why a partition has only limited effectiveness. Generally, a partition must be quite close to the source in order to be effective, and the more low frequency noise is present, the larger the partition must be.

23.8 STRAIGHT-LINE TRANSMISSION

This refers to air absorption of sound rays traveling in a straight line. The sound energy will be reduced in amplitude as it travels. Therefore, distance from the source affects the level of sound.

The inverse square law states that each time you double the distance from the source, you decrease the noise level by 6 dB. In theory, a source that measures 90 dBA at 1 meter, would measure 84 dB at 2 meters, 78 dB at 4 meters, 72 dB at 8 meters, etc. However, this inverse square law is true only for certain conditions. Under ideal conditions, a simple point source, in air in the free field, such as at the top of a tall flag pole, would decrease 6 dB for each doubling of distance. However, this applies only along a radial path from the source and where there are no interfering objects. Most real sources are somewhat directional, however, and therefore the sound levels at equal distances from the source are not independent of the direction.

In a real installation of an electric generating system, many factors appear in the sound field to upset the theoretical predictions. For example, walls, buildings, signs, people, and machinery commonly change the resulting sound field, particularly near such objects. With an obstacle in the sound path, part of the sound will be reflected, part absorbed, and the remainder will be transmitted through the object. In general, the object must be larger than one wavelength in order to significantly disturb the sound. For example, at 10 kHz, the wavelength is 3.4 cm. Even a small object such as a measuring microphone will disturb the sound field. At 100 Hz, where the wavelength is 3.4 meters, sound isolation becomes much more difficult.

The nature of the surface over which sound travels also affects decay of sound with distance. Thick grass in an open field can increase attenuation, particularly at distances greater than 50 meters. A hard, paved surface can increase reflection, and make noise control more difficult.

The inverse square law does not work when measurements are taken in the near field. At distances very close to a noise source, sound propagation is non-linear, and difficult to predict. If you try to measure the source noise of a gen-set up close, the SPL may vary significantly with only a small change in position. Therefore, measurements should not be taken in the near field, which is a distance less than the wavelength of the lowest frequency sound emitted from the set or at less than twice the greatest dimension of the set, whichever is the greater. Most gen-sets emit some sound at the 31.5 Hz frequency. The wavelength of that sound is 340 meters/second divided by 31.5 Hz, or 10.8 meters. Therefore, to be out of the near field when taking measurements, 11 meters distance is a good guide. Unfortunately, many specs and standards call for readings to be made at one meter. While this may be fine for smaller machines with no frequencies lower than 250 Hz, it can cause unpredictable results for large gen-sets. If a manufacturer's data gives sound pressure levels in the near field, we cannot use this data to predict the source sound power levels, or sound pressure levels at other distances.

Other errors may arise if you measure too far away from the gen-set. Depending on the site, reflections from walls and other objects may be just as strong as the direct sound, and correct measurements will not be possible. This is termed the **reverberant field**. Between the reverberant field and near field is the free field, where the inverse square law does work, that is, the level drops 6 dB for a doubling in distance from the source. Measurements of sound pressure level should be made in the free field. It is possible that an installation will have so many reverberant conditions in a small space that no free field exists.

23.9 MEASURING SOUND WORKING IN THE REAL WORLD

When taking noise measurements, the height of the microphone should be consistent. This height should be recorded, along with the distance from the source, direction of the microphone, and all ambient conditions such as temperature, humidity, wind speed, and ambient noise levels. Read the directions accompanying the sound level meter to determine if the microphone should be pointed directly at the noise source, or if the microphone tip should be at a 90° angle so the sound will graze the front of the microphone.

The sound level meter is a delicate instrument, and should be carefully handled in order to give valid results. It should be calibrated before and after a set of readings is made. When making measurements outside, the microphone should be fitted with a windscreen. One frequently overlooked condition is temperature. The microphone must have ample time to adjust to ambient conditions. For example, if you are going to test outdoors in the summertime, do not take the SL meter from an air conditioned office and immediately start taking measurements. By the same token, don't take the instrument from a heated space to a cool one, and expect to get valid readings. It is best to allow several hours for the instrument to adjust to the temperature conditions at the test site before making any readings.

Microphones are so sensitive that the operator's body, and even the meter case, can interfere with readings by reflecting sound waves back to the microphone. It may be advisable to mount the microphone on a tripod and connect it by extension cable to get the meter and operator away from the sound field. Experiments have shown that at frequencies of around 4000 Hz, reflections from the body may cause errors of up to 6 dB when measuring less than one meter from the body.

Another very important factor influencing the accuracy of measurements is the level of background noise compared with the noise signal to be measured. Obviously, the background noise must not drown out the signal of interest. In practice, this means that the level of the signal must be at least 3 dB higher than the background noise. Measure the total noise level with the gen-set running. Measure the background noise level at the same spot (with the gen-set off). If the difference between the two readings is less than 3 dB, the background noise level is too high for an accurate measurement. If it is between 3 and 10 dB, a correction to the reading is necessary. Make the correction by subtracting the number found in the chart used for addition of sound levels. If the difference in levels is 7, subtract 1 from the combined reading.

If the specification calls for octave band readings, ambient noise should be read at each octave band, and corrections made to the gen-set readings as necessary.

Diesel engine noise may be divided into two parts: exhaust noise and structure noise. If the engine is fitted with an effective silencer (see Chapter 18), the structure noise tends to dominate, that is, the noise radiated away from the engine walls. This structure noise consists of noise due to rotation (mechanical) and noise due to combustion. The control of this noise through damping and vibration isolation can be helpful, but in many cases an enclosure is required in order to meet the desired noise level.

One point to note is that when a gen-set is placed in an enclosure, the noise level outside that enclosure will be reduced. However, the noise level inside the enclosure will be higher. The smaller the enclosure is in relation to the size of the set, the noisier it is apt to be inside, and the harder to predict the noise reduction outside the enclosure. Not only are you in the near field, but the irregular surface of the machinery makes up an extremely complex sound source. A close-fitting enclosure is one in which the volume of the noise source is one-third or more of the total enclosure volume. At certain frequencies, the normal mass law transmission loss is negated and the enclosure walls actually become "transparent." These frequencies are related to the resonance frequencies of the enclosure walls and to the standing wave resonances in the air space. The overall effect is a reduction in the overall insertion loss, which would be predicted for a larger enclosure. When using enclosures or barriers for noise control, it is helpful to understand some of the terms commonly used.

Noise reduction (NR) is the difference in sound pressure levels at two specific points inside and outside the enclosure (NR = $SPL_1 - SPL_2$). This is easy to measure, but hard to predict, partly because so many conditions affect how much the noise level of a gen-set will increase when placed in an enclosure.

The *transmission loss* (TL) is $10 \log_{10}$ of the ratio of the incident acoustic intensity on a wall to transmitted acoustic intensity. TL = $10 \log_{10} (I_1 / I_t)$. This is the easiest to predict in theory, and the hardest to measure in practicality.

Insertion loss (IL) is the difference between the sound pressure levels at the same point, with and without an enclosure. This is the easiest to measure, and the hardest to predict theoretically.

Manufacturers sometimes quote values of TL for wall construction, when actually the results are of NR.

The best enclosure made will give disappointing results if there are even small cracks where sound can leak out. At high frequencies, the leak effect is particularly noticeable. A small leak can reduce the TL by over 10 dB. The leak provides an air path for the transmission of energy. When it is necessary to cut holes in an enclosure for pipes or cables, in addition to flexible connectors to isolate any vibration noise, the space between the pipe or cable and the opening must be sealed.

Gaps around doors must be sealed. The thicker the wall, the more seriously a leak will degrade the performance of the enclosure. One square inch of hole can transmit as much energy as a 100 ft. wall. A crack that is 0.01 inch wide × 12 feet long has an area of 1 sq. inch. In a relatively thin wall, such a crack might reduce the transmission loss by 3 dB. If this thin wall is made four times thicker, the transmission loss will be decreased by 12 dB from the same 1 sq. inch crack area.

In supplying air for cooling in a generator set enclosure, the opening must be treated with inlet and outlet mufflers, sound traps, or other devices so the attenuation for those areas is equivalent to the walls.

Leaks will cause "air flanking." It is also common to have mechanical paths that flank energy around a wall and thus reduce its effectiveness. This is also called **structure-borne noise**. In this case, vibration isolation is critical in reducing the noise. Gen-sets should be vibration isolated from the enclosure or else the enclosure walls can act like a sounding board. Also, the enclosure should not be rigidly attached to a floor, a base, or other large surface that is an efficient low frequency sound radiator. Gasketing, such as a medium density closed cell neoprene, should be used in a thickness adequate to isolate the enclosure.

Cooling system fan noise can be a significant contributor to overall gen-set noise. This noise can be reduced by using a systems approach, by increasing radiator efficiency, slowing down the fan, and removing upstream fan obstructions. The fan should be well balanced and mounted on vibration isolators.

For more information on generator set enclosure design see Chapter 22.

23.10 SUMMARY

This chapter has covered a long list of terms and principles. The bottom line is that if noise from electric generating equipment is to be controlled, a systems approach must be applied.

a. The systems design must take into account the many environmental factors that impact the final result.

b. All components should be selected to contribute to the noise control solution. Technical input from the manufacturers of exhaust silencers, radiators, vibration isolators, and enclosures can help the designer achieve maximum results.

c. The package must be assembled and installed with great care and attention to detail in order to avoid degrading the system performance.

The systems approach will enable our industry to meet one more of the challenges of the 1990s.

REFERENCES

[1] Application of Sound Power Level Ratings, for ducted air moving devices, recommended typical dBA calculations, AMCA, Publication 303, 1973.

[2] Crocker, M. J. Machinery Noise Sources and Basic Methods of Control, Lecture 4, from Inter-Noise

Seminar, Techniques of Noise Control, INCE, New York, 1980.

[3] Crocker, M. J. and Price, A. J. Noise and Noise Control, Vol. I, CRC Press, Boca Raton, Florida, 1975, 79.

[4] Lang, W. W. Basic Properties of Noise, from Inter-Noise Seminar, Techniques of Noise Control, INCE, New York, 1980.

[5] Lang, W. W. Noise Criteria, from Inter-Noise Seminar, Techniques of Noise Control, Institute of Noise Control Engineering, New York, 1980.

[6] Measuring Sound, Bruel and Kjaer Instruments, Inc., Cleveland, OH.

[7] Peterson, A. Handbook of Noise Measurement, 9th ed., Gen Rad, Inc., Concord, Mass., 1980.

[8] Sound and Air-Conditioned Space, Carrier Corporation, 1965.

[9] Thumann, A. and Miller, R. Secrets of Noise Control, 2nd edition, Fairmont Press, Atlanta, GA., 1976.

[10] Witt, M. Noise Control — A Guide for Workers and Employees, U.S. Department of Labor, OSHA 3048, Office of Information, 1980.

Batteries and Battery Chargers

Warren Henderson

CHAPTER 24

INTRODUCTION

Analysis of emergency generator set failures shows that battery problems cause at least half of all failures to start. These failures are not necessary. Generator set specifications frequently give detailed attention to the engine, the generator, and peripheral equipment but little more than a voltage for the cranking battery. Reliability of a generator set installation requires a suitable battery and charger and programmed maintenance. Proper specifications and maintenance can prevent almost all battery failures.

Battery failure could be the result of an old battery reaching the end of its expected life. More often the battery dies prematurely because of improper application or charging or lack of a preventive maintenance program. A good preventive maintenance program should catch even the expiring battery. All too often batteries fail to provide the expected life, because the specifier did not recognize and address the conditions that cause failure.

The National Electrical Code recognizes that improper battery charging is a frequent cause of battery failure. Its Article 700 requires an automatic charger and an alarm to indicate when it fails to function. It also requires periodic maintenance.

Engine cranking batteries are available today with a guaranteed life of from 1 to 25 years. Battery chargers are available to assure the maximum life from the battery.

Dependable starting systems require a knowledge of battery characteristics, a detailed analysis of system components, and minimum routine maintenance.

24.1 BATTERY CHARACTERISTICS

A battery is a group of electrochemical cells interconnected to supply a nominal dc voltage to power an electrical load. The number of cells connected in series determines the nominal voltage rating of the battery. In a lead acid battery, each cell produces 2.0 volts. In a nickel cadmium battery, each cell produces 1.2 volts. Thus there are three cells in a 6 volt lead acid battery and five cells in a 6 volt nickel cadmium battery.

The size of any given cell in a battery is the basic factor that determines the discharge current capacity of the entire battery. Discharge capacity of the battery is its ability to supply a given current for a given period of time at a given initial cell temperature while maintaining voltage above a given minimum value. As a cell starts to discharge, there is a decrease in voltage due to the effective internal resistance of the cell. This voltage drop increases with an increase in discharge current, thus lowering the output voltage of the cell by that amount. At a continuous rate of discharge, the voltage gradually becomes lower as the discharge progresses. As the cell nears exhaustion, the voltage drops very rapidly to a value where it is no longer effective. The value at the point of the discharge curve where the voltage begins to fall rapidly is designated as the **final voltage**. It varies, with the rate of discharge being lower with higher ampere rates. For a lead acid cell, it may be as high as 1.85 volts for comparatively low rates of discharge. It may be as low as 1.0 volt at extremely high discharge rates, such as experienced in engine cranking applications. Battery ampere-hour capacity varies with discharge rate. A typical 100 ampere-hour lead acid battery will deliver 12.5 amperes for eight hours to a final voltage of 1.75 volts per cell. The same battery will deliver 25 amperes for three hours, 50 amperes for one hour, and 150 amperes for one minute, corresponding to effective ampere-hour capacities of 75, 50, and 2.5 respectively. See Figure 24-1.

Battery performance varies with ambient temperature. See Figure 24-2. Lowering the battery temperature reduces battery capacity and, therefore, the available cranking time. The loss in capacity is greater at higher discharge rates. Raising the battery temperature increases the battery capacity but decreases the expected battery life. Batteries do not like temperatures in excess of 90°F and the temperature of the electrolyte should never exceed 110°F. Battery capacity is rated at 77°F (25°C). Since 70°F (21°C) is common for generator set applications, a correction may be required.

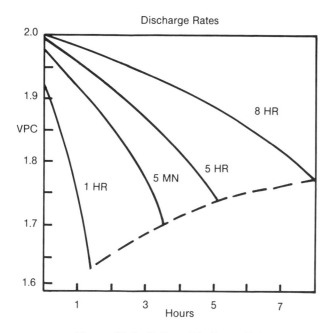

Figure 24-1. Battery Discharge Rates

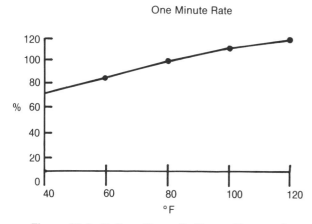

Figure 24-2. Battery Capacity Versus Temperature

24.2 BATTERY CHARGING

A battery charger changes alternating current (ac) to direct current (dc) to restore the charge in storage batteries. The term *charge* applies to the conversion of electrical energy to chemical energy within the battery. Battery chargers normally used in engine starting systems are solid state, constant potential chargers. A constant potential charger is a charger in which the voltage at the output terminals of the charger is held to a constant value. Actually, batteries require two precise voltages: float and equalize.

24.2.1 Float voltage. Float voltage is the minimum charger voltage required to overcome local chemical action between trace impurities and the negative plates of the cell. Local action is the reason a battery becomes discharged during storage. Most manufacturers recommend wet batteries be placed on charge within three months from the time of shipment from the factory, if lead-antimony, or six months, if lead-calcium.

In any service, a battery should receive the correct amount of charge to fully charge it and maintain it in that condition, but no more. In other words, undercharge or overcharge should be avoided. An insufficient amount of charge, even to a small degree, will cause gradual sulfation of the negative plates, with eventual loss of capacity and reduction of battery life. An excessive amount of charge will tend to corrode the grids of the positive plates, thus weakening them physically and increasing their electrical resistance. If the overcharging is at comparatively high rates, gassing will be excessive and will tend to wash active material from the plates. For these reasons float voltage should be maintained within ±1%.

24.2.2 Equalize voltage. Equalize and or recharge voltage is the voltage given to a storage battery to ensure complete restoration of the active material on the plates of all the cells. The purpose of the periodic equalizing charge is to ensure that all of the battery's cells are at full charge level. Battery cells are complex assemblies with many variables that affect cell performance. No two cells are exactly alike. During float charging, which keeps most of the battery's cells at full charge, some cells may gradually lose a little voltage. These low cells can be upgraded or equalized with all the other cells by periodically supplying a prolonged charge at the specified equalize voltage. Usually manufacturers recommend equalizing lead-antimony batteries every one to three months. Many state that lead-calcium batteries floated at 2.25 volts per cell do not require equalizing charges. A battery, however, should be equalized when the float voltage for any cell is below 2.13 volts or more than 0.04 volts below the average for the battery. Equalization is also required when the temperature corrected specific gravity of any cell is more than 10 points below its full charge value.

Equalize charges are usually given at 2.33 volts per cell for lead-acid batteries. Equalize voltages should be maintained at ±2% of the nominal setting.

24.3 DESIGN OF FLOAT SYSTEMS

In a float system the charger, battery, and load are permanently interconnected, as shown in Figure 24-3. When ac power is present, the charger regulates the voltage supplied to the load and the battery. Recommended float and equalize voltages for typical batteries are given in Table 24-1.

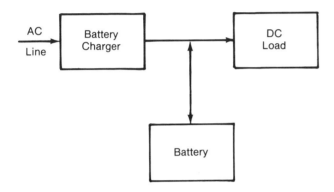

Figure 24-3. Float System

Table 24-1. Voltage Ranges Per Cell at 77°F (25°C)

Battery Type	Nominal Voltage*	Float Voltage**	Equalize Voltage**
Lead-antimony	2.0	2.15–2.19	2.25–2.35
Lead-calcium	2.0	2.15–2.25	2.30–2.40
Nickel-cadmium	1.2	1.40–1.45	1.50–1.65

*The nominal value is an arbitrary reference value selected to establish equipment ratings.
**Recommended float and equalize voltages are a function of electrolyte specific gravity and temperature. They increase with elevated specific gravity. They decrease with increasing temperature.

Lead-acid and nickel-cadmium batteries require different float and equalize voltages. For example, a 24 volt nominal, 12 cell lead-acid battery may be equalized at a voltage as high as 28.2 volts [12 × 2.35 vpc (volts per cell)] and a 24 volt nominal, 20 cell nickel-cadmium battery may be equalized at a voltage as high as 32 volts (20 × 1.6 vpc). If, for example, one of the loads in the system cannot exceed 30 volts, it is obvious that only 18 nickel-cadmium cells can be used. The maximum number of cells and the minimum voltage per cell required by the starting system are given by equations 24-1 and 24-2.

$$\text{Maximum number of cells} = \frac{\text{Maximum system voltage}}{\text{Recommended vpc}}$$

(Equation 24-1)

$$\text{Minimum voltage per cell} = \frac{\text{Minimum allowable voltage}}{\text{Number of cells}}$$

(Equation 24-2)

24.4 BATTERY SELECTION

Battery selection requires a knowledge of the starting current draw, the duration of the starting operation, the number of cells, and the minimum voltage per cell. Starting currents are often available from the engine displacement tables. These values, however, must be corrected for actual starting conditions, such as actual engine oil temperature and viscosity. Lowering engine temperature increases the power required to crank the engine. This results in higher starting current being required from the battery. Battery selection also depends upon the specified life and type.

24.5 TYPES OF BATTERIES

Batteries are either primary or secondary. We are concerned here with secondary batteries. Secondary batteries are storage batteries and are rechargeable. The two common types of rechargeable batteries are lead-acid and nickel-cadmium. In the lead-acid battery, electricity is produced by chemical reactions between the plates and the sulphuric acid electrolyte, producing approximately 2.0 volts in each cell. The specific gravity of the acid is a good measure of the charge in the battery. In the nickel-cadmium battery, 1.2 volts per cell is produced between nickel hydroxide and the cadmium plates. Since the potassium hydroxide electrolyte does not enter into the reaction, specific gravity is not a measure of the charge of this battery. The different types of batteries in each of the two chemical types result from details of construction.

24.5.1 Starting, lighting and ignition (SLI).
The SLI battery is a lead-acid automotive or truck battery. It has thin plates, high energy density, and a relatively short life of 1-3 years. It has been designed to be charged by an automotive alternator at 2.4 volts per cell. It is assumed that the battery will be charged 2-4 hours each day. Typical emergency generator set operation is considerably different. Overcharging will result in rapid water loss and shortened life. Undercharging will result in permanent loss of capacity. Voltage checks during high ampere discharge should be made yearly or more often. If the battery has deteriorated, it should be replaced before a failure to crank occurs at some critical time.

24.5.2 Stationary.
Stationary lead-acid batteries have much thicker plates, longer life, and higher cost. They were designed for utility and telephone applications. When used as cranking batteries, the manufacturer should be consulted for suitable high discharge rates. The guaranteed life may be from 1–20 years.

24.5.3 Valve regulated, recombinate.
Valve regulated recombinate lead-acid batteries are constructed and pressurized to produce water by recombining hydrogen and oxygen gases produced during charging. Pressure relief is provided. Water cannot be added to this type battery. Accurate charging is required to avoid loss of electrolyte. The guaranteed life is 1-20 years.

24.5.4 Pocket plate nickel-cadmium.
Pocket plate nickel-cadmium batteries have pockets made of nickel plated steel strips, which are finely perforated and formed into channels. The active material is placed in these pockets and then covered by a steel strip. The pockets greatly reduce the shedding of material, giving long life (up to 25 years).

24.5.5 Fiber plate nickel-cadmium.
Fiber plate nickel-cadmium batteries use a very porous fiber substance to hold the active material.

24.6 BATTERY CHARGER SELECTION

In order to specify a battery charger, it is necessary to know the ac input voltage, dc output voltage, and dc output amperes. Engine starting battery chargers are usually 120 volts 60 hertz, or 220 volts 50 hertz, single phase. As shown previously, the charger output voltage rating is dictated by the type of battery and the number of cells being charged. For this reason, it is preferable to specify dc output voltage as a specified number of cells of a particular battery type, such as 12 cells of lead-acid instead of 24 volts dc. The dc output is determined by equation 24-3.

$$A = \frac{kAH}{R} + L \qquad \text{(Equation 24-3)}$$

Where:

 A = ampere capacity of the charger
 k = 1.1 for lead-acid or 1.4 for nickel-cadmium
 AH = ampere-hours removed from the battery
 R = recharge time in hours
 L = continuous load on the charger and battery during charging

It is important that the charger rating be set by the battery capacity and the continuous load. Oversizing the charger is not desirable since excessive charging current will overheat the battery. Undersizing is equally undesirable. If the charging current is too low, it will not drive off the crystalline lead sulphate adhering to the plate and capacity will be reduced.

All rectifiers and chargers have ripple produced by the nature of the rectified ac input. Ripple is injurious to electronic equipment attached to a battery. Excessive ripple will shorten battery life. Large batteries, however, are excellent filters of ripple.

24.7 TYPES OF CHARGERS

There are many types of battery chargers, the simplest consisting of a step down transformer and a single diode. The long periods of inactivity of emergency generator set operation require a charger specifically designed for this service. Trickle and single rate chargers are not satisfactory. Three types of satisfactory battery chargers are currently in use with engine generator sets: silicon controlled rectifier, mag-amp and ferroresonant.

24.7.1 Silicon controlled rectifier (SCR) type.
This type of charger uses SCRs instead of diodes for rectification. See Figure 24-4. SCRs are semiconductors similar to diodes but have an additional gate connection. They conduct in only one direction, but conduction starts only after a signal is applied to the gate. Conduction stops only when forward current is nearly zero or bias is reversed.

As a battery is discharged its voltage drops. This voltage is compared with the desired voltage in an amplifier, which causes SCR firing circuits to output a signal to both SCR gates. The output current is compared with the desired current limit in another amplifier. If the current exceeds the current limit value, the current limit amplifier overrides the voltage amplifier. Control of the SCR current can be accomplished by changing the conduction angle for each half cycle, or by time proportioning for a number of cycles to obtain the output needed.

24.7.2 Mag-amp type.
The mag-amp charger, Figure 24-5, consists of a step down transformer, silicon diodes, a saturable reactor, and a control board. Battery voltage is sensed by a zener diode.

As the battery discharges, current flows through the dc coil of the reactor, producing saturation. Saturation of the reactor raises the transformer primary voltage and the charging current. When the battery reaches zener voltage, the zener conducts, causing the transistors to bypass the reactor, desaturating the reactor, and lowering the transformer voltage and charging current.

24.7.3 Ferroresonant type.
This charger consists of a ferroresonant transformer, silicon diodes, and a resonating capacitor. See Figure 24-6. Regulation and current limiting are achieved by saturating the resonant winding.

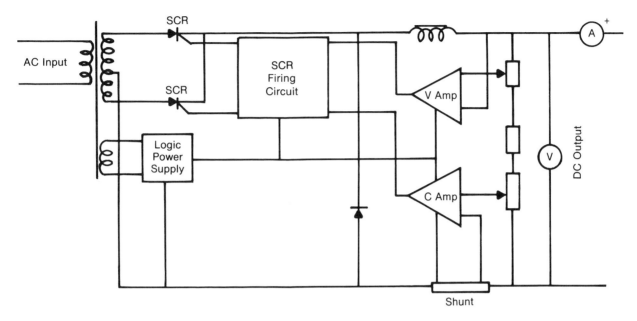

Figure 24-4. SCR Charger

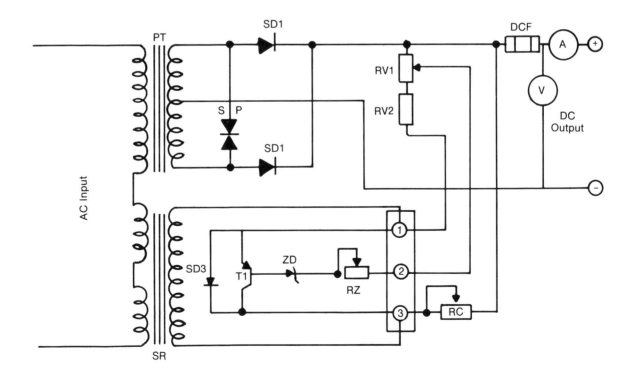

Figure 24-5. Controlled Mag Amp Charger

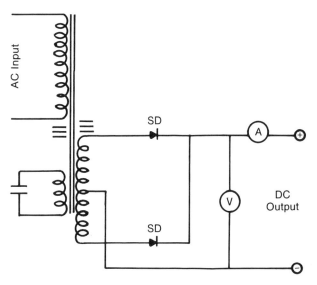

Figure 24-6. Ferroresonant Charger

24.8 BATTERY CHARGER REQUIREMENTS

24.8.1 Float/equalize output. A switch to change the output voltage from float to equalize is needed to allow for regular high rate (equalize) charging to correct any inequalities among individual cells, and to allow faster recharging of the battery after testing or emergency use.

24.8.2 Automatic ac line compensation. Automatic ac line compensation assures that regardless of ac input voltage swings, the float and equalize voltages will be maintained within limits so that the battery will not be subject to under or overcharging.

24.8.3 Battery discharge protection. Battery discharge protection is required so the battery will not discharge through the charger with the loss of ac supply.

24.8.4 Charger protection. Charger protection requires the battery charger to be self-protected and the output current to be limited to a safe value under cranking loads. A circuit breaker or fuse is needed to protect the ac primary circuit. Charger diodes need ferroresonant or other protection from line and load transients.

24.8.5 Isolation. Isolation of ac input and dc output eliminates the possibility of fuses blowing on grounded equipment.

24.8.6 Output indication. Visual indication of both charge rate and system voltage should be provided by a dc ammeter and voltmeter. Operation at equalize voltage should be indicated by a light.

24.8.7 Interference. The charger should not produce RFI or EMI noise that would affect nearby electronic or communication equipment.

24.9 OPTIONAL CHARGER ACCESSORIES

Typical optional accessories are:

a. 0–24 hour equalize timer

b. Automatic equalize timer

c. Low dc voltage alarm relay

d. High dc voltage alarm relay

e. Current failure relay

f. AC power failure relay

g. Ground fault relay

24.9.1 0-24 hour equalize timer. The 0–24 hour equalize timer replaces the float equalize switch. The time is set to the number of hours of equalize time desired. The timer eliminates the need to manually return the charger to the float setting.

24.9.2 Automatic equalize timer. The automatic equalize timer replaces the standard float-equalize switch. When power is applied to the unit, the charger will charge at the high rate. The timer times out to automatically return the charger to the float rate. If an ac power failure occurs at the charger's ac input for over five seconds, the timer will automatically return to the high rate when ac power is returned. It will also automatically return to the high rate when the generator set is cranked. If there is no ac power failure or the engine generator set is not tested, there will be no equalizing.

24.9.3 Low voltage alarm. The low voltage alarm can be used to indicate the system is on batteries or to protect the battery from potential discharge damage in combination with a disconnect device. The relay is adjustable from float voltage to 1.75 volts per cell for lead-acid or 1.0 for nickel-cadmium.

24.9.4 High voltage alarm. The high voltage alarm can be used to indicate the system is at equalize voltage or can be used to shut down a system. The relay is adjustable from equalize voltage to 2.6 volts per cell.

24.9.5 Current failure relay. The current failure relay is used to indicate loss of charger output. This fulfills the National Electrical Code, Article 700-7(c), requirement for indication of charger failure to charge.

24.9.6 AC input power failure relay. The ac power failure relay activates on loss of the ac input.

24.9.7 Ground fault relay. The ground fault relay protects against electrical shock.

24.9.8 Diagnostic lights. A green light for AC-ON, a yellow light for equalizer, a red light for LVDC (low voltage dc), and a red light for HVDC (high voltage dc) can greatly help operators diagnose battery problems.

24.10 MAINTENANCE

As stated in the introduction, periodic maintenance is necessary for any trouble free battery installation. Any effective maintenance program requires a schedule for each item and a written record of results, including figures such as volts per cell or specific gravity.

24.10.1 Water additions. Except for the valve regulated type, batteries will use water. Battery charging current, especially overcharging, acts to decompose water into hydrogen and oxygen. The most important maintenance function is periodic water replacement. Replacement water should be free of minerals. Under no circumstances should the electrolyte level be allowed to drop below the top of the separator in a cell. Not only would this reduce the battery's discharge capacity, but the battery plates could suffer permanent damage. Any excessive water usage by any or all cells should be noted and the cause determined.

24.10.2 Connections. Battery connections tend to lose tightening torque. Terminals should be checked frequently for heating under load and should be torque checked once a year. Any discoloration or other evidence of heating should be noted and investigated.

24.10.3 Cleaning. Batteries must be kept clean. Dirt can cause leakage across or between cells. With clean batteries, evidence of electrolyte leakage, overcharging, gassing, and loose terminals is quickly noticeable.

24.10.4 Cell balance. At least once per year cell voltage or specific gravity should be checked for equality. Cells that fail to charge equally or use excessive water are a sign of trouble.

24.10.5 Capacity. At least once per year, more often as a battery nears the end of its expected life, the capacity under load should be checked. Replacement of weak batteries avoids the inevitable time when a battery fails to crank the engine.

24.11 OTHER CONSIDERATIONS

24.11.1 Equipment operating voltage. With the loss of ac power, the battery voltage declines during discharge. Thus, equipment powered by batteries must be designed to operate over a fairly broad range of voltage supply. Although starting motors are usually designed to operate at voltages as low as 50% of nominal, other auxiliary equipment, such as governors, fuel valves, and lights may not operate satisfactorily on this low voltage. In that case, a separate battery will be required for the auxiliary equipment.

24.11.2 Cable resistance. Batteries must be located as close to the starter motor as possible. The total resistance of the positive and negative cables should not exceed 0.0015 ohms for 12 volt, 0.002 for 24 volt, and 0.0025 for 32 volt dc systems.

24.11.3 Point of charger supply. For most applications, the charger should be supplied from the normal supply. This assures continuation of charging if the load circuit is left unenergized.

24.11.4 Codes and standards. System designs must meet industry codes and standards including The National Electrical Code 1993 and NFPA 110 1993, Emergency and Standby Systems.

24.11.5 Cold temperatures. Batteries have a negative temperature coefficient for voltage, -2.5 mV per cell lead acid and -3.0 mV per cell nickel cadmium per degree Celsius. Batteries in a cold environment need to have heaters. All valve-regulated batteries require a temperature-compensated battery charger for optimum battery life. A temperature sensor is attached to the battery case and is connected to the sensing circuit of the battery charger. With increasing battery temperature, the charging voltage is decreased.

BIBLIOGRAPHY

[1] Asco Facts, Vol. 3, nos. 6 and 7, Batteries and Battery Chargers Used with Engine-Generator Sets, Automatic Switch Company.

[2] EGSA Standard 100B 1977, Performance Standard for Use With Engine Generating Sets, Electrical Generating Systems Association.

[3] EGSA Standard 100C 1980, Performance Standard, Battery Chargers for Engine Starting and Control Batteries, Electrical Generating Systems Association.

[4] Henderson, Warren. Battery and Charger Fundamentals, Presentation to EGSA Chicago Conference, June 1989, Electrical Generating Systems Association.

[5] Migliaro, Marco. Maintaining Maintenance Free Batteries, Institute of Electrical and Electronic Engineers, Industrial and Commercial Power Systems Conference Paper, May 1989.

[6] Preventive Maintenance for Batteries, Diesel Progress North American, March 1986.

[7] Ripley, Jack. Batteries for Gen Set Cranking, Presentation to Power Fair '89, May 1989, Electrical Generating Systems Association.

Appendix

INTRODUCTION

This appendix is merely a convenience to those who use this book. The items printed here are not original but are useful material that members of the EGSA Education Committee submitted. These are items strictly pertaining to on-site power generation that to our knowledge have not been assembled in a convenient place. A reference book of this type is an appropriate place to find this information.

The major emphasis is upon conversion factors. The international nature of EGSA makes us aware of the need for easy conversion from one system to another. The omnibus trade bill of 1988 states that all United States federal agencies must convert to the SI system by 1992. Our readers may welcome the SI conversion factors printed herein. The SI system stands for the international system of measurement used by all international standards organizations and almost all countries worldwide. The United States is the only significant exception. The SI system is based upon the meter, kilogram, second, ampere, and kelvin. It is incorrect to refer to the SI system as the metric system. Such metric system units as the calorie and the dyne are not a part of the SI system. Even the Celsius scale of temperature measurement is derived from the basic kelvin. As Jim Wright pointed out in Chapter 2, the kelvin is a unit, not a degree kelvin as in a degree Celsius. It is simply a unit as ampere is a unit.

This appendix also includes tables of factual material such as the properties of fuels and a few useful nomograms for estimating quantities that may be difficult to calculate. An interesting section that will no doubt be popular is Rules of Thumb. We hope that readers will not replace accurate calculation methods with any of these rules of thumb. Rules of thumb give "ballpark estimates" but should not be used for design work.

A1. SI CONVERSIONS

Using Table A-1 multiply English units in the first column by the factor in the third column to obtain SI units of the second column.

Table A-1. Conversions to SI

To convert from	to	multiply by
ampere-hour	coulomb (C)	3600
Btu	joule (J)	1055
degree Celsius	kelvin (K)	tK = degree C + 273.15
degree Fahrenheit	degree Celsius	degree C = (degree F − 32)/ 1.8
degree Fahrenheit	kelvin (K)	tK = (degree F + 459.67)/1.8
footcandle	lux (lx)	1.076
gallon (US liquid)	litre	3.785
horsepower	watt (W)	746
inch	meter (m)	.0254
inch of Hg	pascal (Pa)	3376.85
inch of water	pascal (Pa)	248.84
kilowatt hour	joule (J)	3 600 000
pound	kilogram (kg)	.3732

A2. POWER CONVERSION FACTORS

Table A-2 is a convenient chart for converting between metric horsepower, horsepower, kilowatts, Btu per minute, and foot pounds per second. Multiply the quantity in the left hand column by the appropriate factor to obtain the quantity in the upper horizontal column.

A3. PRESSURE CONVERSION CHART

Using Table A-3, convert the quantities in the vertical left hand column to those in the top horizontal column by multiplying by the factor shown. This chart can be used for all pressures, including atmospheric.

Table A-2. Approximate Power Conversion Factors

From	Into				
	C.V., P.S., Metric HP	Horsepower (HP)	Kilowatt (kW)	$\frac{BTU}{Min}$	$\frac{FT\text{-}LBS}{Sec.}$
	Multiply By				
C.V., P.S., Metric HP	1.000	.9863	.7355	41.8377	542.5
Horsepower (HP)	1.0139	1.000	.7457	42.418	550.0
Kilowatt (kW)	1.3597	1.341	1.000	56.896	737.6
$\frac{BTU}{Min}$	.0239	.0236	.0176	1.000	12.966
$\frac{FT\text{-}LBS}{Sec.}$	.001843	.001818	.001356	.07712	1.000

Table A-3. Pressure Conversion Chart

From		= psi	= in H_2O	= in Hg	= bar	= mm Hg	= cm H_2O	= kgm/cm^2	= kPa	= atm
					Multiply By					
psi	X		27.68	2.04	6.89×10^{-2}	51.7	70.03	7.03×10^{-2}	6.90	6.80×10^{-2}
in H_2O	X	3.61×10^{-2}		7.36×10^{-2}	2.49×10^{-3}	1.87	2.54	2.54×10^{-3}	0.249	2.46×10^{-3}
in Hg	X	0.491	13.6		3.39×10^{-2}	25.4	34.5	3.45×10^{-2}	3.39	3.34×10^{-2}
bar	X	14.5	401.4	29.53		750	1020	1.02	100	0.987
mm Hg	X	1.934×10^{-2}	0.535	3.937×10^{-2}	1.33×10^{-3}		1.36	1.36×10^{-3}	0.133	1.316×10^{-3}
cm H_2O	X	1.42×10^{-2}	0.394	2.89×10^{-2}	9.81×10^{-4}	0.736		1.0×10^{-3}	9.81×10^{-2}	9.7×10^{-4}
kgm/cm^2	X	14.22	393.7	28.95	0.981	735.3	1000		98.1	0.968
kPa	X	0.145	4.01	0.295	0.0100	7.5	10.20	10.2×10^{-2}		9.87×10^{-3}
atmosphere	X	14.70	406.8	29.92	1.013	760	1033	1.033	101.3	

1 torr = 1 mm Hg   1 k N/m^2 = 1 mm Hg

A4. BTU ENERGY EQUIVALENTS

Refer to Table A-4.

Table A-4. Energy Equivalents

	Btu Equivalent
Kilowatt hours	3413
Ton of coal	25,000,000
Barrel of crude oil	5,800,000
Gallon of gasoline	125,000
Gallon of No. 2 fuel oil	140,000
Gallon of propane (vaporized)	91,500
Cu. ft. of propane	2,520
Cu. ft. of natural gas	1,031
Therm	100,000
Cord of wood	20,000,000

A5. BARREL OF OIL EQUIVALENT ENERGY

Refer to Table A-5.

Table A-5. Barrel of Oil Equivalent Energy

1 Barrel of oil = 5,800,000 Btu
5.604 cu. ft. natural gas
0.22 tons of bituminous coal
1700 kWh of electricity

A6. LIQUID MEASURE EQUIVALENT VOLUMES

Refer to Table A-6.

Table A-6. Liquid Measure Equivalent Volumes

Barrel	42 gallons
Gallon	231 cu. in.
Cubic foot	7.48 gallons
Cubic meter	35.3 cu. ft.
Metric ton	7.35 barrels

A7. APPROXIMATE WEIGHTS OF VARIOUS LIQUIDS

Refer to Table A-7.

Table A-7. Weights of Liquids

	Pounds per U.S. Gallon	Specific Gravity
Diesel fuel	6.88 to 7.46	.825 to .895
Ethylene glycol	9.3 to 9.8	1.12 to 1.15
Furnace oil	6.7 to 7.9	.80 to .95
Gasoline	5.8 to 6.3	.67 to .75
Kerosene	6.25 to 7.1	.90 to .92
Lubricating oil	7.5 to 7.7	.90 to .92
Water	8.34	1.00

A8. INTERNATIONAL STANDARD NUMBER PREFIXES

Refer to Table A-8. Note there is no billion or trillion. The capital M stands for millions, not for one thousand as in Roman numerals. A million watts is MW and a thousand watts is kW (lower case k). A space is left between each three numerals without a comma. Even four digit numbers are written with the space; for example, 1 000.

Table A-8. International Standard (SI) Numerical Prefixes

Multiplication factors	Prefix	Symbol
1 000 000 000 000 = 10^{12} (exponent)	tera	T
1 000 000 000 = 10^{9}	giga	G
1 000 000 = 10^{6}	mega	M
1 000 = 10^{3}	kilo	k
100 = 10^{2}	hecto*	h
10 = 10^{1}	deca*	da
0.1 = 10^{-1}	deci*	d
0.01 = 10^{-2}	centi*	c
0.001 = 10^{-3}	milli	m
0.000 001 = 10^{-6}	micro	μ
0.000 000 001 = 10^{-9}	nano	n
0.000 000 000 001 = 10^{-12}	pico	p
0.000 000 000 000 001 = 10^{-15}	femto	f
0.000 000 000 000 000 001 = 10^{-18}	atto	a

*To be avoided where possible. For example, use 100 millimeters instead of 10 centimeters.

A9. PROPANE BULK TANK TABLE

One of the major problems in using propane (boiling point $-44°F$) is obtaining sufficient vapor at cold temperatures. Table A-9 gives vaporization data over a range of temperatures and withdrawal rates. The table assumes that the tank at least half full.

A10. USEFUL EQUATIONS

A10.1 Brake mean effective pressure (BMEP).
BMEP cannot be measured and is not a true measure of an engine's capability. If it were, engines would be getting worse rather than better. It still appears in specifications.

4 cycle engine

$$\text{BMEP} = \frac{\text{BHP} \times 792\,000}{\text{engine displacement} \times \text{RPM}} \quad \text{(Equation A-1)}$$

2 cycle engine

$$\text{BMEP} = \frac{\text{BHP} \times 396\,000}{\text{engine displacement} \times \text{RPM}} \quad \text{(Equation A-2)}$$

Where:

Engine displacement is in cubic inches

$$\text{BHP} = \frac{\text{kW}}{0.746 \times \text{generator efficiency}} + \text{fan horsepower}$$

A10.2 Torque.

$$\text{Torque in pound feet} = 5250\,\frac{\text{HP}}{\text{RPM}} \quad \text{(Equation A-3)}$$

A10.3 Peak of sine wave.

Peak of sine wave = 1.414 rms value (Equation A-4)

A10.4 Rectification formulae.
Equations to determine the ac voltage input for dc voltage output into resistance load:

Half wave

    Single phase Vac = 2.22 Vdc (Equation A-5)

    Three phase Vac = .6757 Vdc (Equation A-6)

Full wave bridge

    Single phase Vac = 1.11 Vdc (Equation A-7)

    Three phase Vac = 0.74 Vdc (Equation A-8)

A11. EXHAUST BACK PRESSURE NOMOGRAPHS

Figures A-1 and A-2 are nomographs for determining the back pressure in lengths of pipe. The total back pressure includes the pipe, flexible connection, all elbows, and the engine silencer. The nomographs are based upon the basic Equation A-9.

$$\text{Delta P} = \frac{1.8 \times 10^{-5} \times Q^2}{d^{5.18}} \quad \text{(Equation A-9)}$$

Table A-9. Propane Bulk Tank Table
(Tank kept at least 1/2 full)

	Temperatures						
	32°F	20°F	10°F	0°F	-10°F	-20°F	-30°F
Withdrawal Rate	Size Tank Required						
50 CFH-125,000 Btu/Hr.	115 Gal.	115 Gal.	115 Gal.	250 Gal.	250 Gal.	400 Gal.	600 Gal.
100 CFH-250,000 Btu/Hr.	250 Gal.	250 Gal.	250 Gal.	400 Gal.	500 Gal.	1000 Gal.	1500 Gal.
150 CFH-375,000 Btu/Hr.	300 Gal.	400 Gal.	500 Gal.	500 Gal.	1000 Gal.	1500 Gal.	2500 Gal.
200 CFH-500,000 Btu/Hr.	400 Gal.	500 Gal.	750 Gal.	1000 Gal.	1200 Gal.	2000 Gal.	3500 Gal.
300 CFH-750,000 Btu/Hr.	750 Gal.	1000 Gal.	1500 Gal.	2000 Gal.	2500 Gal.	4000 Gal.	5000 Gal.

Notes:
1. Btu/Gal. vaporized   91,500
2. Btu/Pound vaporized   21,560
3. Btu/Cu. Ft. of vapor   2,520

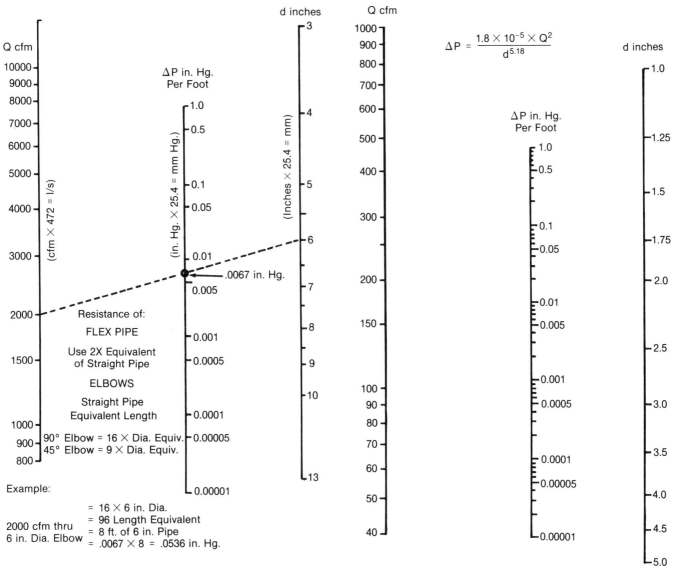

Figure A-1. Back Pressure Nomograph

Figure A-2. Back Pressure Nomograph

A12. RULES OF THUMB

Everyone loves those short cuts! Rules of thumb are good for a quick estimate or a check to see if a multiplier or decimal place may have been missed. Rules of thumb are strictly short cuts. They ignore correction factors, round off significant figures, and pay no attention to individual machine performance. Using them for engineering design calculations must be avoided.

A12.1 Engine horsepower.

Engine net horsepower = $1.5 \times$ kW of generator

Assumes approximately 0.90 generator efficiency

A12.2 Motor starting.

Gen-set maximum across-the-line motor start = 0.5 HP per kW

Ignores differences in generators, motor code letters and types of load.

A12.3 Total motor load.

Total motor load a generator can power = $1.14 \times$ kW

Generator kW to power motor load =

$0.877 \times$ total motor HP

Assumes approximately 0.85 motor efficiency and 0.8 power factor or better.

A12.4 Engine derating (naturally aspirated engines only).

Altitude derating = 4% per 1000 ft altitude above 1000 ft.

Temperature derating = 1% per 10°F above 70°F.

A12.5 Generator heat rejection to cooling air.

Generator rejected heat = $340 \times$ kW = Btu/hr

Assumes approximately 0.90 efficiency. Subject to large error for different efficiencies. If the generator efficiency is known, a better approximation is the following.

Generator rejected heat =

$3413 \times [(1/\text{eff.})-1] \times$ kW = Btu/hr

or

Generator rejected heat =

$$\frac{3413 \times [(1/\text{eff.})-1] \times \text{kW}}{60} = \text{Btu/min.}$$

A12.6 Engine heat radiated to cooling air, liquid cooled engines.

Dry exhaust manifold.

Engine radiated heat = 0.10 of fuel consumption in Btu.

Fuel consumption in Btu/min =

$$\frac{\text{Btu content of fuel}}{\text{Fuel consumption/hr/60}}$$

See section A4 for approximate fuel Btu content. This calculation gives only the heat radiated to the cooling air. The heat rejected to the coolant is a much larger figure.

A12.7 Engine energy distribution at rated load.

Total Btu consumed =

1/3 to exhaust + 1/3 to coolant + 1/3 useful work

A12.8 Elevator feedback.

Regenerative power for full load down =

0.4 power full load up

Regenerative power for braking full load down =

0.5 power full load up

A12.9 Sound attenuation.
Noise level reduces 6 dBA each time distance from the generator set is doubled.

A13. EFFECT OF VOLTAGE AND FREQUENCY VARIATION ON INDUCTION MOTORS

Many smaller generator sets have frequency variations much greater than generally found on utility lines. Some may also have voltage variations up to 10% of rated. In some cases, motors are operated voltages lower or higher than their rating. Figures A-3 and A-4 give a picture of what happens to motor performance with voltage and frequency variations from the motor rating. The charts are general and cannot be guaranteed.

In general, motors will operate successfully under the following conditions. They will not necessarily meet the standards established for normal rating.

a. Where the variation of voltage does not exceed 10% above or below normal.

b. Where the frequency variation does not exceed 5% above or below normal.

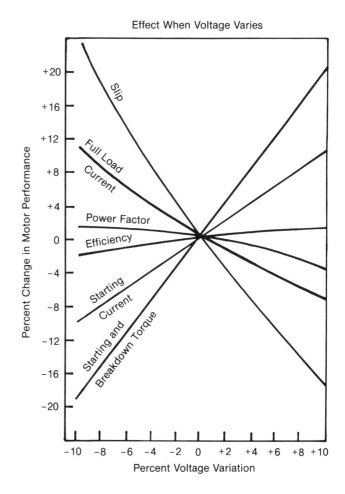

Figure A-3. General Effect of Voltage Variation on Induction Motor Characteristics

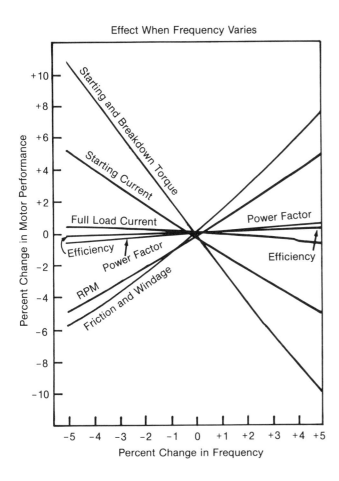

Figure A-4. General Effect of Frequency Variation on Induction Motor Characteristics

Table A-10. Multipliers to Convert Three Phase kVA to Amperes and Amperes to Three Phase kVA

Voltage	Multiplier		Voltage
	3 Phase kVA to Amperes	Amperes to 3 Phase kVA	
190	3.039	.3291	190
200	2.887	.3464	200
208	2.776	.3603	208
220	2.624	.3810	220
225	2.566	.3897	225
230	2.510	.3984	230
240	2.406	.4157	240
380	1.519	.6582	380
400	1.443	.6928	400
416	1.388	.7205	416
440	1.312	.7621	440
450	1.283	.7794	450
460	1.255	.7967	460
480	1.203	.8314	480
575	1.004	.9957	575
600	.9623	1.039	600
2000	.2887	3.464	2000
2100	.2750	3.637	2100
2400	.2406	4.157	2400
3300	.1750	5.716	3300
4160	.1388	7.205	4160

Example 1.
A three phase induction motor has a nameplated full load running current of 40 amperes, and a nameplate voltage of 208 volts. What is the full load running kVA of this motor?

The multiplier to convert amperes to kVA @ 208 volts is .3603. 40 amperes × .3603 = 14.41 running kVA.

Example 2.
A three phase generator is rated at 125 kVA at 480 volts ac. What is the full load running current of this generator?

The multiplier to convert kVA to amperes @ 208 volts is 2.776. 125 kVA × 2.776 = 347.0 full load amperes.

c. Where both voltage and frequency vary from normal, both quantities low or high are compensating factors, within limits. If one is low and the other high, however, the total variation of voltage and frequency should not exceed 10%, with frequency not more than 5%.

A14. AMPERE CHARTS

A14.1 kVA. Table A-10 gives multiplication factors for three-phase amperes to kVA and kVA to amperes.

A14.2 kW at 0.80 power factor. Table A-11 gives ampere ratings for three-phase 0.80 power factor generator sets.

A15. MOTOR RATINGS

Table A-12 gives typical running kW, running kVA and, starting kVA for NEMA design B, C, and D motors. Starting power factors vary from approximately 0.5 for small motors to 0.3 for 250 HP motors. Power factors increase as motors accelerate.

A.16 POWER FACTOR IMPROVEMENT

Table A-13 gives kVAR of capacitors per kW required to improve various power factors lagging. The preferred location for power factor correction capacitors is in a position where they will be switched with the load being corrected.

Table A-11. Generator Set Ampere Ratings — Three Phase 0.80 P.F.

KW	kVA	208V	216V	230V	240V	416V	460V	480V	600V	2400V	4160V
50	62.5	174	167	157	150	87	78	75	61	—	—
60	75	208	201	188	180	104	94	90	72	—	—
75	93.8	260	251	235	226	130	118	113	90	—	—
100	125	347	334	314	301	174	157	150	120	—	—
125	156	433	417	392	376	217	196	188	150	—	—
150	187	519	500	469	450	260	235	225	180	—	—
175	219	608	585	550	527	304	275	263	211	—	—
200	250	694	668	628	601	347	314	301	241	—	—
225	281.3	781	752	707	677	390	353	339	272	—	—
250	312	866	834	783	751	433	392	375	300	—	—
300	375	1040	1002	941	902	521	471	451	361	90	52
350	438	1216	1171	1100	1054	608	550	527	422	105	61
400	500	1388	1337	1255	1203	694	628	601	481	120	69
450	562.5	1562	1503	1413	1355	781	707	677	545	136	78
500	625	1735	1671	1569	1504	867	785	752	601	150	88
600	750	2082	2005	1883	1804	1041	941	902	722	180	104
700	875	2429	2339	2197	2105	1214	1098	1053	842	211	121
800	1000	2775	2673	2510	2406	1388	1255	1203	962	241	139
900	1125	3123	3007	2824	2706	1561	1412	1353	1083	271	156
1000	1250	3470	3341	3138	3007	1735	1569	1504	1203	301	174
1100	1375	3817	3675	3452	3308	1908	1726	1654	1323	331	191
1200	1500	4163	4009	3765	3609	2082	1883	1804	1443	361	208
1300	1625	4511	4343	4079	3909	2255	2040	1955	1564	391	225
1400	1750	4858	4678	4393	4210	2429	2196	2105	1684	421	243
1500	1875	5204	5012	4707	4511	2602	2353	2255	1804	451	260

Table A-12. Typical Three Phase Motor Characteristics for NEMA Design B, C and D Motors

Typical 3 Phase Motor — Characteristics					NEMA Design B, C and D Motors [1]				
HP	RPM	Running kW	Running kVA	kVA Start	HP	RPM	Running kW	Running kVA	kVA Start
1	3600	1.05	1.3	13	40	3600	33.3	37	221
	1800	1.06	1.4	12		1800	35.2	39	220
	1200	1.02	1.5	12		1200	34.5	39.2	220
						900	34.6	41.3	216
2	3600	1.9	2.2	19					
	1800	1.9	2.3	13	50	3600	43.5	48	276
	1200	2.0	2.7	18		1800	43.5	48	275
						1200	43.2	48	274
3	3600	2.9	3.2	25		900	42.0	49.9	272
	1800	2.8	3.4	24					
	1200	2.8	3.7	24	60	3600	49.5	55	336
						1800	51.5	57	331
5	3600	4.6	5.2	35		1200	51.1	58	330
	1800	4.6	5.4	34		900	51.3	61.2	328
	1200	4.5	5.8	34					
	900	4.5	6.3	33	75	3600	64	71	419
						1800	63	70	416
7-1/2	3600	6.7	7.5	48		1200	63	71.5	417
	1800	6.9	7.9	46		900	66	77	414
	1200	6.8	8.2	45					
	900	7.0	9.3	44	100	3600	85	94	552
						1800	84	93	556
10	3600	8.8	9.8	62		1200	84.5	96	555
	1800	8.8	10.1	60		900	86	99	552
	1200	8.7	10.5	58					
	900	9.4	12.3	56	125	3600	108	119	698
						1800	106	117	696
15	3600	13.2	14.7	88		1200	109	123	695
	1800	13.0	14.7	84		900	113	127	690
	1200	12.9	15.2	82					
	900	13.7	17.4	81	150	3600	127	139	836
						1800	125	136	830
20	3600	16.7	18.6	112		1200	131	148	828
	1800	17.2	19.4	112		900	136	153	820
	1200	17.4	20.3	112					
	900	17.4	21.6	110	200	3600	167	183	1110
						1800	164	180	1105
25	3600	20.5	22.8	139		1200	168	195	1100
	1800	21.6	24.3	138		900	178	201	1060
	1200	22.0	25.5	138					
	900	22.0	26.2	136	250	3600	204	224	1380
						1800	200	220	1370
30	3600	25.2	28	166		1200	205	232	1360
	1800	25.5	28.6	165		900	213	239	1345
	1200	25.0	28.6	165					
	900	25.5	31.1	161					

[1] Design A motor may have starting kVA values that are as much as 50% higher. Many 3600 rpm motors are designed A.

Table A-13. kW Multipliers for Determining Capacitor Kilovars

Desired Power Factor in Percentage

Original Power Factor in Percentage		80	81	82	83	84	85	86	87	88	89	90	91	92	93	94	95	96	97
	50	0.982	1.008	1.034	1.060	1.086	1.112	1.139	1.165	1.192	1.220	1.248	1.276	1.306	1.337	1.369	1.403	1.440	1.481
	51	0.937	0.962	0.989	1.015	1.041	1.067	1.094	1.120	1.147	1.175	1.203	1.231	1.261	1.292	1.324	1.358	1.395	1.436
	52	0.893	0.919	0.945	0.971	0.997	1.023	1.050	1.076	1.103	1.131	1.159	1.187	1.217	1.248	1.280	1.314	1.351	1.392
	53	0.850	0.876	0.902	0.928	0.954	0.980	1.007	1.033	1.060	1.088	1.116	1.144	1.174	1.205	1.237	1.271	1.308	1.349
	54	0.809	0.835	0.861	0.887	0.913	0.939	0.966	0.992	1.019	1.047	1.075	1.103	1.133	1.164	1.196	1.230	1.267	1.308
	55	0.769	0.795	0.821	0.847	0.873	0.899	0.926	0.952	0.979	1.007	1.035	1.063	1.093	1.124	1.156	1.190	1.227	1.268
	56	0.730	0.756	0.782	0.808	0.834	0.860	0.887	0.913	0.940	0.968	0.996	1.024	1.054	1.085	1.117	1.151	1.188	1.229
	57	0.692	0.718	0.744	0.770	0.796	0.822	0.849	0.875	0.902	0.930	0.958	0.986	1.016	1.047	1.079	1.113	1.150	1.191
	58	0.655	0.681	0.707	0.733	0.759	0.785	0.812	0.838	0.865	0.893	0.921	0.949	0.979	1.010	1.042	1.076	1.113	1.154
	59	0.619	0.645	0.671	0.697	0.723	0.749	0.776	0.802	0.829	0.857	0.885	0.913	0.943	0.974	1.006	1.040	1.077	1.118
	60	0.583	0.609	0.635	0.661	0.687	0.713	0.740	0.766	0.793	0.821	0.849	0.877	0.907	0.938	0.970	1.004	1.041	1.082
	61	0.549	0.575	0.601	0.627	0.653	0.679	0.706	0.732	0.759	0.787	0.815	0.843	0.873	0.904	0.936	0.970	1.007	1.048
	62	0.516	0.542	0.568	0.594	0.620	0.646	0.673	0.699	0.725	0.754	0.782	0.810	0.840	0.871	0.903	0.937	0.974	1.015
	63	0.483	0.509	0.535	0.561	0.587	0.613	0.640	0.666	0.693	0.721	0.749	0.777	0.807	0.838	0.870	0.904	0.941	0.982
	64	0.451	0.474	0.503	0.529	0.555	0.581	0.608	0.634	0.661	0.689	0.717	0.745	0.775	0.806	0.838	0.872	0.909	0.950
	65	0.419	0.445	0.471	0.497	0.523	0.549	0.576	0.602	0.629	0.657	0.685	0.713	0.743	0.774	0.806	0.840	0.877	0.918
	66	0.388	0.414	0.440	0.466	0.492	0.518	0.545	0.571	0.598	0.626	0.654	0.682	0.712	0.743	0.775	0.809	0.846	0.887
	67	0.358	0.384	0.410	0.436	0.462	0.488	0.515	0.541	0.568	0.596	0.624	0.652	0.682	0.713	0.745	0.779	0.816	0.857
	68	0.328	0.354	0.380	0.406	0.432	0.458	0.485	0.511	0.538	0.566	0.594	0.622	0.652	0.683	0.715	0.749	0.786	0.827
	69	0.299	0.325	0.351	0.377	0.403	0.429	0.456	0.482	0.509	0.537	0.565	0.593	0.623	0.654	0.686	0.720	0.757	0.798
	70	0.270	0.296	0.322	0.348	0.374	0.400	0.427	0.453	0.480	0.508	0.536	0.564	0.594	0.625	0.657	0.691	0.728	0.769
	71	0.242	0.268	0.294	0.320	0.346	0.372	0.399	0.425	0.452	0.480	0.508	0.536	0.566	0.597	0.629	0.663	0.700	0.741
	72	0.214	0.240	0.266	0.292	0.318	0.344	0.371	0.397	0.424	0.452	0.480	0.508	0.538	0.569	0.601	0.635	0.672	0.713
	73	0.186	0.212	0.238	0.264	0.290	0.316	0.343	0.369	0.396	0.424	0.452	0.480	0.510	0.541	0.573	0.607	0.644	0.685
	74	0.159	0.185	0.211	0.237	0.263	0.289	0.316	0.342	0.369	0.397	0.425	0.453	0.483	0.514	0.546	0.580	0.617	0.658
	75	0.132	0.158	0.184	0.210	0.236	0.262	0.289	0.315	0.342	0.370	0.398	0.426	0.456	0.487	0.519	0.553	0.590	0.631
	76	0.105	0.131	0.157	0.183	0.209	0.235	0.262	0.288	0.315	0.343	0.371	0.399	0.429	0.460	0.492	0.526	0.563	0.604
	77	0.079	0.105	0.131	0.157	0.183	0.209	0.236	0.262	0.289	0.317	0.345	0.373	0.403	0.434	0.466	0.500	0.537	0.578
	78	0.052	0.078	0.104	0.130	0.156	0.182	0.209	0.235	0.262	0.290	0.318	0.346	0.376	0.407	0.439	0.473	0.510	0.551
	79	0.026	0.052	0.078	0.104	0.130	0.156	0.183	0.209	0.236	0.264	0.292	0.320	0.350	0.381	0.413	0.447	0.484	0.525
	80	0.000	0.026	0.052	0.078	0.104	0.130	0.157	0.183	0.210	0.238	0.266	0.294	0.324	0.355	0.387	0.421	0.458	0.499
	81		0.000	0.026	0.052	0.078	0.104	0.131	0.157	0.184	0.212	0.240	0.268	0.298	0.329	0.361	0.395	0.432	0.473
	82			0.000	0.026	0.052	0.078	0.105	0.131	0.158	0.186	0.214	0.242	0.272	0.303	0.335	0.369	0.406	0.447
	83				0.000	0.026	0.052	0.079	0.105	0.132	0.160	0.188	0.216	0.246	0.277	0.309	0.343	0.380	0.421
	84					0.000	0.026	0.053	0.079	0.106	0.134	0.162	0.190	0.220	0.251	0.283	0.317	0.354	0.395
	85						0.000	0.027	0.053	0.080	0.108	0.136	0.164	0.194	0.225	0.257	0.291	0.328	0.369
	86							0.000	0.026	0.053	0.081	0.109	0.137	0.167	0.198	0.230	0.264	0.301	0.342
	87								0.000	0.027	0.055	0.083	0.111	0.141	0.172	0.204	0.238	0.275	0.316
	88									0.000	0.028	0.056	0.084	0.114	0.145	0.177	0.211	0.248	0.289
	89										0.000	0.028	0.056	0.086	0.117	0.149	0.183	0.220	0.261
	90											0.000	0.028	0.058	0.089	0.121	0.155	0.192	0.233

Example: Total kW input of load from wattmeter reading 100 kW at a power factor of 70%. The leading reactive kVAR necessary to raise the power factor to 95% is found by multiplying the 100 kW by the factor found in the table, which is .691. Then 100 kW × .691 equals 69.1 kVAR. Use 70 kVAR.

Acknowledgments

EGSA could not have produced this book without the efforts contributed by member companies. Those who furnished materials for illustrations and provided time for the authors are:

Allison Gas Turbine Div., General Motors Corp.
Automatic Switch Co.
Basler Electric Co.
Charles Industries, Ltd.
Detroit Diesel Corp.
Hercules Engines, Inc.
Kato Engineering Co.
Korfund Dynamics Co.
Lister-Petter Inc.
Kohler Co.
Marathon Electric Mfg. Co.
Modine Mfg. Co.
Frank W. Murphy Mfgr.
Nelson Industries, Inc.
Power & Energy International, Inc.
Prichard-Brown Div., C. King, Inc.
Pryco, Inc.
Selkirk Metalbestos Div., Household International
Waukesha Engine Div., Dresser Industries, Inc.
Westinghouse Electric Corp.
Woodward Governor Co.

Index

AC voltage generation, 2.1.5
Adjustable speed motor drives, 2.8.3
Air gap definition, 2.1.1
Air intake cleaner dry type, 8.4
Air starting, 11.9
Air-cooled diesel engine
   advantages, 11.1
   air intake cleaner, 11.7
   aluminum vs. cast iron, 11.3
   disadvantages, 11.2
Air-cooled spark ignited engine
   advantages, 9.1
   aluminum vs. cast iron, 9.4
   cylinder configurations, 9.10
   disadvantages, 9.2
   horizontal vs. vertical shaft, 9.9
   two vs. four-stroke cycle, 9.4
Alternating current (ac), 1.3
Alternator, synchronous generator
   altitude, 2.5.2 (f)
   armature, 2.2
   delta connected, 2.3.3
   effect of environment, 2.5.3
   electrical loads, 2.6.2
   four lead delta connected, 2.3.3 (a)
   four lead, wye connected, 2.3.2 (e)
   lead identification, 2.2.7
   loading, 2.6
   load bank testing, 2.6.4
   permanent magnet, 2.4.2 (c)
   single phase, 2.3.1
   sizing for motor starting, 2.7.2
   temperature, 2.5
   ten lead, wye connected alternator, 2.3.2 (f)
   three phase, 2.3.2
   transient voltage dip, 2.7.2 (b)
   twelve lead, 2.3.4
   wye connected, 2.3.2
Altitude considerations
   alternator, 2.5.2 (f)
Ambient noise, 23.2.7
Amortisseur bars definition, 2.1.1
Amortisseur windings, 2.8.6
Ampere definition, 1.1

Apparent power, 1.8.2 (a)
Arc extinguishing, 5.2.3, 6.5.2, 6.5.6, 6.11.7
Armature
   definition, 2.1.1
   winding, 2.2.2 (b), 2.2.3, 2.2.4, 2.2.5, 2.2.6
   alternator, 2.2
Available fault current, 6.5.4
Battery
   characteristics, 24.1
   charger requirements, 24.8
   charger selection, 24.6
   charger types, 24.7
   charging, 24.2
   equalize voltage, 24.2.2
   float voltage, 24.2.1
   maintenance, 24.10.2
   performance, 24.1
   selection, 24.4
Bearing bracket definition, 2.1.1
Bridge rectifier, 1.4.1 (b)
Brush
   definition, 2.1.1
   holder definition, 2.1.1
   type exciters, 2.1.4 (a)
Brushless exciters, 2.1.4 (c)
Btu energy equivalents, A4
Butane, 15.2
Bypass-isolation switch, 6.10.9
Capacitance, 1.7, 1.7.3
Capacitors, 1.7
Circuit definition, 1.1
Circuit breaker
   continuous current rating, 5.4.
   definition, 5.1
   Interrupting rating, 5.4.3
   low voltage power, 7.2.1 (a)
   mechanical interlock, 5.3.6
   medium voltage, 7.2.2
   molded case, 5.1, 5.2.1, 5.2.4 (f), 5.4.6
   motor operator, 5.3.5
   standards, 5.1.2
   with transfer switches, 6.9.5
Cogeneration, 7.3.4, 10.7, 12.2.5
Collector ring definition, 2.1.1

Combustion chamber design, 10.3
Commutator definition, 2.1.1
Conductor, 1.7.1
Continuous duty, 2.5.2 (d)
Continuous load, 6.5.3
Cooling system
    air-cooled diesel engine, 11.4
    air-cooled spark ignited engine, 9.5
    air-to-boil (ATB), 10.6.4
    engine mounted radiator, 16.2.3
    gas turbine, 12.6
    heat rejection to coolant, 10.6.6
    hot well/cold well system, 16.4
    liquid cooled diesel, 10.6
    quanity of coolant, 10.6.6
    radiator cooling air, 10.6.3
    radiator, 16.2
    remote radiator, 16.2.2
    spark ignited liquid cooled engine, 8.1
    water cooled systems, 16.1
Core (magnetic) definition, 2.1.1
Damper bars, 2.1.1
Deaeration, 16.8
Decibel, 23.2.3
Diesel engine
    combustion chamber design, 10.3
    compression ratio, 10.1
    four-stroke cycle, 10.2.1
    fuel injection system, 10.4
    open chamber system, 10.3.1
    performance, 10.5
    pre-combustion system, 10.3.2
    principle, 10.1
    selection of cooling equipment, 10.6
    two-stroke cycle, 10.2.2
Diffraction, sound, 23.7
Digester (sewage) gas fuel systems, 15.3
Diodes, 1.4.1
Direct current (dc), 1.2
Electric power, 1.8
Electrical current definition, 1.1
Electromagnetic field, 2.1.3
Electromagnetic induction, 1.6
Electromagnetism and electromagnets, 1.5.2
Electromagnetism, 4.1
Electromotive force (EMF) definition, 1.1
Electron definition, 1.1
Emergency power, 6.1
Emergency system maintenance, 6.11
Enclosure
    design considerations, 22.1
    environment considerations, 22.3
    materials, 22.2
    noise control, 22.4

    temperature control, 22.5
End bell definition, 2.1.1
Engine electrical system, 8.7, 11.9
Engine flange load and SAE flywheel designs, 21.8.2
Engine performance ratings, 10.5
Engine safety, 11.12
Excitation
    field forcing, 4.2
    induction generator, 3.1.6
    manual, 4.2, 4.3
    torsional, 21.4.1
Exciter
    definition, 2.1.1
Exciter
    armature definition, 2.1.1
    field definition, 2.1.1
    rotor definition, 2.1.1
    stator definition, 2.1.1
Exhaust gas heat exchangers, 16.6
Exhaust system
    air-cooled diesel engine, 11.8
    dry, 17.2
    piping, 17.1
    pressure tight pipe, 17.3
    spark ignited liquid cooled engine, 8.6
Faraday, 1.6
Fault current support, 4.9
Field coil definition, 2.1.1
Field pole definition, 2.1.1
Fire pump loads, 6.9.6
Foundation considerations, 21.8.6
Frame, alternator definition, 2.1.1
Frequency, 2.4.1
    compensation, 4.8
    regulation, 2.4.1
Front Engine Mounts, 21.8.3
Fuel injection
    distributor or rotary pump system, 10.4 (e)
    electronic unit injector, 10.4 (d)
    multiple plunger pump system, 10.4 (a)
    pressure-time system, 10.4 (b)
    unit injector system, 10.4 (c)
Fuel system
    air-cooled diesel, 11.6
    air-cooled spark ignited engine, 9.7
    carburetor, 8.3
    electrical control module, 14.3
    fuel transfer pump, 8.3
    pump head, 14.2.3
    pump lift, 14.2.2
    pump prime, 14.2.4
    pumping system, 14.2
    testing, 14.3.3
Full wave rectification, 1.4.1 (b)

Gas mixing, 15.1.5
Gas turbine
    advantages, 12.2
    air inlet and exhaust systems, 12.9
    ambient sensitivity, 12.3.4
    combustion, 12.1.2
    compressor, 12.1.1
    disadvantages, 12.3
    fuels and combustion, 12.8
    performance, 12.2.1
    power turbine, 12.1.3
    principle of operation, 12.1
    regenerative cycle, 12.10 (a)
    single shaft and split shaft, 12.4
    specific fuel consumption (SFC), 12.3.4
Gaseous fuel
    fuel treatment, 15.1.4
    pressure regulators, 15.1.1
Gen-set crankshaft evaluation criteria, 21.4.4
Governor
    air-cooled diesel, 11.10
    centrifugal ballhead, 13.1.6
    digital governing, 13.16
    droop-droop parallel operation, 13.14
    droop speed control, 13.12
    electrical vs. mechanical, 13.9
    hydraulic servo, 13.4
    hydraulic, 13.5
    hydromechanical governor basics, 13.3
    isochronous-droop parallel operation, 13.13
    isochronous load sharing, 13.15
    isochronous speed control, 13.11
    isochronous, 13.7
    load sharing, 13.10
    loop dynamics, 13.8.1
    mechanical governor basics, 13.2
    proportional type electric, 13.8
    servomotors, 13.4
    spark ignited liquid-cooled engine, 8.8
    speed droop, 13.6
    speed sensing techniques, 13.1
Ground fault protection, 6.8
Grounding
    delta connected alternator, 2.3.4 (b)
    wye connected alternator, 2.3.2 (b)
Half wave rectification, 1.4.1 (a)
Harmonic
    content, 2.8
    distortion, 2.8
HD-5 propane
    carburetors, 15.2.5
    filter, 15.2.1
    fuel system, 15.2
    fuel tank, 15.2.3

    vaporizer, 15.2.2
Horsepower
    equations, 2.6.2 (c) (3)
Hydraulic starting, 11.9
Ignition system
    air-cooled spark ignited engine, 9.8
    spark ignited liquid-cooled engine, 8.5
Imbalance, 21.7.3
Impedance, 1.7.4, 4.1
Inductance, 1.7, 1.7.2
Induction generator
    construction, 3.1.5
    excitation, 3.1.6
    frequency and voltage, 3.1.7
    performance, 3.1.4
    protection, 3.2.2
    pushover torque, 3.2.2
Induction motor loads, 2.6 (a)
Inductive reactance (XL), 1.7.2.d
Inrush current
    motor load, 6.5.1
    tungsten lamp load, 6.5.1
Insulating materials
    armature, 2.2.5
    classes of, 2.5.2
Insulator, insulation definition, 1.1
Interrupting current, 6.5.2
Inverse square law, 23.8
Kilovolt-amperes reactive kvar, 1.8.3 (a)
Laminated core definition, 2.1.1
Lateral vibration, 21.3
Lead identification, alternator, 2.2.7
Lubricating oil system
    air-cooled diesel engine, 11.5.1
    dry sump, 11.5.2
    extended run, 11.5.2
    full flow filter, 8.2
    gas turbine, 12.7
    pressure regulator valve, 8.2, 11.5.1
    pressure relief valve, 8.2
    spark ignited liquid-cooled engine, 8.2
    wet sump, 11.5.2 (4)
Magnetism, 1.5
Main bearing distress, 21.7.5
Mass law, sound transmission loss, 23.2.8
Mass-elastic shaft system, 21.2.1
Motor ratings, A15
Motor starting, 2.7, 4.1
    code letters, 2.7.1 (a)
    form, 2.7.2 (c)
    full voltage, 2.7.1 (c) (d)
    reduced voltage, 2.7.1 (b) (e)
Natural gas engine, 15.1
    carburetors, 15.1.2

Index

turbocharged, 15.1.1
Non-linear loads, 2.8
Octave bands, 18.2, 23.2.5
Oersted, 1.5.2
Ohm definition, 1.1
Ohm's Law, 1.7.1 (b), 1.7.4, 1.8.1, 4.1
Operational amplifier (op amp), 4.4
Overhang and web deflection, 21.7.4
Overlapping neutral transfer contacts, 6.8
Overload capability, alternator, 2.5.2 (e)
Paralleling, 4.10, 7.3.3, 7.4.3, 13.13, 13.14, 13.15
    import-export control, 7.3.4
    var/power controller, 7.3.4
    with utility, 7.3.4
Permanent magnet (pilot) exciter, 2.1.4 (d)
Permanent magnet field, 2.1.3
Permanent magnets, 1.5.1
Permeability, 1.5.2
Power, 1.8
    apparent, real, reactive, 1.8.3 (a)
    power right triangle, 1.8.3 (c)
Power factor
    correction, 2.6.1 (d)
    lagging, 2.4.2 (e) (2)
    leading, 2.4.2 (e) (3)
Protective controls
    alarms, 19.5
    cooling, 19.2
    lubrication, 19.1
    overspeed, 19.3
    shutdown, 19.6.1
Protective relaying, 7.3.4
Recovery time, 2.6.1 (b) (3)
Rectification, 1.4
Rectifier, 1.4.1, 2.1.1
Refraction, sound, 23.6
Residual magnetism, 2.1.4 (c)
Resistance, 1.7, 1.7.1
Resonance, natural mode of vibration, 23.2.9
Revolving armature alternators, 2.1.2
Revolving field alternators, 2.1.2
Right hand rule, 1.5.2
Rotor definition, 2.1.1
Safety, engine, 8.9
Separately derived system, 6.8
Shaft
    axial vibration, 21.3.3
    bending failures, 21.7.1
    misalignment, 21.7.2
    shock, from dropping, 20.1.1
Short circuit (fault) current support, 4.9
Short circuit protection, 5.2.4 (c) (d)
Shunt trip, 5.3.1
SI conversions, A1

Silencer
    back pressure, 18.6
    configuration, 18.7
    factors affecting silencing, 18.5
    manufacturer needed information, 18.4
    material, 18.7
    types, 18.4
Sine wave generation, 2.1.5
Single phase ac, 2.2.1 (a)
Skid and Mountings, 21.2.2
Skid design, 21.8
Slip, negative, 3.1
Sound
    absorbtion, 23.4
    basics, 23.2
    definitions, 18.1
    measureing, 23.9
    octaves, 18.2, 23.2.5
    power level, 23.2.1
    pressure level, 23.2.2
    transmission, 23.3
Stability
    voltage regulator, 4.5
Standby duty, 2.5.2 (d)
Static exciters, 2.1.4 (b)
Stator definition, 2.1.1
Switchgear
    bus bars, 7.2.1, 7.2.3
    generator control panel, 7.3.1
    low voltage metal-enclosed, 7.2.1
    metal-clad, 7.2.2
    metal-enclosed interrupter, 7.2.3
    outdoor, 7.2.1
    random access paralleling, 7.3.3
    sequential paralleling, 7.3.3
    three source priority load system, 7.3.2
    two source system, 7.3.2
    voltage classifications, 7.1
    synchronizing, 7.3.3
Synchronous speed, 3.1.2
Temperature rise, 2.5.2 (b)
Torsional analysis
    harmonic synthesis method, 21.5
    dampers, 21.6
    excitation input, 21.4.1
Total system load, 6.5.1 (b), 6.9.1
Transfer switch
    arc gap, 6.5.2
    arcing contacts, 6.5.6, 6.11.7
    arcing tip, 6.5.6
    blow-on contacts, 6.5.4 (a)
    controls, 6.6
    manual, 6.3
    types of loads, 6.9.1

Transfer (day) tank
    connections, 14.1.2
    sizing, 14.1
Transferring motors
    closed transition, 6.7.1 (d)
    inphase transfer, 6.7.1 (a)
    motor load disconnect, 6.7.1 (b)
    motor load shedding delayed reconnection, 6.7.2
    timed center off, 6.7.1 (c)
Transient voltage, 2.6.1
Trip elements
    electromechanical, 5.2.4 (a)
    magnetic, 5.2.4 (c)
    solid state, 5.2.4 (f)
    thermal overload, 5.2.4 (b)
Uninterruptible power supplies, 2.8.4
Variable frequency drives (VFD), 6.7.3
Vibration
    sources of, 20.1
    theory, 20.2
Vibration isolators
    elastomeric, 20.4.3.2
    helical wirerope, 20.4.3.3
    pneumatic, 20.4.3.4
    spring, 20.4.3.1
    stiffness and load-deflection (LD) curves, 20.4.2
Volt definition, 1.1
Volt-amperes reactive VAR, 1.8.3 (a)
Voltage dip, 2.6.1 (b) (4)
Voltage regulator
    frequency compensation, 4.8
    power input, 4.7
Voltage regulation, 2.4.2
Voltage, induced, 1.6
Volts per hertz, 4.1
    frequency compensation, 4.8
Weights of liquids, A7
Withstand current rating, 6.5.4, 6.9.4
X/R ratio, 6.5.4, 6.9.4